# CHANGING TROPICAL FORESTS

## HISTORICAL PERSPECTIVES ON TODAY'S
## CHALLENGES IN CENTRAL & SOUTH AMERICA

### EDITED BY HAROLD K. STEEN & RICHARD P. TUCKER

PROCEEDINGS OF A CONFERENCE SPONSORED BY
THE FOREST HISTORY SOCIETY AND IUFRO FOREST HISTORY GROUP

The Forest History Society is a nonprofit, educational institution
dedicated to the advancement of historical understanding of man's
interaction with the forest environment. It was
established in 1946. Interpretations and conclusions in FHS
publications are those of the authors; the institution takes
responsibility for the selection of topics, the competence of the
authors, and their freedom of inquiry.

Work on this book and its publication were supported by
grants to the Forest History Society.

ISBN 0-8223-1247-6

# TABLE OF CONTENTS

SECTION II: MEXICO

SECTION III: BRAZIL AND AMAZONIA

# Introduction

The International Union of Forestry Research Organizations (IUFRO) formed a Forest History Group in 1961. In 1981 the Forest History Group formed a tropical subgroup, which sponsored its first conference in Canberra, Australia, in 1988. This tropical conference in San José was the second, quickly followed by the third in Honolulu, and plans are evolving for a fourth in Africa. These four conferences will complete an agenda that includes making both the tropical regions and the Southern Hemisphere an active part of the IUFRO Forest History Group, far too long dominated by northern and temperate issues.

The Forest History Society is headquartered at Duke University, as is the Organization for Tropical Studies with its impressive facilities in Costa Rica. This proximity made it convenient to site the Latin American conference in San José; OTS's experience and infrastructure greatly enhanced both planning and implementation. Of course, generous financial support from the Rockefeller Foundation was very helpful, making it possible for Latin American scholars to attend, providing simultaneous English-Spanish/Spanish-English translation of all papers, and publishing these proceedings.

Richard Tucker, University of Michigan and then the chairman of the tropical section of the Forest History Group, took on major responsibility for overall architecture - inviting specific scholars and designing sessions. Elinor Melville, York University, lined up papers on Mexico. As chairman of the Forest History Group and executive director of the Forest History Society, I focussed on administrative matters--financial support, housing, transportation, assembling these proceedings, and similar tasks. Catherine Christen, graduate student at Johns Hopkins University but in residence in San José, provided yeoman service with local arrangements and by attracting local conference participants.

The twenty-three papers included in these proceedings range in scope and geography from broad overviews to detailed accounts of specific sites. Some investigators reported on forest conditions, others on grasslands, and still others on the people and institutions that caused change. All in fact dealt with change-- change through time and change brought about by different causes. Ecosystems are not static but are changing through natural forces or through human intervention. This conference and its predecessor in Australia and its successors in Hawaii and Africa look at these changes.

February is the dry season in Central America, and the weather behaved appropriately, and delightfully. Conference activities began with a welcoming breakfast at OTS, and then on to the sessions at nearby University of Costa Rica opened by Keynoter Gerardo Budowski, University for Peace, San José. Following three days of thoughtful--even provocative--papers, we boarded a bus and traveled northeast from San José for an overnight excursion into the humid lowlands and the OTS field station at La Selva.

We did more than look out of bus windows! OTS staff had packed our schedule that ended with a 10:00 p.m. "jungle walk" and began with a 6:00 a.m. bird watching tour. In between, naturalists led small groups of us on treks to experience directly tropical flora and fauna. The trip back included a stop at Irazú, a steaming volcano just two hours north of San José. The final, formal event was a farewell banquet with Dr. Alvaro Ugalde, director of the Office of International Environmental Relations of the Costa Rican government, as after-dinner speaker.

There is much more to a conference than the important papers; distant colleagues become acquainted and collaborations become feasible. Thus, the primary purposes of IUFRO--exchange of information and increased collaboration--are realized.

Harold K. Steen
Durham, North Carolina

Perceptions of Deforestation in Tropical America:
The Last 50 Years

Gerardo Budowski
University for Peace

Data for the last 50 years have shown an accelerated rate of deforestation, at a speed never before witnessed in the whole region. But during this period the perception about deforestation has also drastically changed especially in the last 10-15 years. One wonders why there is such a disparity.

Until about 2 decades ago the prevalent attitude, transmitted since conquest and colonization, was that tropical forests are unlimited, that deforestation and conversion to other land uses was a symbol, in fact a proof of "progress", sanctioned by laws that recognized that such "improvements" were required to claim property rights. Such attitude is still prevalent in some countries, and it is deeply rooted in rural areas.

For decades, in fact centuries, the official policy was to "conquer" the forest, "reaffirm sovereignty" and promote spontaneous or planned colonization schemes on forest land. With few exceptions, Indians and their ways of life were little considered; rather policies toward their "acculturation" prevailed.

By far the main conversion was from forest to pasture, a trend that still prevails, sometimes directly, sometimes with a short intermediate phase of food crops and often preceded by logging of the few commercial species. Another significant factor has been land speculation to claim ownership over large forestry estates, a matter that was greatly facilitated by road construction and the availability of generous credits and other incentives. This reached a peak in the 1960s and 1970s.

Large expenditures were allowed to promote migration schemes by landless farmers from populated and often degraded areas to "depopulated" forest lands, with considerable support from international financing agencies. Many of such schemes failed, since the new areas opened were often significantly different from the original habitat of settlers because of climate, soils, health conditions, and marketing of products. Also, local failures resulted from conflicts with native populations, such as indigenous groups or extractors of rubber, nuts, chicle (*sapodilla*), and other forest products.

1

The efforts of conservation organizations, particularly from outside the region, were often received with anger by local governments as well as by the affected populations.  But with increasing local conservation awareness and the emergence of local environmentally conscious leaders, as well as increase in cooperation projects from abroad, this has changed gradually and in the last 5-10 years quite drastically.  There is now a widespread recognition about the beneficial influence of forests, since concern with species depletion or extinction and even carbon dioxide sequestration by trees is often in the news.

Even forest science has evolved in universities in the last decades.  From 2-3 forest schools in 1940, the number now is well over 50 and continuously increasing.  Also, there is much more emphasis in social forestry, particularly fuelwood production, agroforestry, management of wildlands and buffer zones, forest hydrology, and even ecotourism derived from forest areas. The amazing growth of national parks and other protected areas, both in numbers and in size, is a reflection of such trend, even if many of the areas have no management plans and are poorly protected.

In spite of these favorable trends, it appears very difficult to break the dynamics of deforestation, itself exacerbated by population growth and scarcity of available land for cultivation, at least in some countries.  This often results in the invasion of what were hitherto considered marginal lands, such as areas with steep slopes, very high or very low rainfall, swampy conditions, highly acid soils or with aluminum toxicity, even upper mountains where repeated frosts are a limiting factor.  This has of course disastrous consequences, often leading to catastrophes.

Forest exploitation also has evolved, but not necessarily in a favorable way. Highly priced timbers like mahogany and Spanish cedar are coming to an end while new timbers, ignored 2 or 3 decades ago, now appear on the market.  But in broad lines, the sawmill industry is still archaic with enormous wastes in the forest and at the mill, although here too there are some well managed enterprises.  With the exception of Trinidad, there are no sizeable examples of sustainably managed natural forests in tropical America.  In many countries there have been attempts to correct such situations by enacting stringent laws including incentives and disincentives, so as to favor sustainable production, protection and reforestation, but the results have rarely met the initial expectations.

In some countries forest plantations have witnessed a spectacular success, especially with fast growing species like *Eucalyptus* spp., *Pinus* spp., Teak (*Tectona grandis*), Melina (*Gmelina arborea*), laurel (*Cordia alliodora*) and neem (*Azadirachta indica*).

Technical assistance programs have exercised a tremendous influence notably through the United Nations system as well as other governmental and non-governmental groups.  Also during the last 15 years there has been a tremendous surge of local conservation entities.  Other recent developments include:

- A better appreciation of endogenous technologies for using natural forests, leading to the designation of extractive reserves and a better recognition of the natives' rights, particularly Indians and long term river settlers.

- Creation of official structures such as Secretariats of Natural Resources or for the Environment.  A worthwhile step was the recent creation of the National Institute for Biodiversity in Costa Rica.

- The interesting trend that allows parts of the foreign debt to be condoned in favor of conservation action frequently related to the preservation of biodiversity in forested areas.

- Presence of an increased number of Latin American leaders in different branches of forestry; many of them occupy influent positions in international aspects.

- Increasing interest of the rest of the world in tropical American forestry issues frequently accompanied by the desire to get involved in the search for solutions.

## Conclusion

After a period of wanton and generalized tropical forest destruction for conversion to other uses, there is since about 10-15 years, a strong reaction against such a tendency and a strong will to defend nature.  This is particularly evident among certain official sectors, financing agencies and donors from outside the region, and a number of local organizations, as well as the better educated public.

But the weight of centuries of forest clearance and the inherent culture that goes with it, as well as population pressure, landless farmers, increased demands for timber and other forest products, all combine to create a gap between wishful thinking and reality.  Against such a dynamic situation, officials dealing with forest protection as well as foreign assistance programs have generally been impotent and frustrated.  This has lead to an increasing number of catastrophes, some quite visible, like erosion, change in water regime, disappearance of wildlife, while others like climatic change loom in a near future.  This in turn generates improvisations or enacting legal and other measures that usually turn out to be ineffective and rarely solve the problem and its origin.

The discussions coordinated by FAO, leading ultimately to the elaboration of the tropical forestry action plan for each country during in the last 3 years, have undoubtedly had an impact to counterbalance the described trends, but it is still an open question whether such planning can be converted into real achievements in the field. The same can be said for the theoretical frameworks towards a strategy for sustainable development that has been going on in some countries with great success.

— A better understanding of the many perceptions of deforestation as perceived by different population groups could undoubtedly help design more realistic action plans involving educational activities, as well as realistic field projects and promoting a better use of available funds. This in turn would help to forestall stop catastrophes and help to create a more satisfying attitude toward the remaining forests. It would also stimulate more natural and man-made reforestation and develop the real potential of tropical American forests as a source for a multitude of products and services.

The Tasks of Latin American Environmental History

Warren Dean
New York University

This paper will consider the possibility that some version of environmentalist views of past human experience might become generally accepted among Latin American historians and that some significant number of them might come to specialize in environmental history. The discussion will assume that objects of historical analysis and conceptions of their significance are determined by contemporary social and political constellations that orient and stimulate intellectual inquiry. Therefore I shall outline the historical traditions that have been cultivated in Latin America and the intellectual and social milieux from which they have emerged. I shall then discuss some of the writings of historians who have examined the environmental past in order to understand their purposes in dealing with these subjects. Finally, I shall consider some of the principal subjects of contemporary concern that might lead historians to an examination of origins and solutions.

Latin American historians have always written engaged history. Their task, their burden, has been to elaborate images of the past that defend their polities, their cultures, and their economies and point them in chosen directions. Earlier generations of historians, members of a lettered upper class, sought to legitimize the oppressive and exclusionary political systems that had emerged from the independence struggles, endowing them with idealized and ill-fitting attributes of the European nation-states. Their efforts were part of a broader oligarchical program that replaced institutions and social arrangements imposed by European colonialism with other, newer European forms that they declared to be politically liberal even though their effect was to conserve and broaden elite privileges. These historians, backed by funds extracted by the central governments from the burgeoning export trade in raw materials and foodstuffs, were founders or improvers of public universities, presses, archives, and historical societies.

Spanish and Portuguese colonialism and nineteenth-century oligarchical liberalism offered striking subjects for conservationist critiques,and indeed visiting foreign field researchers had raised such questions intermittently since the late eighteenth century. Latin America had attracted many of the leaders in European natural sciences, from Humboldt to Darwin to Warming, and some of them worked for long years there. The export trade had provided funds as well for a few centers of scientific investigation that built upon these illustrious beginnings. Many of their directors were European and North American scientists who had emigrated to Latin

America under contract. Their research usually had practical and immediate applications in such fields as public health, forestry, geology, agriculture, and veterinary science. Herpetological studies at the Butanta Institute in Sao Paulo, for example, were intended to provide anti-toxins for the hundreds of thousands of Italians introduced into the state to tend coffee. These institutions not only directed their efforts to the control of diseases and the improvement of rural productivity, they also generated important critiques of wasteful timber cutting, erosion, polluted water supplies, and the destruction of native populations. Indeed they were the principal source of these critiques. In Argentina, for example, the director of the museum of natural history, Florentino Ameghino, was an advocate of watershed management and tree planting on the pampas.[1]

Lay public advocates of these causes were not entirely lacking. There were a few prominent Latin Americans, amateur enthusiasts in the natural sciences or journalists, lawyers, or businessmen: André Rebouças, for example, a distinguished Brazilian civil engineer and timber merchant, was the first to propose, in 1878, a network of national parks. In the same decade the Mexican government passed its first forest laws, possibly influenced by the Mexican Natural History Society, founded in 1868. Francisco P. Moreno, Argentina's expert in its border disputes with Chile, donated to his government in 1904 the land he had been awarded in payment of his services, in order to form that country's first national park. In the case of Alberto Torres, a Brazilian political philosopher, the link between neocolonial capitalist development and environmental degradation was explicitly elaborated.[2]

Although some of these environmental issues found from time to time a response among urban middle-class sectors, it cannot be said that conservationism, an appreciation of nature, or a cult of the wilderness were either widely diffused or profoundly nourished in Latin American societies, not, at least in its Europeanized sectors. It is possible to offer historical explanations for this lack of a defensive mechanism that might have curbed the capitalist onslaught on their resource base. One that might be valid is that the city had been the privileged residence of the Europeanized elites since the conquest; they held in contempt the peasantry, the slave work force, and tribal peoples beyond the pale. They defined their second-hand urbanity as civilization and the folk knowledge of the exploited interior as barbarism. Their climates, geography, and native flora and fauna were often seen as factors that had caused the degeneration of their rural countrymen. Thus even as they dreamed of repopulating their countryside with European immigrants--yet another wave of strangers to the land--they hopefully introduced exotics like trout, wild boar, house sparrows, red deer, and rock pigeons to their landscapes.[3]

It is not so surprising, therefore, that Latin American historians of the period very rarely integrated the concerns of their scientist colleagues in their own writings. One does observe, however, in the 1920s and 1930s, as a consequence of a maturing disenchantment with Europe, a nativist introversion that raised historical themes closer to the physical realities of their countryside: the origins of mestizo cultural and social forms, folk beliefs, customs, and usages, and the penetration and settlement of the frontiers. Joao Capistrano de Abreu and Sérgio Buarque de Holanda, in Brazil, were exemplars of these tendencies. Scientific appreciation of the native environment appears to have accelerated as well in this time of economic crisis and renovating political movements, as a first generation of Latin Americans come to direct existing scientific institutions and to found new ones. In Mexico, for example, Enrique Beltran refounded the Natural History Society and directed the Mexican Institute of Natural and Renewable Resources. In Brazil, among other initiatives, a forest service was created, the constituent assembly passed codes of forestry and fish and game, and a national symposium on conservation took place in Rio de Janeiro in 1934, organized by Alberto José Sampaio.[4]

During this interwar period European and North American social scientists, among them historians and historical geographers, began to carry out field research in Latin America. Of the North Americans the most influential among environmental historians in the U.S. was Carl Sauer, whose interests ranged over all the Western Hemisphere. His speculations concerning native population densities, agricultural origins, plant and animal dispersal, and the anthropic origins of ecosystems were underlain by a profound vision of the relation of the human species with all others, a vision that transcended political boundaries and disciplines.[5]

The involvement of foreign social scientists concerned with environmental issues presents complex historical interactions. Although earlier levies of natural scientists often seized upon opportunities to inject alien political and social perspectives into their reports and no doubt were influential in local political as well as scientific circles, the social sciences were constantly interactive on a political level. These influences would be worth tracing. A sensational example: the Nazis sent to the Brazilian state of Espirito Santo a geographer to study the fate of German colonists who had emigrated there in the 1870s, the purpose being to discover if the "Aryan race" maintained its superiority in tropical climes. The dismayed report of the geographer, who found a desperate and fever-ridden peasantry, was suppressed until after the war. One wonders what were the influences of this and other German geographers, several of whom were active in Brazil, upon issues as important as this. The flow of environmental specialists did increase during World War II, as part of U.S. government's campaign to enhance its image in Latin America and to increase the production of war-related exports.[6]

In the postwar world, social and environmental sciences in Latin America gained momentum through adherence to the ideology of economic developmentalism. The region's public policy makers had been impressed at the effectiveness of central planning in creating the war machines of the great powers and in reconstructing the civilian economies at war's end. The neglect to which they found Latin America relegated, while the U.S. bestowed vast funds on its former enemies and while the imperial powers tried to restore their damaged links with their colonies, stimulated them to independent efforts to close the gap with the industrialized nations. Concepts of underdevelopment and autarky, elaborated by UNESCO's Economic Commission for Latin America, provided theoretical justification for accelerated growth policies, while the critical state of the Latin American economies provided urgent practical motives. These goals appeared to the military to be essential to national security, and counted with their support. When, by 1960, the U.S. government became at last convinced that economic development was necessary to avoid further Cuban-style revolutions in the hemisphere, it abandoned the laissez-faire preachments it had directed to the region and also supported the spread of multinational manufacturing plants, thereby intensifying the drive to development.

Ecosystems came under closer scrutiny, as "nature" came to be reconceived as "natural resources" or "renewable resources." Institutes were formed to "develop" and "exploit" them. Thus not only economists, but geographers, geologists, physicists, foresters, agronomists, sociologists, and biomedical scientists had roles to play in this grand enterprise. The mobilization of public opinion in favor of development planning, which appears only in retrospect an easy task, was also the work of specialists inside and outside government.[7]

It might be hypothesized that developmentalism became the central object of public policy and the principal issue of public debate as the result of a underlying trans-formation that can most validly be regarded as environmental: populations were suddenly growing very rapidly and most of this growth was becoming concentrated in urban agglomerations. For the directing elites, this new reality had immediate political implications, since the urban poor were sufficiently concentrated and conveniently located to topple governments. Unlike rural populations, they had to be supplied uninterruptedly with food, water, and transport. They had to be provided jobs and they had to be manipulated so as not to threaten existing hierar-chical social arrangements. But in a more durable and material sense, the rapid growth and concentration of human numbers in cities imposed vastly greater demands upon the ecosystems upon which they proliferated, like fungi, upon a living substrate. The capital resources needed to bring food, fuel, water, and raw materials to the cities were immense. These products had to be extracted from the countryside even if there were no goods to offer in exchange. In real economic terms the cities were to some greater or lesser degree parasitical, or so were they becoming in ecological terms, since the metropolises of the mid-twentieth century preempted larger and larger life zones, recycled less and less of their inputs, and increasingly toxified their effluents.

It is clear that the direction of public policy in the 1950s and 1960s created an ambience that was unfavorable to consideration of issues that would later be come to be called environmentalist. The politically strategic middle sectors were especially favored by the measures that were chosen, indeed rapid development offered a premium for their skills. Large-scale projects with evident negative impact upon ecosystems were therefore defined exclusively as liberation from despised economic backwardness. The building of penetration roads into the Amazon basin, for example, was at the time hailed with pride by the several national governments and was, it ought also to be noted, greeted with admiration in the foreign press. It is true that, of all the social scientists whose fields of study were affected by rapid planned development, anthropologists were least likely to be recruited, and were the first to raise criticisms of these projects. Even they, however, were often enough set to implementing rescue measures whose real purpose was to increase the chances of success of official development projects, in the fading hope of preserving their charges, in some measure, from greater harm.[8]

Critiques of the development programs arose largely from the political left. Their criticisms did not question the necessity or desirability of industrialization, which indeed they wished for as a necessary prerequisite to a desired radical social transformation. But they easily and early recognized the ideological character of these programs, and pointed out their undesirable political and economic consequences. The products of the new factories were most commonly affordable only by the middle class, who had, nevertheless, to be subsidized with cheap consumer credit to permit them to buy them in sufficient quantity. Only a minority of the urban working class was absorbed by these factories, or indeed by any formal employment. Wage levels did not keep pace with improvements in productivity and sometimes not even with inflation. Thus even proletarianization was incomplete and precarious. Although reduced income and wealth concentration had been promised, instead it intensified. The inflow of foreign capital was viewed by the left as an absolute evil, alienating to external centers of decision, often government-backed cartels, the ultimate control of investment, research and development, and surrendering profits achieved in the most advanced sectors of the economy. That Cuba had in the meantime chosen an alternate path to development, which, it was supposed, would be more efficacious as well as more just, considerably fortified these arguments.[9]

Much of the historiography of the era was directly or indirectly preoccupied with economic development and its political impact. The most urgent task of the historian was to offer explanations for the backwardness of their societies and their economies. Concepts of "cultural diffusion" and "obstacles to growth," elaborated by anthropologists, sociologists, and political scientists, were applied retrospectively. The impact of imperialism and foreign capital upon historical patterns of development was gauged. A theory of "dependency" was elaborated that ascribed the underdevelopment of the "periphery" to the internalization of structures and ideologies designed to perpetuate subordination to the neo-imperialist forces of the advanced, industrialized "center," whence profits exclusively flowed. These interpretations provided justification for the statist and autarkical policies that were being pursued, but they did not suggest radically different paths of growth nor question their advisability.[10]

The emergence of environmentalist critiques at the international level, most notably at the first environmental congress at Stockholm in 1972, therefore encountered unfavorable political and ideological reception by the directing elites of Latin America, whose mandates, to the extent that they possessed or sought any at the time, were based upon economic developmentalism. The appeal to give consideration to environmental costs was therefore entirely unwelcome. It was suspected by some on the left as well as the right, and indeed this suspicion has not entirely been laid to rest, that environmentalism was being promoted by the advanced countries in order to forestall the rapid development of Latin America or to divert attention from social conflicts. [11]

Despite initial hostility, environmentalism has been taken up as an intellectual and political issue in several countries. Numerous environmentalist organizations, ranging from national political parties to neighborhood groups, have been formed, and nature conservation organizations founded earlier have become more activist and committed to political action. Development economists have exchanged concerns with natural scientists and natural scientists have found their voice as activists. The media have responded with coverage of environmental problems, and there has grown in consequence a widespread popular consciousness that has been expressed in political debates. These concerns have also taken hold among at least some members of the possessing classes. Earlier attitudes of resistance to foreign interference in sovereign affairs have been to a degree replaced by mutual willingness to engage in genuinely internationalist collaboration with first-world environmentalist groups. Governments have therefore felt obliged to respond with legislation, administrative regulation and increased funding.[12]

It is not surprising that these expressions are still precarious and largely ineffectual indications of the relatively diffuse aspirations of some sectors of the public, considering that these countries have passed through 10 years of the gradual collapse of dictatorial regimes and their replacement by weak and ill-coordinated civil governments, followed by another decade of economic disorganization and depression. Nevertheless, this rapid transformation of the political climate is significant, in that it mirrors real changes in social relations, developments in technical and scientific networks, and the re-evaluation of collective experiences in the human relation to the environment.

The governments of Latin America, casting about in the early 1970s for responses to foreseeable international pressure, discovered that they had at hand cadres of natural scientists, employed in public research centers and universities, who had been elaborating critiques of development projects, despite the developmentalist goals of their institutions. These scientists, some of whom had been suffering isolation or reprisals for their unwelcome positions, might now be put to research that, the government no doubt hoped, would whitewash its development projects. To the institutes of "natural resources" could be appended the label "environmental sciences," so as to signify the intended reorientation. As it became apparent that ministerial level environmental protection agencies would have to be formed, the maverick scientists could also be recruited to staff them.[13]

The reconceptualization of the human relation to the natural world that these public demonstrations of concern imply clearly call for retrospective analysis. The most important source of historians to take up this task are the existing cadres of natural scientists who are already engaged in reshaping public policy. They are relatively numerous, they already possess the necessary scientific and technical background to explicate environmental questions, they regularly deal in the historical aspects of their own discipline, and they see most clearly the political implications of their subject matter. Evidently, they also lack training in historical method, in data collection and interpretation, and perhaps in the larger questions of political and economic history. They also may find it difficult to criticize the policies of their own scientific institutions. Perhaps the ideal solution would be the formation of mixed research teams of historians and scientists, and the enrollment of apprentices from both fields, to engage in cross-disciplinary training at the doctoral or post-doctoral level. There may also be some recruitment of historians into the consulting firms that are mandated by new laws to carry out environmental impact statements.

Will there be many scientists or technicians who will want to take up retrospective analysis, considering how great is the need for confronting contemporary environmental problems, and how great will be the career rewards and satisfactions for studying the past? Forest history has suffered the disadvantage that it has offered few opportunities for employment: schools of forestry, unlike schools of engineering, medicine, and other technological and scientific disciplines, have employed few historians. As these schools convert to training more general specialists in environmental sciences, they should be encouraged to recruit historians. Perhaps no more than a few will be enlisted for this task, but they will influence others, if their analyses and interpretations should prove compelling. One suspects that a significant number of these new historical scholars will be women, since some of the natural science disciplines still fail to reward women scientists according to their merits, and they may well find academic history marginally more congenial.

Meanwhile, it is significant that social scientists in Latin America, whose positions have historically been leftist, have gradually incorporated environmentalist arguments in their critiques of existing political and social arrangements. The reestablishment of civilian governments that have been obliged to try to co-opt peasant movements and to give an accounting of their treatment of tribal peoples has exposed the environmental dimension of the oppression to which these groups have been subjected and has revealed that peasant and tribal methods of environmental exploitation are to some degree conservationist. Cross-class solidarity has been achieved through these exposés, which have evoked sympathy, and through consumerist concerns, for example, the uncontrolled application of biocides in agriculture. Radical critics have also revived economic nationalism by pointing out that foreign companies often engaged in environmentally destructive activities which were forbidden in their home countries.[14]

The imposition of inappropriate technologies by incompetent public authorities, inevitably more dangerous for the urban lower class, has also provoked political reactions: the tragedies of Cubatão, where industrial pollution from state-owned companies has caused loss of life as well as destruction of neighboring forests, and Goiânia, where abandoned nuclear wastes ravaged a lower-class neighborhood. State-owned electric companies have had to defend themselves against campaigns that have opposed their attempts to build hydroelectric plants, not only because of the displacement of peasants and Indians, but also because of the destruction of forests, rivers, and natural formations. These political debates have gained clarity and significance through the occasional contribution of historians, or of social or natural scientists who have sought to define pre-existing environmental conditions or reveal the interests that underlay developmental goals.[15]

It seems likely that these political concerns will generate further inquiries into the origins of the multiple environmental crises to which the economic development of this century has brought Latin America. This engagement undoubtedly will be useful in mitigating the impact of future planning. And yet it will not realize the full potential of the environmental view of history. Conceiving of the environmental dimension of the human condition as a side-effect of other, supposedly more decisive activities, such as the class struggle, capital accumulation, the spread of imperialism, the triumph of science and technology, or the subjection of women, is to trivialize a historical reality of immensely greater, enduring consequence. I consider it to be the ultimate intellectual task of environmental historians not merely to persuade liberal economic historians that what they regard as externalities must be included in their equations of economic growth and development, or to persuade Marxian historians that the degradation of the environment is one of the forms that the oppression of the proletariat customarily takes, but to make environment the dominant paradigm, demonstrating that these externalities and this degradation are in fact the most consequential of human activities and that human interaction with the environment must be regarded as the central issue in human history.

This is admittedly an inherently tragic outlook and therefore a most thankless task for academic historians, who depend on the good will of public authorities, private donors, and parents, all of whom expect meliorist views of human experience to be served up to impressionable youth. Those whose interpretations might induce despair are likely to be invited to drink hemlock. And yet, the imposition of environmentalism upon history may be entirely compatible with the reigning academic post-modernism, one of whose tenets, at least insofar as I can interpret it, is that some of the quandaries of human existence are simply not amenable to human rectification. Even the most engaged of historical inquiries, derived from activist sympathies, may finally lead to this: to a grim, but realistic school of history, one that may finally and decisively turns all historical inquiry away from vain insistence on the idea of progress and dispel the hollow triumphalism of all former interpretations of our common and conflictive past.

## Notes

1. Antonio Elio Brailovsky, "Politica ambiental de la generacion del 80," in Nora L. Siegrist de Gentile, *et al.*, *Tres estudios argentinos* (Buenos Aires, 1982), 287-364. On one of the scientific institutions, see Silvia F. de M. Figueirôa, *Um Século de Pesquisas em Geociências* (Sao Paulo, 1985).

2. Rebouças, *Excursao ao Salto do Guayra* (Rio de Janeiro, 1876); José Sarukhan, "Mexico," in E. J. Kormondy and J. F. McCormick, *Handbook of Contemporary Developments in World Ecology* (Westport, CT, 1981); Moreno, *Reminiscencias* (2d ed., Buenos Aires, 1979); Torres, Alberto, *As Fontes da Vida no Brazil* (Rio de Janeiro, 1915).

3. See for example Helmut Sick, "A Invasao da América Latina pelo Pardal," *Boletim Nacional, Zoologia* [Rio de Janeiro] (1959) No. 207.

4. Joao Capistrano de Abreu, *Caminhos Antigos e Povoamento do Brasil* (Rio de Janeiro, 1930); Buarque de Holanda, *Monções* (Rio de Janeiro, 1945). Ricardo Ramirez, *La condicion legal*

*de los bosques y su conservacion* (Mexico, 1900); Tereso Reyes, *Mexico, Esta en peligro de perecer* (Mexico, 1932). On Brazilian conservationism, see Warren Dean, "Forest Conservation in Southeastern Brazil," *Environmental Review*, 9 (Spring 1985).

5. Sauer's most extensive historical work was *The Early Spanish Main* (Berkeley, CA, 1966); see his bibliography in Sauer, *Selected Essays, 1963-1975* (Berkeley, CA, 1981). Another American with penetrating views was the forestry specialist Roy Nash--see his *The Conquest of Brazil* (New York, 1926); and "Brazilian Forest Policy," *Bulletin of the Pan-American Union*, 58 (July 1924), 688-706.

6. W. A. Engler, "A Zona Pioneira ao Norte do Rio Doce," *Revista Brasileira de Geografia*, 13 (April-June 1951), 223-264.

7. Emanuel Adler, *The Power of Ideology: The Quest for Technological Autonomy in Argentina and Brazil* (Berkeley, CA, 1987); José Leite Lopes, *Ciência e Libertaçao* (Rio de Janeiro, 1969). On geographers, see Nilson Cortez Crocia de Barros, "Trabalho . . . Historia do Conservacionismo no Brasil" (Unpublished term paper, Department of History, University of Sao Paulo, May 1985).

8. Environmentalism, although weak, was never extinguished at the political level. In 1949, conferences sponsored by UNESCO and the International Union for the Protection of Nature drew together organizations engaged in nature protection, and stimulated local campaigns. See *Campana latino-american para la proteccion de la naturaleza y la conservacion de los recursos naturales* (Buenos Aires, 1952).

9. Celso Furtado, *O Mito do Desenvolvimento Econômico* (Rio de Janeiro, 1974).

10. Two of the most widely read of the historical studies were Celso Furtado, *The Economic Growth of Brazil: A Survey from Colonial to Modern Times* (Berkeley, CA, 1963 [1959]); Aldo Ferrer, *La economia argentina; etapas de su desarrollo y problemas actuales* (Buenos Aires, 1963). The classic text of dependency is Fernando Henrique Cardoso and Enzo Faletti, *Dependency and Development in Latin America* (Berkeley, CA, 1979 [1969]).

11. Thomas G. Sanders *Development and Environment: Brazil and the Stockholm Conference* (Hanover, NH, 1973). See also Warren Dean, "Confronting the Environmental Crisis in Latin America," *ILAIS Occasional Paper* [NYU] (1973). An interesting attempt at framing questions of US-Latin American relations in environmental terms is Andrew Maguire and Janet Welsh Brown, eds., *Bordering on Trouble: Resources and Politics in Latin America* (Bethesda, MD, 1986), published for the World Resources Institute. Suspicion of foreign hidden motives is of much earlier date than 1972: When, in 1946, for example, it was proposed that the scientific study of the Amazon be made the responsibility of an institute under UNESCO auspices, the Brazilian government reacted with outrage, and in response created the National Institute for Amazon Research (INPA) in Manaus.

12. Carlos Augusto de Figueiredo Monteiro, *A Questao Ambiental no Brasil, 1960-1980* (Sao Paulo, 1981); Eduardo Viola, "O Movimento Ecologica no Brazil (1974-1986): Do Ambientalismo a Ecopolitica," in José Augusto Padua, org., *Ecologia e Politica no Brasil* (Rio de Janeiro, 1987), pp. 63-109. For encounters of environmentalists and development specialists, see Manuel Correia de Andrade, *et al.*, *Meio-Ambiente, Desenvolvimento e Subdesenvolvimento* (Sao Paulo, 1975); Osvaldo Sunkel, coord., *Estilos de desarrollo y*

*medio ambiente en América Latina* (Mexico, 1980-81). See also Enrique Leff, *Ecologia y capital; hacia una perspectiva ambiental del desarrollo* (Mecico, 1986). An eminent agricultural specialist converted to environmentalism and recruited to government service is José Lutzenberger, see his *Ecologia; do Jardim ao Poder* (Porto Alegre, 1985). For a report on Brazilian elite sympathetic to environmentalism, see José Ruy Gandra, "A mancha verde se espalha," *Exame-VIP* (27 June 1990), pp. 26-31. Evidence of internationalist sympathies appears in Fernando Gabeira's history of Greenpeace: *Greenpeace; Verde Guerrilha da Paz* (2d ed.; Sao Paulo, 1988).

13. Among early studies defending government policies, see Nelson Geigel Lope-Bello, *La experiencia venezolana en proteccion ambiental* (Caracas, 1974); Clara Martins Pandolfo, *A atuaçao da SUDAM na preservaçao do patrimônio florestal da Amazônia* (Belém, 1972).

14. For appreciations of peasant conservation, see François Lartigue, *Indios y bosques: politicas forestales y comunales en la Sierra Tarahumara* (Mexico, 1983); Juan Ansion, *El arbol y el bosque en la sociedad andina*, prologue by Chris E. van Dam (Lima, 1986). On consumerist reforms, see Antonio Elio Brailovsky, *Negocio de envenenar* (Buenos Aires, 1988). For critiques of foreign companies' operations, see Cuauhtemoc Gonzalez Pacheco, *Capital extrangero en la selva de Chiapas, 1863-1982* (Mexico, 1983); Rafael Virasoro, *La Forestal Argentina* (Buenos Aires, 1972). For a general synthesis, see Luis Vitale, *Hacia una historia del ambiente en América Latina* (Caracas 1983).

15. See for example Fernando Gabeira, *Goiânia, rua 57: O Nuclear na Terra do Sol* (Rio de Janeiro, 1987); Francisco Graziano Neto, *Questao Agraria e Ecologia; Critica da Moderna Agricultura* (Sao Paulo, 1982).

Microfossils and Forest History in Costa Rica

Sally P. Horn
University of Tennessee

Documentary and field evidence of forest history can be supplemented and extended in some areas by analyzing pollen grains, charcoal fragments, diatoms, and other plant microfossils preserved in terrestrial and marine sediments. These fossils, and the sediments in which they are preserved, can provide information on pre-settlement forest composition and fire regimes, on the impact of climatic change on forest ecosystems, and on forest clearance, crop production, agricultural burning, soil erosion, water pollution, and other anthropogenic disturbances (Deevey, 1978; Berglund, 1986; Delcourt, 1987; Binford and Leyden, 1987; Byrne and Horn, 1989; Roberts, 1989).

Paleoecological records derived from the study of sedimentary microfossils are available from a number of sites in Latin America, but the data network is spatially and temporally incomplete (Livingstone and van der Hammen, 1978; Flenley, 1979; Bradbury, 1982; Colinvaux, 1987; Markgraf, 1989). This paper focuses on Quaternary paleoecological studies in Costa Rica.[1] The excellent system of national parks and reserves in Costa Rica has made the nation a key site for tropical ecological research, and we know a great deal about the present distributions and dynamics of its varied plant and animal life (Janzen, 1983a; Gómez, 1986). But we have only just begun to investigate the longer-term history of the rich Costa Rican biota. In this paper I review the pollen and charcoal records now available for Costa Rica, and then present the preliminary results of ongoing paleoecological studies in the central Atlantic lowlands. As background for my review of Costa Rican paleoecology, I first provide a brief overview of present environmental conditions.

**Environmental Setting**

The mountainous backbone of Central America divides Costa Rica into 3 broad physiographic provinces: the Pacific lowlands, the Atlantic lowlands, and the Central highlands (Figure 1). The Pacific lowlands include a series of low coastal hills and valleys and 3 prominent peninsulas. The Atlantic lowlands comprise the lowermost eastern slopes of the Central highlands and the adjacent Atlantic coastal plain. The Atlantic coastline is significantly shorter and straighter than the Pacific coastline, and the topography inland is more uniform.

16

Figure 1.  Physiography of Costa Rica and locations mentioned in text.

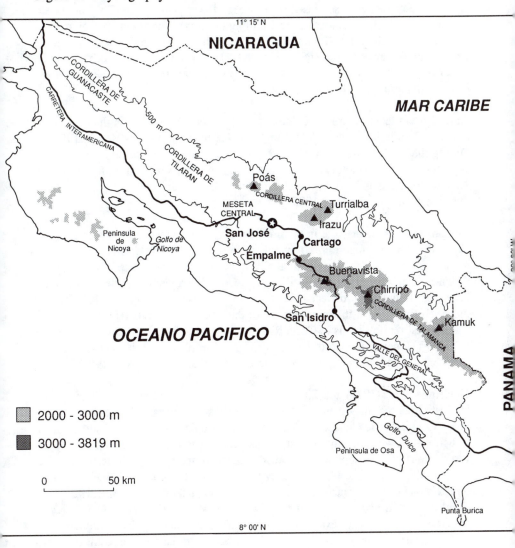

The Central highlands are formed by 2 great mountain chains. The northern chain begins near the Nicaraguan border with the volcanic peaks of the Cordillera de Guanacaste and extends southeastward through lower peaks in the Cordillera de Tilaran to the high volcanoes of the Cordillera Central. The southern mountain chain is the Cordillera de Talamanca, a rugged plutonic range that begins near the geographic center of Costa Rica and extends slightly beyond the border with Panama. Between these 2 mountain chains lies the populous Meseta Central, the heart of modern Costa Rica.

Annual mean temperatures in Costa Rica range from over 27° C in the central and northern Pacific lowlands to less than 10° C on the highest peaks (Coen, 1983). During the northern hemisphere winter the Pacific lowlands and much of the adjacent highland region experience a 3 to 6 month dry season. Rainfall in the Atlantic lowlands is more evenly distributed, although dry spells of shorter duration do occur (Herrera, 1985).

The province of Guanacaste in the northern Pacific lowlands is the driest area of Costa Rica; here annual rainfall may be as low as 1300 mm (Herrera, 1985). The number of *wet* months, and the total annual rainfall, increases with increasing distance southward along the Pacific coast. On the Osa peninsula the rainy season lasts from mid-March to December, and annual precipitation reaches 4000 mm or more, rivaling that in the rainiest areas of the Atlantic lowlands (Herwitz, 1979). The highest rainfall totals in Costa Rica occur at mid-elevations on the windward (Atlantic) slopes of the Cordillera Central and Cordillera de Talamanca, where moist trade winds dump up to 6000 mm of rainfall annually (Coen, 1983).

Holdridge *et al.* (1971) and Tosi (1969) have identified and mapped 12 life zones in Costa Rica that they equate with distinct vegetation types. Three zones are recognized in the lowlands: tropical dry forest, tropical moist forest, and tropical wet forest. Lowland Guanacaste and adjacent areas fall within the tropical dry forest life zone, and support semi-deciduous forest and savanna associations. The remnant deciduous forests found today in this area are low and relatively open, with 2 strata of trees (Hartshorn, 1983). Common genera include *Bursera*, *Bombacopsis*, *Cochlospermum*, *Enterolobium*, *Guazuma*, *Luehea*, and *Tabebuia* (Hartshorn, 1983). The savannas of Guanacaste are dominated by introduced African grasses (Parsons, 1970). Today's savannas are largely the result of clearing and burning for pasture, and where protected from fire are being reinvaded by woody plants (Janzen, 1986). But similar associations may have existed on shallow soils and in swampy areas prior to human settlement (Holdridge, 1953).

The Atlantic lowlands, and the wetter central and southern Pacific lowlands, fall within the tropical moist and tropical wet forest life zones. Tropical moist forests are tall, multistratal, semi-deciduous or evergreen forests (Hartshorn, 1983). Characteristic genera include *Carapa, Copaifera, Cordia, Dialium, Dipteryx, Anacardium, Guarea, Eschweilera, Manilkara, Lecythis, Ochroma, Virola, Vitex,* and *Terminalia* (Goodland, 1971).

Tropical wet forests are tall, multistratal evergreen forests characterized by very high species diversity (Hartshorn, 1983). This diversity makes the identification of "characteristic" genera difficult. Goodland (1971) lists the following taxa as common: *Pentaclethra, Nectandra, Ocotea, Coussapua, Licania, Manilkara, Virola, Vochysia, Dialium, Aspidosperma, Hura, Guatteria, Anacardium, Sideroxylon, Ceiba, Castilla, Swartzia, Vitex, Pouteria, Sterculia, Astrocaryum, Iriartea, Socratea, Euterpe.*

Mangrove forests are found along protected coasts within all 3 life zones; genera represented are *Rhizophora, Laguncularia, Avicennia, Conocarpus,* and *Pelliciera* (Simberloff, 1983).

With increasing altitude the tropical forests of the Pacific and Atlantic lowlands give way to premontane and montane formations in which temperate taxa, particularly *Quercus,* assume greater importance. Montane oak forests were once extensive in the Meseta Central, but have long since been cleared for agriculture and charcoal production. Today the largest stands occur at higher elevation in the more remote areas of the Cordillera de Talamanca (Standley, 1937-38; Burger, 1977).

In the premontane rainforests of mid-elevations, common oak associates are members of the Lauraceae (especially *Nectandra, Ocotea,* and *Persea*), *Cedrela, Sapium, Ulmus, Zanthoxylum, Oreomunnia, Alfaroa,* and *Weinmannia.* With increasing elevation, *Drimys, Podocarpus,* and *Alnus* become important (Standley, 1937-38). Epiphytic and terrestrial ferns, including tree ferns, are particularly abundant in these cool montane forests.

The genus *Pinus* is not represented in the montane forests of Costa Rica, but pine pollen is found in lake and marine sediments due to long distance transport by winds and currents. Pines form extensive stands in the drier interior highlands of southern Mexico, Guatemala, Honduras, and Nicaragua, and also occur on the Atlantic coastal plain of Nicaragua and Honduras and on several Caribbean islands (Parsons, 1955).

The highest peaks of the Cordillera de Talamanca extend above timberline, and support open grass- and shrub-dominated páramo vegetation. A characteristic component of the Costa Rican páramos is the bamboo *Chusquea* (*Swallenochloa*) *subtessellata*, which in many areas occurs in nearly monospecific stands (Janzen, 1983b; Clark, 1989). Elsewhere the bamboo grows intermixed with a variety of evergreen shrubs, among which the genera *Hypericum*, *Vaccinium*, *Pernettia*, *Senecio*, and *Escallonia* are especially common (Horn, 1989a). Whether or not this páramo vegetation represents the natural plant cover of the high peaks has attracted some interest. The Buenavista páramo along the Inter-American highway route is included within the montane rainforest life zone in the Holdridge system (Tosi, 1969), and Janzen (1973, 1983b) believes that the area did support low forest, rather than páramo, prior to extensive human disturbance.[2] The higher and more remote Chirripó massif is classified within the subalpine rain páramo life zone in the Holdridge system (Tosi, 1969), and here the treeless vegetation is usually considered natural (Hartshorn, 1983).

## Microfossil Records from Costa Rica

The first paleoecologist to work in Costa Rica was Paul Martin, who in 1959 recovered a series of cores from montane and lowland mires (Martin, 1960). The longest core, and the only one ever analyzed, was from a montane bog at the Parque Vicente Lachner site (2400 m) in the Cordillera de Talamanca (Martin, 1964). The present vegetation of this site is lower montane rain forest (*sensu* Holdridge *et al.* 1971) dominated by *Quercus*. In a 13 m core, Martin found 2 sections in which nonarboreal pollen (particularly Gramineae, Umbelliferae, and Compositae) and spores of *Lycopodium* and *Isoetes* were abundant, and pollen of *Quercus* and other montane forest trees was scarce. Martin postulated that these zones corresponded to periods in which there was a regional shift in vegetation from montane forest to páramo as a result of climatic cooling during the Late Pleistocene. The depositional environment had changed along with the vegetation, with a marsh or peat bog occupying the site when it was surrounded by forest, and a lake occupying the site when it was surrounded by páramo. The uppermost "páramo zone" was dated at 20,750 yr B.P. The lower "páramo zone" was too old to date by conventional radiocarbon methods; extrapolating from sedimentation rates calculated for the upper part of the core suggests that it may have spanned the period from ca. 110,000 yr B.P. to ca. 48,000 yr B.P.

No further coring took place in Costa Rica for over 20 years, but in the mid-1970s Kesel (1983) collected surface exposures for paleoecological analyses on the lower slopes of the Cordillera de Talamanca in the Valle del General (Figure 1). Even before the post-1940 wave of deforestation in the valley there were numerous large savannas on the lower Talamancan slopes; Barrantes (1965) has attributed these to the clearing and burning of forests by pre-Columbian occupants. Kesel's preliminary pollen data suggest that open savanna vegetation might in some areas predate human arrival. The earliest human artifacts yet recovered in Costa Rica are Clovis and Magellan spear points typical of those used by Paleo-Indian hunters 10,000 to 12,000 years ago (Snarskis, 1981). But 2 samples of lacustrine or low-energy fluvial sediments dated at 17,050 yr B.P. and 12,830 yr B.P. contained pollen spectra suggesting open grass savanna with scattered trees. Additional surface exposures, and cores from nearby locations, are now being analyzed by Kam-biu Liu of the Louisiana State University (Liu, pers. comm. 1989). His results should shed further light on questions of forest and savanna history in the Valle del General.

Pollen analysis of marine sediments from the Pacific continental shelf and slope of Costa Rica began in the early 1980s. Samples from the upper 24 m of sediment recovered at Deep Sea Drilling Project Site 565, located 42 km west of the Nicoya Peninsula, revealed shifts in pollen frequencies that seem generally consistent with Martin's record of altitudinal shifts in montane vegetation belts (Horn, 1985a). Four pollen zones were recognized in the section, which was estimated to span ca. 145,000 years. In the basal zone, lowland tropical pollen types (Urticales, *Piper, Alchornea*) were common and montane taxa were scarce. The 2 middle zones were dominated by montane/temperate types such as *Quercus, Podocarpus*, and *Pinus*. Pollen spectra in the upper zone were variable, perhaps because of drilling disturbance.

Provisional pollen diagrams have been prepared for 2 short cores from Golfo Dulce in southwestern Costa Rica (Horn, 1985b). The diagrams show changes in the frequency of lowland tropical pollen types (*Cecropia* and other *Urticales*, and *Piper*) that may reflect forest clearance, but these records have yet to be dated.

Clary (1986) has examined the pollen content of coastal marine sediments recovered from a mangrove forest on the Nicoya Peninsula. The samples were from an undated sediment core (c. 2 m long) recovered downstream from the Nacascolo archaeological site. Most of the samples contained too few pollen grains for a standard count, so results must be considered tentative. No marked changes in plant composition within the mangrove forest were apparent in the pollen record. Charcoal concentrations were more variable, suggesting periods of intensified agricultural activity in upstream areas. One sample from near the base of the core contained 2 grains of maize pollen.

The first lake sediment cores recovered in Costa Rica were 2 short cores raised in 1985 from Lago Chirripó (3520 m), a glacial lake located within the Chirripó páramo in the Cordillera de Talamanca. The cores were recovered from water depths of 6 and 9 meters using plastic tubes fitted with rubber pistons. Analysis of the charcoal content of the nearshore core (110 cm long; collected 20 m offshore) revealed that natural or human-set fires have affected the Chirripó highlands throughout the deposition of the ca. 4000 yr sedimentary record (Horn, 1989c). The fires may have been set by lightning or people within the Chirripó páramo, or they may have burned into this area following the ignition of surrounding forests.

Both short cores contained dispersed macroscopic charcoal, but the nearshore core from Lago Chirripó also preserved 2 distinct layers of macroscopic charcoal that may have been deposited during a lower lake stand. Radiocarbon determinations on bulk sediment samples that encompassed these layers yielded age estimates that are coeval with radiocarbon determinations made on macroscopic charcoal recently unearthed beneath mature tropical wet forest at the La Selva Biological Station in the Atlantic lowlands (Horn and Sanford, in prep.). This correspondence could be a meaningless coincidence, but it might also be a paleoclimatic signal, perhaps reflecting short-term regional rainfall anomalies such as El Niño, or longer-term shifts in regional climate. Other core analyses planned and in progress in Costa Rica should shed light on the issue.

The charcoal record in the short cores from Lago Chirripó has been confirmed and extended through the analysis of a longer core recovered in 1989 from a lake in the adjacent Valle de las Morrenas. The Morrenas core has also provided the first information available on the timing of deglaciation in Costa Rica (Horn, 1990). The pollen diagram shows that páramo-like vegetation has covered the high peaks since deglaciation some 10,000 years ago (Horn, in prep.). Some or all of the fires evident in the charcoal record may have been human-set, and these fires may have affected timberline position on the massif. But the Chirripó páramo does not appear to be an area in which human action has carved grassland from forest.

Other cores from the Cordillera de Talamanca are now under study by Antoine Cleef and associates at the University of Amsterdam (Cleef, written comm. 1990). Their longest core (11 m) was recovered from the La Chonta bog (2310 m) near El Empalme. An initial abstract on the pollen stratigraphy (Cleef *et al.* 1990) suggests general similarity with Martin's (1964) record from the nearby Lachner bog. No radiocarbon dates were available when the abstract was prepared, but the authors estimated that the record covered the last ca. 30,000 years. Páramo and open scrub vegetation surrounded the La Chonta site during the deposition of the lowest 7 m of the section, which is thought to date to the last glacial period. Montane forests migrated upslope during the deposition of the upper (Holocene?) section. According to Cleef *et al.* (1990), some discrepancies with the Lachner record exist, but these have yet to be elaborated.

Clary (1990) has recently investigated the pollen content of soil samples from archaeological sites in the Cordillera de Tilarán that were studied as part of the

Arenal prehistory project coordinated by Payson Sheets. Pollen preservation was generally poor. A few pollen grains of maize and other possible cultigens were found at a site corresponding to the Silencio phase (A.D. 600-1200), and 1 possible maize pollen grain (along with maize phytoliths and macrofossils) was recovered from an older, Tronadora phase site (2000 B.C.-500 B.C.). The Arenal samples from human habitation sites provide important clues on human subsistence, but even better preservation would not have made them suitable for reconstructing local or regional forest history. For that purpose, sediment cores from lakes or mires are more appropriate.

**Ongoing Paleoecological Research in the Atlantic Lowlands**

To provide records of forest history and human-environment interactions within Costa Rican rainforests, I am collaborating with Dr. Gilbert Vargas of the University of Costa Rica and Dr. Kurt Haberyan of Troy State University, Alabama, in a study of the microfossil content of sediment cores from a series of lakes and swamps in the central Atlantic lowlands. During the preColumbian period this area was an important trade zone and a buffer between Mesoamerican and South American influences (Stone, 1977; Snarskis, 1981). Today it is a key site for tropical ecological research, with activity centered at the La Selva Biological Station. Mature forests here and in adjacent areas are often described as "virgin" (Hartshorn, 1983; Organization for Tropical Studies, 1989), implying little previous human impact. But recent excavations have revealed evidence of prehistoric human occupation and fires at La Selva (Quintanilla, 1990), suggesting a more disturbed history.

Our paleoecological work in the Atlantic lowlands will provide an important historical perspective for studies of contemporary ecology in this region (Hamburg and Sanford, 1986). We anticipate that our results will also provide tests for various hypotheses relating to prehistoric human subsistence, climatic history, and rainforest disturbance and diversity.

The first sites selected for study were Laguna María Aguilar and Laguna Hule on the lower northern slope of Volcán Poás in the Cordillera Central. The lakes were cored during the summer of 1988, using a square-rod piston sampler operated from a raft made by lashing plywood to 2 rubber rafts. The 6 cores recovered constitute the first long (piston) sediment cores recovered from Costa Rican lakes. The environmental setting of the lakes, and the nature and age of the lake sediments, are briefly described here.

Laguna María Aguilar (770 m) is a small (<5 ha.) natural lake with a maximum depth of about 7 m. The origin of the lake is uncertain; it does not appear to occupy a volcanic crater, but it may be dammed by a debris flow or lahar (Gerardo Soto, written comm. 1990). The lake is surrounded by cattle pastures and remnant areas of tropical wet forest (premontane transition, as mapped by Tosi, 1969). Three cores were recovered, of which the longest is 7.4 m. The base of the deposits consists of wood (tree trunks or branches) and gravel.

Laguna Hule (720 m) is the largest and deepest lake in a series of 3 lakes within an explosion crater (Alvarado, 1989; Soto and Alvarado, 1990). The lake has a surface area of about 55 ha. and a maximum depth of 26.5 m (Gerardo Umaña, written comm., 1990). The steep interior walls of the crater are covered with tropical wet forest (premontane transition) vegetation. From this lake we also recovered 3 cores, the longest of which is ca. 4 m. All cores reached the volcanic bedrock that underlies the Hule lake sediments.

The piston cores from Laguna María Aguilar and Laguna Hule were extruded in the field, wrapped in saran wrap and foil, and packaged in PVC tubes for return to the United States. The sediments consist of dark, organic-rich lake muds with interbedded volcanic ash layers. Preliminary loss-on-ignition values indicate that the sediments contain 30-60 percent organic matter, calculated on a dry weight basis. The inorganic fraction is especially rich in diatoms.

Sediment and wood samples from the cores were submitted to the University of Arizona Isotope Laboratory and Beta Analytic, Inc., for radiocarbon determination. The most reliable date, on wood at the base of the 7.4 m Laguna María Aguilar core, is 4000 yr B.P. (all dates reported are uncalibrated). Bulk sediment samples from the bases of 2 of the Laguna Hule cores yielded radiocarbon age estimates of 7580 yr B.P. and 6910 yr B.P. However, surficial sediments from this lake yielded an apparent age of 3230 yr B.P., suggesting contamination with old carbon. It appears that radiocarbon-free carbon dioxide of volcanic origin, or groundwater carbon with an apparent age, is entering the lake and affecting the isotopic composition of organisms that are later incorporated within the lake muds. Similar incidences of anomalously old radiocarbon dates from volcanic areas have been described by Olsson (1986) and Byrne and Horn (1989).

Subtracting the surface discrepancy from the youngest basal radiocarbon date suggests that the Hule cores might span ca. 3680 years. "Correcting" the basal date in this fashion may not be justified, in that the dating error may have varied over time (Brubaker, 1975). Nevertheless, the "corrected" basal date does appear consistent with the results of a study of volcanic deposits and soils in road cuts near Laguna Hule (Melson *et al.*, 1988). A carbonized tree branch or small tree found beneath volcanic tephra deposited during the eruption or eruptions that formed Laguna Hule yielded an uncalibrated age estimate of 5140 yr B.P, indicating that the lake can be no older than this. Based on the degree of soil formation on top of the Hule tephra, Melson *et al.* (1988) estimated that the lake was formed only about 3000 years old.

To improve the chronology of the Laguna Hule and Laguna María Aguilar cores, we plan to submit near-surface samples for Pb-210 analyses. The results should provide acceptable time control for the uppermost sediments, and in this way facilitate our comparison of recent pollen spectra and recent forest history. The Pb-210 dates can then be extrapolated downcore to provide an approximate chronology for the lower sediments.

It may be possible to further refine the chronology of the cores through analysis of volcanic ash samples (Einarsson, 1986) and through AMS dating of pollen residues (Brown *et al.*, 1989). By dating just the pollen, rather than also the sediments that preserve the pollen, it may be possible to eliminate or minimize the error caused by the introduction of isotopically inactive carbon (Stephen Hall, written comm. 1990).

Several dozen samples from the Laguna Hule and Laguna María Aguilar cores have been processed for pollen analysis, but counts are still in progress. Pollen preservation in excellent, and samples from both cores contain sufficient pollen for analysis. The pollen residues also contain charcoal particles. Some of the pollen types in the surficial sediments represent long-distance transport from montane forests (*Podocarpus, Quercus, Alnus, Ulmus*), but most of the pollen grains are from taxa characteristic of lowland forests (*Alchornea, Urticales, Piper, Bombacaceae, Palmae*). This contrasts with the situation in the Mayan region of northern Central America, where pollen spectra from lowland lakes are often dominated by montane taxa (Tsukada and Deevey, 1967).

To improve our basis for interpreting the pollen, charcoal, and diatom records from Laguna María Aguilar, Laguna Hule, and other sites in the Atlantic lowlands, we are collecting and analyzing surface sediments and plankton and water samples from a series of natural and artificial lakes in Costa Rica. Our goal is to elucidate relationships between modern microfossil distributions and the geography, water chemistry, vegetation, climate, and disturbance history of lakes and surrounding areas. This information will allow us to reconstruct in greater detail the environmental history of the central Atlantic lowlands. The need for contemporary microfossil data is especially great in the case of diatoms, which have yet to receive systematic research attention in Costa Rica.

## Conclusion

The tropical forests of Costa Rica are among the best known examples of such forests in the world. An extensive literature now exists on the ecology of these forests, but their longer-term history remains obscure. What was the nature of precontact vegetation? How did prehistoric and early historic human activity affect Costa Rican forests? How did plant communities respond to climatic change during the Pleistocene and Holocene? Answers to these and other questions are beginning to emerge though the study of pollen grains and other microfossils in terrestrial and marine sediments.

## Acknowledgments

The recovery and initial dating of the Laguna Hule and Laguna María Aguilar sediments were supported by a Professional Development Award and other funds from the University of Tennessee. For field and logistical assistance in Costa Rica I thank Gilbert Vargas, Carolyn Hall, Douglas Baird, Jr., Isabel Avendaño, Jairo Mora, Olger Duran, and Carlos Zuñiga. Figure 1 was drafted by Thomas Wallin under the direction of William Fontanez in the Cartographic Services Laboratory of the University of Tennessee.

## Notes

1. See Graham (1987) and references therein for information on Tertiary paleoecology in Costa Rica and neighboring areas.

2. Available historical accounts suggest that any such wholesale conversion of forest to páramo must have predated the 20th century. The construction of the Inter-American highway during the 1940s greatly altered oak forests *below* timberline, but did not lead to a significant expansion of páramo vegetation on the high Buenavista peaks (Horn, 1989b).

# References

Alvarado I., G.E. (1989). *Los Volcanes de Costa Rica*. EUNED, San José, Costa Rica.

Barrantes, M. (1965). *Las Sabanas en el Sureste del País*. Instituto Geográfico Nacional, San José, Costa Rica.

Berglund, B.E., Ed. (1986). *Handbook of Holocene Palaeoecology and Palaeohydrology*. Wiley, Chichester.

Binford, M.W. and Leyden, B. (1987). Ecosystems, paleoecology and human disturbance in subtropical and tropical America. *Quaternary Science Reviews* 6, 115-128.

Bradbury, J.P. (1982). Holocene chronostratigraphy of Mexico and Central America. *Striae* 16, 46-48.

Brown, T.A., Nelson, D.E., Mathewes, R. W., Vogel, J. S., and Southon, J. R. (1989). Radiocarbon dating of pollen by accelerator mass spectrometry. *Quaternary Research* 32, 205-212.

Brubaker, L.B. (1975). Postglacial forest patterns associated with till and outwash in northcentral upper Michigan. *Quaternary Research* 5, 499-527.

Burger, W. (1977). Fagaceae. *Fieldiana Botany*, n.s., 40, 59-82.

Byrne, R., and Horn, S. (1989). Prehistoric agriculture and forest clearance in the Sierra de los Tuxtlas, Veracruz, Mexico. *Palynology* 13, 181-193.

Clark, L.G. (1989). Systematics of Chusquea Section Swallenochloa, Section Verticillatae, Section Serpentes, and Section Longifoliae (Poaceae-Bambusoideae). *Systematic Botany Monographs*, 27, 1-127.

Clary, K.H. (1986). An analysis of pollen from core 2, from the Nacascolo archaeological area, on the Nicoya Peninsula, northwestern Costa Rica. Castetter Laboratory for Ethnobotanical Studies, Technical Series #175. Department of Biology, University of New Mexico, Albuquerque.

Clary, K.H. (1990). Pollen evidence for prehistoric environment and human subsistence, the Arenal project. Submitted for inclusion in the final report on the Arenal prehistory project, edited by P.D. Sheets.

Cleef, A.M., Hooghiemstra, H., Noldus, G., and Kappelle, M. (1990). Historia del clima y la vegetación del último glaciar holoceno de la turbera la Chonta (2310 m alt.), Cordillera de Talamanca, Costa Rica (abstract). Quinto Congreso Latinoamericano de Botánica, Havana, Cuba, June 1990.

Coen, E. (1983). Climate. *In Costa Rican Natural History* (D. Janzen, Ed.), pp. 35-46. University of Chicago Press, Chicago.

Colinvaux, P. (1987). Amazon diversity in light of the paleoecological record. *Quaternary Science Reviews* 6, 93-114.

Deevey, E.S. (1978). Holocene forests and Maya disturbance near Quexil Lake, Petén, Guatemala. *Polskie Archiwum Hydrobiologii* 25, 117-129.

Delcourt, H.R. (1987). The impact of prehistoric agriculture and land occupation on natural vegetation. *Trends in Ecology and Evolution* 2, 39-44.

Einarsson, T. (1986). Tephrochronology. *In Handbook of Holocene Palaeoecology and Palaeohydrology.* Wiley, Chichester.

Flenley, J.R. (1979). *The Equatorial Rainforest: A Geological History.* Butterworths, London.

Gómez, L.D. (1986). *Vegetación de Costa Rica.* EUNED, San José, Costa Rica.

Goodland, R. (1971). Introduction to the principal vegetation types of Costa Rica. *In Handbook for Tropical Biology in Costa Rica* (C. Schnell, Comp.), n.p. Organization for Tropical Studies, San José, Costa Rica.

Graham, A. (1987). Miocene communities and paleoenvironments of southern Costa Rica. *American Journal of Botany* 74, 1501-1518.

Hamburg, S.P. and Sanford, R.L., Jr. (1986). Disturbance, *Homo sapiens*, and ecology. *Bulletin of the Ecological Society of America*, 67, 169-171.

Hartshorn, G. (1983). Plants: Introduction. *In Costa Rican Natural History* (D. Janzen, Ed.), pp. 118-157. University of Chicago Press, Chicago.

Herrera, W. (1985). *Clima de Costa Rica.* EUNED, San José, Costa Rica.

Herwitz, S.R. (1979). *The Regeneration of Selected Tropical Wet Forest Species in Corcovado National Park, Costa Rica.* M.A. thesis, University of California, Berkeley.

Holdridge, L.R. (1953). La vegetación de Costa Rica. *In Atlas Estadístico de Costa Rica* (E.W. Trejos and A. Archer, Eds.), pp. 32-33. Ministerio de Economía y Hacienda y Dirección General de Estadística y Censos, San José.

Holdridge, L.R., Grenke, W.C., Hatheway, W.H., Liang, T., and Tosi, J.A., Jr. (1971). *Forest Environments in Tropical Life Zones: A Pilot Study.* Pergamon Press, Oxford.

Horn, S.P. (1985a). Preliminary pollen analysis of Quaternary sediments from Deep Sea Drilling Project Site 565, western Costa Rica. *Initial Reports of the Deep Sea Drilling Project* 84, 533-547.

Horn, S.P. (1985b). Estudio palinológico preliminar de dos núcleos cortos del Golfo Dulce, Costa Rica. *Anales de la Escuela Nacional de Ciencias Biológicas* 29, 57-70.

Horn, S.P. (1989a). Postfire vegetation development in the Costa Rican páramos. *Madroño* 36, 93-114.

Horn, S.P. (1989b). The Inter-American highway and human disturbance of páramo vegetation in Costa Rica. *Yearbook of the Conference of Latin Americanist Geographers* 15, 13-22.

Horn, S.P. (1989c). Prehistoric fires in the Chirripó highlands of Costa Rica: Sedimentary charcoal evidence. *Revista de Biología Tropical* 37, 139-148.

Horn, S.P. (1990). Timing of deglaciation in the Cordillera de Talamanca, Costa Rica. *Climate Research* 1, 81-83.

Janzen, D.H. (1973). Sweep samples of tropical foliage insects: description of study sites, with data on species abundances and size distributions. *Ecology* 54, 659-678.

Janzen, D.H., Ed. (1983a). *Costa Rican Natural History*. University of Chicago Press, Chicago.

Janzen, D.H. (1983b). *Swallenochloa subtessellata* (Chusquea, Batamba, Matamba). *In Costa Rican Natural History* (D. Janzen, Ed.), pp. 330-331. University of Chicago Press, Chicago.

Janzen, D.H. (1986). *Guanacaste National Park: Tropical Ecological and Cultural Restoration*. EUNED, San José, Costa Rica.

Kesel, R.H. (1983). Quaternary history of the Río General Valley, Costa Rica. *National Geographic Research Reports* 15, 339-358.

Livingstone, D.A. and van der Hammen, T. (1978). Palaeogeography and Palaeoclimatology. *In Tropical Forest Ecosystems: A State-of-Knowledge Report*. UNESCO, Paris.

Markgraf, V. 1989. Paleoclimates in Central and South America since 18,000 B.P. based on pollen and lake level records. *Quaternary Science Reviews* 8, 1-24.

Martin, P.S. (1960). Effect of Pleistocene climatic change on biotic zones near the equator. *Yearbook of the American Philosophical Society*, 1960, 265-267.

Martin, P.S. (1964). Paleoclimatology and a tropical pollen profile. *In Report of the VIth International Congress on Quaternary, Warsaw 1961*. Volume II, 319-323.

Melson, W.G., Sáenz R., R., Barquero H., J., and Fernández S., E. (1988). Edad relativa de las erupciones del Cerro Congo y Laguna Hule. *Boletín de Vulcanología* 19, 8-10.

Olsson, I.U. (1986). Radiometric dating. *In Handbook of Holocene Palaeoecology and Palaeohydrology* (B. Berglund, Ed.), pp. 273-312. Wiley, Chichester.

Organization for Tropical Studies. (1989). La Selva Biological Station: Information for Visitors (mimeo). Organization for Tropical Studies, San Pedro de Montes de Oca, Costa Rica.

Parsons, J.J. (1955). The Miskito pine savanna of Nicaragua and Honduras. *Annals of the Association of American Geographers* 45, 36-63.

Parsons, J.J. (1970). The "Africanization" of the new world tropical grasslands. *Tubingen Geographische Studien* 34, 141-153.

Quintanilla, I. (1990). "Occupaciones PreColumbinas en el Bosque Tropical Lluvioso: Evaluación Arqueológica de la Estación Biológica La Selva." Unpublished report on research conducted with the support of the Andrew Mellon Foundation and the Organization for Tropical Studies.

Roberts, N. (1989). *The Holocene: An Environmental History.* Basil Blackwell, Oxford.

Simberloff, D.S. (1983). Mangroves (Manglares, Mangroves). *In Costa Rican Natural History* (D. Janzen, Ed.), pp. 273-276. University of Chicago Press, Chicago.

Snarskis, M.J. (1981). The archaeology of Costa Rica. *In Between Continents/Between Seas: Precolumbian Art of Costa Rica* (S. Abel-Vidor, R. L. Bishop, W. Bray, E.K. Easby, L. Ferrero A., O. Fonseca Z., H. Gamboa P., L.D. Gómez P., M.M. Graham, F.W. Lange, M.J. Snarskis, L. van Zelst), pp. 15-84. Harry N. Abrams, New York.

Soto, G.J. and G.E. Alvarado I. (1990). Procesos hidrovolcánicos ejemplificados en volcanes de Costa Rica (abstract). *In* Programas y Resúmenes de Ponencias, VII Congreso Geológico de América Central, 19-23 1990, San José, Costa Rica.

Standley, P.C. (1937-38). Flora of Costa Rica. *Field Museum of Natural History Botany Series* 18, 1-1571.

Stone, D. (1977). *Pre-Columbian Man in Costa Rica.* Peabody Museum Press, Cambridge, Massachusetts.

Tosi, J.A. (1969). *República de Costa Rica: Mapa Ecológico.* 1:750,000. Centro Científico Tropical, San José.

Tsukada, M. and Deevey, E.S. (1967). Pollen analyses from 4 lakes in the southern Maya areas of Guatemala and El Salvador. *In Quaternary Paleoecology* (E.J. Cushing and H.E. Wright, Eds.), pp. 303-331. Yale University Press, Connecticut.

Exploitation of Natural Resources in Colonial Central America:
Indian and Spanish Approaches

Murdo J. MacLeod
University of Florida

The standard view of the relationships between humans and the natural environment in precolumbian Mesoamerica has been, in general, a romantic and implausible one. Humans lived in ecological balance with their *ambiente*, which they respected and revered. Intricate geographical complementarity, judicious mixes of foraging, hunting, and agriculture, plus land management which included soil and crop rotation, irrigation, and individual care for cultivated plants, made of Mesoamerica, and indeed other parts of Amerindia, a place as close to an ecological paradise as humans are likely to come.[1]

Into this Eden came the grasping, exploitative, ruthless European invaders, the early capitalists of the sixteenth to eighteenth centuries, accompanied by their devastating pandemics from Eurasia, and their flocks of large, destructive domesticated animals. Even the weeds which they brought with them, in the guts of their cattle and equines, were imperialistic, and drove out native grasses.[2]

When overdrawn, this picture, at least to a skeptical historian, seems to be a soap opera of "good guys and bad guys." It attributes blame to historical and epidemiological processes of which humans at the time had no understanding. Above all, without any apparent recognition of what these views mean theoretically such writers advocate extreme cultural determinism in the anthropological sense. They allow no place for historical facts, individual and group action, or demographic and economic trends. (This is not to deny the interrelationships between all these historical phenomena and culture, of course). What we do know is that many parts of Mesoamerica, notably the Mayan area before the European invasions, experienced their own Malthusian cycles, with recurrent eras of soil exhaustion, overpopulation, systemic crises and other consequent abuses of the natural environment.[3]

After all, if these states were not quite imperial in the modern sense, nevertheless Mesoamerica had its own centralizing states, and these imposed severe burdens on their subject peasantries, both in their lives (as soldiers or sacrificial victims) and in their agricultural and other products, all of which, among other things, led to intensive use, and occasional overuse, of the natural environment.[4]

Having put aside what may be considered as the debris of romantic cultural determinism, this essay can now turn to what may be a more balanced view of cultural and economic differences between Amerindian and what came to be called Spanish or *ladino* societies in post-conquest Central America. (The essay will discuss the first 2 centuries after the European invasions - the post-conquest sixteenth, the entire seventeenth, and the early eighteenth centuries).

Land use by Amerindians varied by region in precolumbian and colonial Central America. Some of these differences may be attributed to cultural factors but even more, perhaps, to demographics. There seems to have been a continuum between foraging and hunting, with little agriculture, such as was found in parts of Mosquitia and the east Caribbean coast, to intensive agriculture and little use of foraging found in a few monocultural spots such as Izalcos (Sonsonate).

Most of Central America of interest to the invading Spaniards, that is the mountain and Pacific zones, geographically about half of the isthmus, fell somewhere in the middle of this continuum and saw various mixtures of fixed intensive agriculture, dominated by the great agricultural trilogy of maize, beans, and squash, and commercial and individual hunting and foraging in what ladinos called the *monte*, that is land not used for agriculture and usually overgrown by first or second growth. Spaniards had little understanding of such mixed land use. Moreover, as the Amerindian population fell catastrophically under the attack of Eurasian diseases, and as shortages of basic staples such as maize and wheat drove up prices in the cities, Spaniards began to look more closely at land use, especially in areas where Amerindians seemed to have abandoned, or had abandoned, all apparent land use. As population and agriculture fell extensive grazing replaced intensive agriculture, domesticated or semi-feral Eurasian animals replaced humans, and meat - a product of less efficient soil use but requiring far less labor - gained on agricultural produce in the diets of *ladinos* and even, although to a lesser extent, of Amerindians. To the Spanish failure to understand blends of agriculture and foraging was added their impatience, which they considered to be entirely rational, to acquire grazing for cows, mules and horses, and their need for more land for agriculture to support Spanish cities and the increasing number of *ladinos* living on *chacras* or early haciendas in the countryside.

One finds this conflict of cultures and economic needs in court cases over land. In many typical *pleitos* even sympathetic judges simply failed to understand the economic and ecological bases of Indian village life in a period of decline. In such litigation a Spaniard would petition for land, claiming that no one lived on it or used it. Local *ladinos*, called in as witnesses, would corroborate these statements. The area was *despoblado* and had been that way for years. The petitioner would then add that exploitation of such unused stretches of land would increase the food supply, the population, and the *vida política y cristiana* of the colony.

Indian villagers would reply that the land in question had been theirs "since time immemorial." In former days, when the village had been larger, these lands had been cultivated on a rotating basis. Nowadays, although it appeared abandoned, the monte was of use to the *pueblo*, and supplied wood, thatch, wild fruits and vegetables, game, and dyes. If a Spanish farm were set up, even for a few grazing animals, the woods and brush would be burned or cut down, the game would disappear, and the villagers would have to pay to use the area. Spanish judges knew about village *ejidos* or *fondos legales*, the common lands owned by each community, but for them this did not include large stretches of overgrown and seemingly unused land, which lacked proper management. So the Indian monte was usurped, the land was cleared, and cattle took over.[5]

In Guatemala much of this battle was waged between the 1570s and the 1630s. Although there were individual cases of this kind throughout the colonial period, and indeed during the Liberal confiscations of the 19th century, by 1640 the Central American Amerindians had lost in principle.

Nevertheless, extra official Indian foraging continued to reappear, especially under extreme conditions. While the sixteenth and early seventeenth centuries, periods of falling populations, ironically rarely suffered from food shortages, the epidemics, droughts, or floods of the late seventeenth and especially the eighteenth century caused frequent famines. Amerindians reacted in various ways. Some stuck it out in the village and lived or died there. Others fled to nearby Spanish towns or haciendas, begging for food or work. These were perhaps the most acculturated. Still others took to foraging in what was left of the monte. The food items which they found there in times of famine have been called "starvation foods," but there is some evidence that they were more than that, and represented, in many cases, inherited knowledge of the various wild foods to be found. In some cases astonished and repelled Spanish officials made lists of what the foraging Indians were eating in times of dearth - a diet, incidentally, which the Spanish officials found repulsive - and these lists contain an amazing variety of gathered edibles. Thus, in a paradoxical way, hard times in Indian villages delayed the seemingly ineluctable process of "ladinoization," at least for some of the more traditional or marginal Amerindians.[6]

So far we have seen how Spanish misunderstandings of Amerindian utilization of natural resources, plus Spanish need for land in times of declining supplies, led to an erosion of Amerindian types of land use, which, in spite of the stubbornness of some in times of famine, led to a weakening of Amerindian cultures. We now turn to the whole process of extraction, which so dominated Spanish-Amerindian relations over much of colonial Central America.

Central America did have some years of colonial boom. Cacao, then indigo, became very dominant crops. Cacao in Izalcos in the sixteenth century, and indigo in San Salvador in the eighteenth became valuable export crops to such an extent that Spaniards took over much of the land and means of production. In these areas Amerindian agriculture became marginal or non-existent.[7]

In most of Central America throughout much of the colonial period, however, there was lack of the great export products which Europeans sought and consumed. The conquering elites and their descendants lived by curiously similar economic means to the elites who preceded them in precolumbian days. The dominant mode of extraction was what Marx and others have called "the tributary mode of production." That is, the rural population was left to produce without great interference in the work place, and its surpluses were then removed via tribute or other forms of extortion.

Before Europeans arrived, however, Amerindian ruling houses and noble elites looked to their peasant producers for almost all their needs. The peasantry supplied not only maize, salt, and cacao, but also the items of rarity and prestige which differentiated the elites from ordinary people, and thus set them apart by providing the visible signs of their superiority. Thus Amerindians tribute before the Spanish invasion of Central America, included such items as jade, obsidian, jaguar skins, and quetzal feathers.[8]

At first the Spaniards changed the tributary system very little, but they quickly realized that they had very little use for feathers or obsidian. The basic problem was a perennial one in colonies. Valuable raw materials were exported to the mother country, and finished items of prestige could not be of colonial or local origin - colonial products were *ipso facto* less prestigious and were thus not worth having for display or conspicuous consumption. Colonists derive their prestige, in part, by consuming and displaying rare and expensive goods from the mother country. Because the best silks and cottons, the best wines and brandies, in short, the luxuries consumed by the Central American colonial elites, all came from outside, it followed that these elites did not need rare and expensive items, except gold and silver of course, from their tributary populations. What was needed from them was staples such as maize, beans, and wheat, and basic materials such as wood for construction and charcoal, or hay for horses. Gradually, then, Spaniards simplified production. They did introduce such items as wheat, animal hides, and wool, and by the seventeenth century Amerindians in some places were paying their head taxes in such items, but in general, as the sixteenth century advanced,

Spanish officials and *encomenderos* simplified Indian tribute. Nearly all villages paid at least some of their tribute in maize, wheat, beans, cotton cloth, and some silver money.[9]

This, of course, had an impact of village labor and production. To begin with many villages were simply ordered to put aside a certain amount of common land, often little more than a large plot, to grow the new products, usually wheat, to which they were not accustomed, so that they would be able to pay their tribute.[10] Some areas around the major Spanish cities, and especially near Santiago de Guatemala, paid most of their tributes in wheat or flour by the late seventeenth century. Even those who paid little or nothing in wheat felt the levelling effect of the standardized simplified tribute system. Indian society, much more than before the conquest, became monocultural, more dependent that ever on the great trilogy of maize, beans and squash.[11]

There are numerous examples of this. Tegucigalpa, now the capital of Honduras, but for most of the colonial period a small, tumultuous silver mining town or *villa*, went through a series of seventeenth-century shortages and look drastic measures to deal with them. Two of the most common were to prohibit all trade in grains or livestock until the needs of the villa were satisfied, and to requisition maize from Indian villages at a fixed price, often a poor price well below open market levels.[12]

The village of Lepaterique, today not far from Tegucigalpa, in the early seventeenth century had been a pueblo of general self-sufficiency with small surpluses in various goods which it paid or traded to Tegucigalpa and to small village fairs in highland Honduras. By May of 1712, when yet another severe shortage hit the villa of Tegucigalpa, things had changed. The *alcalde mayor* of the villa requisitioned 185 *fanegas* (bushels) of maize at 1 1/4 *medios* per *real*, a low price for a time of shortage. The villagers employed a brilliant defensive strategy, an ingenious series of maneuvers which provide an excellent illustration of James Scott's *Weapons of the Weak*, and ended up by selling most of their maize for a fair price on the open market. Three of the village leaders had to spend a few days in the Tegucigalpa city jail, not a pleasant experience, then or now. They also suffered the indignity of taking all the blame and of apologizing abjectly in public in front of the furious hoodwinked alcalde mayor. The point here, however, is that Lepaterique's economy had totally changed, both in fact and in the minds of the villagers. It was now, by its own recognition, a supplier of maize and nothing else to the villa. The village leaders were able to get out of jail by pleading that if they were not quickly released, the plantings of next year's maize crop would be endangered and then where would Tegucigalpa be?[13]

The tribute had another major effect, which reinforced some of the tendencies just mentioned. Almost always, by the early seventeenth century, it forced nearly every Indian village to pay some of its tribute in coinage. This was very finely tuned in most cases, after some early tinkering which had produced unwanted results for Spanish society, especially its merchants.[14] The tribute was designed to monetize the Indian economy, but by keeping most of the payments in maize and a few other staples, it did not monetize it too much. The end result, as had been designed by the authorities, was a society where most Amerindians stayed in their villages as basic subsistence agriculturalists and did not become full-time petty traders, potential rivals to Spanish and *casta* merchants. Nevertheless they did have to enter the labor or trading markets at least briefly to earn their tribute money. Often this meant seasonal labor, as officialdom had intended again. As early as the middle of the sixteenth century Amerindians from far-away Verapaz were working seasonally in the cacao groves of Izalcos just to earn their tribute. Others from the highlands of Chiapas and Huehuetenango went to Soconusco throughout the sixteenth and seventeenth centuries for the same reason.[15]

Many Indians dreaded the Spanish labor market however. *Encomienda* labor had been murderous in its early days, and the *tandas*, or rotational drafts of labor for town or farm work were onerous, and defrauded many of the draftees.[16] There had been no great tradition of free middle distance trade before the arrival of the Spaniards. Trade had been very carefully supervised, organized by the tribute dominated needs of the elites so much that it was hardly trade at all. Only the great towns had large fairs. The Spanish monetized tribute changed all of this. After the population began to recover in the late seventeenth and early eighteenth centuries, Amerindians in Central America took to the trails, and a whole series of regional fairs, often disguised as saints' *fiestas*, sprang up in the highlands of Chiapas and Guatemala.

Some of these exchanges were by barter or in substitute coinages such as cacao. Most, however, involved petty cash transactions, what Sol Tax was later to call "Penny Capitalism." What quickly emerged was regional specialization. Individual villages or groups of villages no longer produced the full panoply of crops and artifacts which would more or less supply their needs. In addition to maize and beans they produced 1 or 2 local specialties, charcoal, tiles, baskets, salt, and fish on the coast, cotton, cloth, and leather straps were some of them, and traded them at the regional level to earn cash for tribute payments and other purposes.[17]

Again, the effect was to drastically reduce variety, especially in agricultural goods, and to produce a peasantry which provided a few staples and a very few basic regional specialties. Once more, what we see disappearing in intricacy, tiny, multi-faceted adjustments to and exploitation of micro-regional ecological niches and advantages.

Spaniards and Indians brought different materials as well as cultures to their exploitation of the natural world. Large domesticated animals, new cultigens, and new implements such as the plough were material factors which had deep influences, whatever the dominant cultural attitudes. Leaving these aside, however, what can be said about attitudes?

There was a considerable difference between what can be termed Amerindian and Spanish cultural attitudes towards the natural world and how to exploit it. This is not to say that Central American Amerindians before the conquest all lived under the same ecological regime. Some inhabited areas of dense population, monoculture, and heavy strains on the environment. Others, especially those of the Caribbean lowland rainforests, relied heavily on hunting and foraging. But most of those in the areas the Spaniards occupied followed patterns of ecological exploitation somewhere in the middle of this continuum, and this invading and colonial Spaniards often failed to understand. What should be emphasized as a final point, however, is that the clash of cultures is only 1 variable among many when we examine the 2 different approaches to environmental use during the centuries under examination here. Of similar importance were materials and technology, demographic trends and cycles, the imperatives of the political situation created by colonialism, and the rise of new cities and their corresponding hinterlands.

## Notes

1. A recent extreme version of this romantic view of precontact America can be found in Kirkpatrick Sale, *The Conquest of Paradise: Christopher Columbus and the Columbian Legacy.* (New York: Alfred A. Knopf, 1990). A less exaggerated but still idealistic view of a part of pre-contact Central America is that of Dan Stanislawski, *The Transformation of Nicaragua, 1519-1548.* (Berkeley: University of California Press, 1983).

2. The bibliography on these matters is large. Much more dispassionate than the 2 authors above, is Alfred W. Crosby, *Ecological Imperialism: The Biological Expansion of Europe, 900-1900.* (Cambridge: Cambridge University Press, 1986).

3. See, for example, the concluding essay by Gordon R. Willey, "Pre-Hispanic Maya Agriculture: A Contemporary Summation," in *Pre-Hispanic Maya Agriculture.* Peter D. Harrison and B.L. Turner II, eds. (Albuquerque: University of New Mexico Press, 1978), pp. 325-373; and

Linda Schele and David Freidel, *A Forest of Kings: The Untold Story of the Ancient Maya.* (New York: William Morrow and Company, Inc., 1990) pp. 343, 345, 379-80, and the materials in the notes, pp. 433, 488, and 489.

4. See note 3, and for central Mexico, Sherburne F. Cook, *Soil Erosion and Population in Central Mexico.* (Berkeley: University of California Press, 1949).

5. E.g., Archivo Nacional de Centroamérica (ANC) Al. Tierras, 51832, f. 7 (1579); ANC, Al. Tierras, 51867, 5932 (1590); ANC, Al. Tierras, 51961, 5939, ff. 9-11v. (1627).

6. See the documents in note 5. For continued foraging in other parts of Mesoamerica see the illuminating document in Archivo de Instrumentos Públicos (Guadalajara)39-8-513 (1727). Foraging today distinguishes even remnant Indian groups from surrounding ladinos. See Jeffrey L. Gould, *To Lead as Equals: Rural Protest and Political Consciousness in Chinandega, Nicaragua, 1912-1927.* (Chapel Hill: University of North Carolina Press, 1990), pp. 98 and 234.

7. Murdo J. MacLeod, *Spanish Central America, A Socioeconomic History, 1520-1720.* (Berkeley: University of California Press, 1973), especially chapters 5 and 10.

8. See, for example, N. Molins Fábrega, *El Códice Mendocino y la economía de Tenochtitlan.* (Mexico: Libro-Mex. Editores, 1956).

9. Little has been done on the study of colonial tribute in Central America. The best study for Mesoamerica is, José Miranda, *El Tributo Indígena en la Nueva España durante el Siglo XVI.* (Mexico: El Colegio de México, 1952). The simplified tribute may be seen in any of the many tribute lists from ANC or the other Central American archives.

10. John C. Super, *Food, Conquest, and Colonization in Sixteenth-Century Spanish America.* (Albuquerque: University of New Mexico Press, 1988), especially pp. 32-37.

11. For a general statement of this proposition see Miranda, cited above. See also, for Santiago de Guatemala, Archivo General de Indias, Audiencia de Guatemala 24, Francisco de Escobedo to Crown, April 6, 1695.

12. Murdo J. MacLeod, "Indian Family Size in Seventeenth-Century Honduras; Some Implications for Colonial Demographic History," *Estudios del Reino de Guatemala: Homenaje al profesor S.D. Markman.* Duncan Kinkead, ed. (Seville: Escuela de Estudios Hispano-Americanos, 1985), pp. 101-115.

13. Archivo Nacional de Hondural (ANH), caja 27, no. 904 (1712).

14. Miranda, *El tributo indígena...*demonstrates this point.

15. Fr. Francisco, Prior de Viana, Fr. Lucas Gallego, and Fr. Guillermo Cadena, "Relación de la Provincia: tierra de la Vera Paz....desde el año de 1544 hasta éste de 1574," *Guatemala Indígena.* 2 (1962), pp. 141-60; Juan de Pineda, "Decripción de la Provincia de Guatemala, Año de 1594," *Anales del Museo Nacional "David J. Guzmán."* 4 (1952), pp. 46-69.

16. Benno Biermann, "Don Fray Juan Ramírez de Arellano, O.P., und sein Kampf gegen die Unterdrückung der Indaner," *Jahrbuch für Geschichte von Staat, Wirtschaft und Gesellschaft Lateinamerikas.* 4 (1967), pp. 318-47.

17. AGI/AG 140, "Relación y forma...de Bisitar," ff. f-7; Viana *et al.*, Relación..., p. 147.

Environmental History of La Selva Biological Station:
How Colonization and Deforestation of Sarapiquí Canton, Costa Rica,
Have Altered the Ecological Context of the Station

Susan M. Pierce
Colorado State University

Much information about ecological systems is derived from research at biological stations. Conclusions and ideas are often extrapolated to other regions, yet few stations are seldom perceived in the context of their environmental history. What research conclusions are skewed and which present land conditions might be misunderstood due to this oversight? The history of the physical condition of the encompassing landscape can lead to a better understanding of the research site, and in addition, the sociological and political framework from which past and present land use practices are derived can lend insight.

Historical land disturbances are not always apparent to a casual observer due to vegetational succession and other processes of time. A superficial evaluation of the site is often inadequate and it is always useful to probe for evidence of past disturbances. Research methods vary and may entail surveying for charcoal and/or cultural artifacts or examining local knowledge via interviews. Since the demand for land is so prevalent in many countries, especially those considered "developing countries", research stations are often islands of native vegetation surrounded by highly manipulated terrain. In these cases, not only the effects of ecosystem fragmentation must be considered, but in addition, the feasibility of the continuing existence of the reserve within such a context must be evaluated. I have studied the environmental history of La Selva Biological Station, Costa Rica and to a lesser extent the Sarapiquí region in which it is located. (Sarapiquí is a canton or political unit similar to a county). "La Selva" is located in northeastern Costa Rica, approximately 90 km north of San José, Costa Rica. The closest town, Puerto Viejo, is 4 km north of the station. La Selva receives scientists from all over the world whose research contributes significantly to our understanding of tropical ecosystems.

Much of the information about the region was acquired from interviews with researchers, landowners, workers, pertinent agency personnel, and longtime residents of the Puerto Viejo vicinity. The data obtained by personal interviews were verified whenever possible by literature review and study of public records. This region has a particularly dynamic land use and forest history which has certainly influenced conditions at the station itself. In this paper I explore the colonization of the Sarapiquí region, the past and present land use trends and resulting deforestation. Subsequently I focus on an analysis of environmental history of the land that comprises La Selva Biological Station.

## Deforestation in Costa Rica

Originally, only 3.5 percent of the country of Costa Rica was non-forested (Sylvander 1981). The area deforested before the arrival of the Spaniards is estimated at 2 percent (Tosi 1974). Deforestation rates fluctuated during the early colonial period (1500-1750) and it was not until after 1750 that a sharp increase in deforestation occurred. Much of this can be explained by population dynamics and colonial agricultural practices that were inappropriate for the climate (Hall 1985). Deforestation rates throughout history vary according to sources, and perhaps understandably so. Almost no written record exists of early forest types and even after 1940, when aerial photographs became more available, the coverage was not complete and/or frequent enough for precise calculations. Sader and Joyce (1988) estimate the total primary forest remaining in 1940, 1950, 1961, 1977, and 1983 as 67, 56, 45, 32 and 17 percent respectively. In 1976, Budowski (1976) estimated the country's deforestation rate as 50,000 ha per year. By 1981, Sylvander claimed that 70,000 ha per year were being cut, with evidence of a slight decline. The Tropical Science Center (1982) reported that half of the country's deforestation has occurred since 1950.

## The History of Sarapiquí Canton (Pre-Columbian-1945)

The canton of Sarapiquí is located to the northwest of San José, on the southeastern border of Nicaragua and to the west of the province of Limon (Figure 1). It is one of the largest cantons in Costa Rica, with an area of 2,349 $km^2$, representing 4.6 percent of the national territory (CODESA 1983). Sarapiquí was not recognized as a canton until 1970 and was previously considered part of the province of Heredia (Vargas E. 1974). The canton is divided into 3 districts: Puerto Viejo, La Virgen, and Horquetas. The majority of the canton qualifies as tropical moist and wet forest, although there is variation in the vegetation and climate within Sarapiquí (Vargas E. 1974).

The San Juan river, which forms the northern boundary of the canton, has had a great influence over the canton's pre-Columbian and colonial development because of its use as a trade route. The canton of Sarapiquí has been occupied for at least the last 3,000 years, with an increase in population between 300 B.C. and 300

Figure 1.  Sarapiqui Canton and Districts

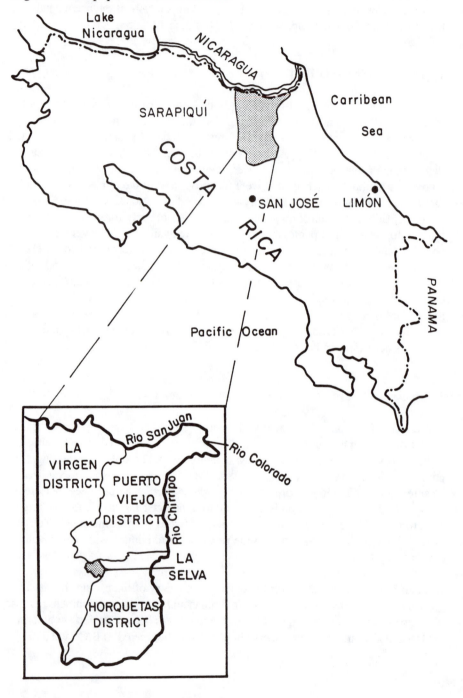

A.D. The pre-Columbian populations practiced agriculture and therefore had an influence on the forests since their arrival (Quintanilla 1990). Several pre-Columbian sites exist in Sarapiquí, including 1 large and complex site located 20 kilometers of La Selva, known as Cubujiqui (Quintanilla 1990).

During the 1700 and 1800s the San Juan River was the major northern entryway into the country for traders and explorers, many of whom were European. Most of the first settlers of Sarapiquí came to the town of Muelle, located just over 4 km north of La Selva Biological Station. Muelle was a significant port during this era and Puerto Viejo de Sarapiquí also received trader traffic (Fernández 1985).

Few people lived in Sarapiquí during the early 1900s. The region's population was reduced dramatically due to a reduction in trade traffic associated with construction of the Panama Canal (CODESA 1983). The first road in Sarapiquí was established in 1925 by Santiago Fernández Hidalgo in his quest to find a coffee exportation route. During this same time, Fernández designated the land between the rivers Sucio and Sarapiquí and the road to Carrillo, as "national wasteland", thus making it available for anyone who was willing to clear the forest (Vargas E. 1974). Bananas were one of the first export crops grown in Sarapiquí from the 1920s through the 1940s. Small farmers established plantations on alluvial soils and sold bananas periodically to companies whose cargo boats would stop at the towns along the Puerto Viejo and Sarapiquí rivers (pers. comm. Victor Paniagua). Timber was another early export. Reports of enormous Spanish cedar (*Cedrela odorata*) lining the river are prevalent among long time residents of Sarapiquí. Spanish cedar, however, was the first species to be targeted for lumber, and it was nearly exterminated by colonists who commuted via waterways (pers. comm. I. Alvarado).

**Events Influencing Sarapiquí's Development: 1940s-Present**

Incentives have been provided by the government of Costa Rica throughout its history (as in nearly every country's history) to "conquer the frontier". One of these incentives included the "Gracias" plan, by which people who supported the existing government in some form or other were rewarded for their efforts by receiving large tracts of land at the frontier. Much of Sarapiquí was parceled out in this manner to people that knew their land only on a map (Vargas E. 1974) In the 1940s, approximately 20,000 ha in Sarapiquí was granted to a group known as "Acción Nacional de Trabajo". This was the result of the government's search for an outlet for the landless people from the capital city region. This land was subsequently divided into 100 ha plots and offered to those willing to clear the forest and plant crops (Butterfield 1988). Another major event that drew people to the Sarapiquí region was the creation of Cinchona, a quinine plantation near the town of Vara Blanca, during World War II. In order to establish Cinchona, the surrounding infrastructure was improved, including the construction of a road from the Central Valley (pers. comm. G. Hartshorn).

In 1952, an epidemic of yellow fever swept through Sarapiquí and killed many people who were far from medical assistance. The fever had its largest impact however on the monkey population. People who were in the region during this episode tell stories of dead monkeys falling from trees and of corpses congregated near water holes. This epidemic was so harsh in fact that monkeys were not seen in the area for 10 years afterwards, and even then it was a rare event (pers. comm. I. Alvarado and J. Mejias). Currently, the voices of Howler monkeys (*Alouatta palliata*), spider monkeys (*Ateles geoffroyi*), and white-faced capuchin (*Cebus capucinus*) are commonly heard in the region's forests.

A second wave of settlers arrived in Sarapiquí during the 1950s. This group consisted largely of people from the Central Valley (the capital city region) with above-average financial means. Most of these people came to establish large cattle ranches, and land in Sarapiquí was still a bargain. These "outsiders" came into the region and bought out the small farmers and then employed these same farmers to work the land. In order to create pastures, large expanses of forest were cut for the first time in Sarapiquí. Nevertheless, only a few of these large ranches were established in the 1950s. It was during the early 1960s that large landholders of the region experimented with another land use, mechanized rice. Much of the land that is relatively flat and near roads was deforested, tilled with heavy machinery, and planted in rice. However, mechanized rice in Sarapiquí lasted only from about 1960-1965. The brevity of this trend was due to climatic conditions, financial expenditure, and rice blast (*Pyricularia oryzae*), a disease that ruined the crops. These rice fields were subsequently converted to cattle pastures, and Sarapiquí began the "cattle boom" era. With a high demand for beef, principally from the United States, credit was made easily available for those wanting to participate in the cattle enterprise, and more forest was cut. Roads were created and improved which resulted in heavy migration, particularly during 1968-1973 when 52 percent of the population older than 5 arrived in Sarapiquí (CODESA 1983). Perhaps the greatest factor contributing to this migration was the establishment of a banana plantation by Standard Fruit Company in 1968, 20 km to the southwest of La Selva in the town of Río Frío (CODESA 1983).

**Current Land Use and Forest Cover in Sarapiquí**

Sarapiquí is an agricultural and beef producing canton with 79.5 percent of the active labor force involved in these activities, as compared to the national average of 37 percent (CODESA 1983). Banana production, centered around Rio Frio, accounts for most of the labor force in the canton (CODESA 1983). It is estimated that over 50 percent of Sarapiquí's territory is in large ranches, with absentee landowners, and is principally used for cattle production (Butterfield 1988). As of 1983, pastures and uncultivated land covered 51 percent of the canton's area, with permanent and temporary agriculture comprising 17 percent, while the remaining 32 percent was forests and swamps (CODESA 1983). The governmental branch in charge of land distribution, Instituto de Desarollo Agraria (IDA), controls 38 percent of the land area in Sarapiquí, where 50 percent of the canton's population resides. IDA farms are typically small parcels with depleted soil conditions that are used primarily for producing subsistence crops, and possibly 1 or 2 export crops (Butterfield 1988). Thus, La Selva Biological Station exists within a truly dynamic environmental context that, in less than 40 years has gone from densely forested and being accessible only by boat or horseback to less than 30 percent forested and one of Costa Rica's most rapidly developing cantons.

**History and Importance of La Selva Biological Station**

La Selva Biological Station (1513 ha) comprises Dr. Leslie Holdridge's original farm (613 ha) and 6 annexes that have been bought since Holdridge's purchase in 1952. These annexes all have extensive land use histories and include the Western Annex (631 ha), the Peje Annex (105 ha), the Las Vegas Annex (68 ha), and Annexes A and B. La Selva is currently owned by the Organization for Tropical Studies (Organización Para Estudios Tropicales) and has become one of the world's foremost tropical ecology research facilities (Figure 2).

The station is located in the Atlantic lowlands, about 4 km upstream from the town of Puerto Viejo de Sarapiquí, between the confluence of Rio Sarapiquí and Río Puerto Viejo (Sludd 1960). The elevation at La Selva is 37-150 m. with an average annual rainfall of 4000-4500 mm. and is classified as Tropical Wet Forest (Holdridge 1971). La Selva is surrounded on 3 sides by cattle pastures and agricultural lands. The station is connected on its southern border to Braulio Carrillo National Park by a land corridor. This corridor is considered to play an important role in the maintenance of La Selva's impressive biological diversity (Clark 1990).

Figure 2.  La Selva Biological Station

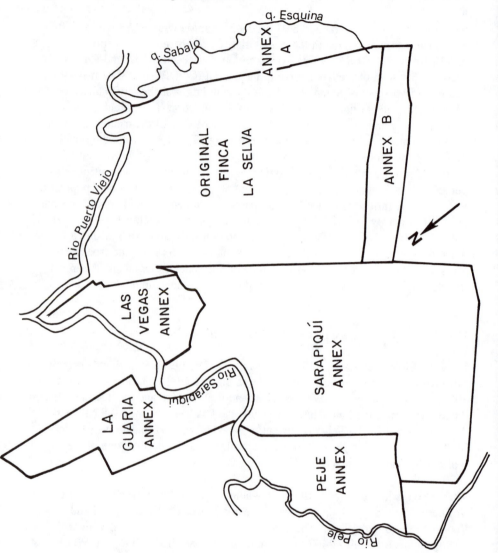

La Selva's flora exceeds 1,900 species, and moths alone number 5,000 species. *Pentraclethra macroloba*, or Gavilan, is considered the dominant tree, though there are 460 other species that are found within La Selva. This rich biological diversity and the well-equipped laboratory and housing facilities available at La Selva attract scientists from all over the world. Over the last 5 years the station usage has increased four-fold and accommodated researchers from 95 institutions (Clark 1990). La Selva Biological Station is central to tropical biology today due to the high proportion of tropical research conducted on this site.

**La Selva's Pre-Columbian History**: Even those forests that are today considered "virgin" or primary at La Selva have been manipulated by humans. Discovery of charcoal within the soils of La Selva initiated inquiries as to how much effect pre-Columbian people had on the surrounding forest (Sanford 1986). Quintanilla recently studied several pre-Columbian sites at La Selva, some dating back 2,000 years (Quintanilla 1990). Seven pre-Columbian sites have been located within the boundaries of the station's grounds. Quintanilla (1990) encountered charcoal deposits, along with large quantities of ceramic artifacts within sites and suggests that fire was used in food preparation and probably for land clearing practices.

**The Original "Finca La Selva"**: In 1950, Associación Compra de Tierra, an association consisting of 7 San José industrial agriculturists, bought a considerable amount of land in the Puerto Viejo region, including the land that comprises present day La Selva (pers. comm. C. Torres). Dr. Leslie Holdridge, a premier tropical dendrologist and one of the founders of the Tropical Science Center in San José, Costa Rica, worked with some of the "Associación's" members in the Ministry of Agriculture. When a few of the association's members resigned, Holdridge was invited to join the land investment group. Holdridge and his counterparts first arrived at the present station locality via horseback and dugout canoe. The land was mostly covered by intact forest, with 1 abandoned hut. Petriceks (1956) noted that much of the recent alluvial soils supported 18-20 year old secondary forest that grew on small abandoned banana plantations. These plantations were most likely abandoned due to a blight of *Fusarium* which destroyed the local banana industry in the 1930s (Hartshorn 1983).

In 1953, Holdridge purchased the 613 ha parcel and named it "Finca La Selva". His original intention was to manage the forest and experiment with tropical forestry and agriculture (pers. comm. Holdridge). However, after a survey conducted by Janis Petriceks in 1956, it was concluded that the farm had a low forestry potential, primarily due to its great species diversity and transportation difficulties. According to Dr. Holdridge, "In those days lumber was very cheap. They only cut Spanish cedar and laurel (*Cordia alliodora*)." Holdridge began to experiment immediately after purchasing La Selva with several kinds of tropical crops. Over a period of 4 to 5 years (1953-1958), he established plantations of cacao (*Theobroma cacao*), pejibaye, or peach palm, (*Bactris gasipaes*), and laurel on the alluvial terraces. Other experimental crops include rubber (*Hevea brasiliensis*), coffee *(Cafea robusta)*, and a Brazilian medicinal plant, *Ipicauana*. Fortunately, Dr. Holdridge and several of his colleagues were avid naturalists and quickly realized the rich biological diversity present within his property. Holdridge states, "there were people like (Paul) Sludd who were so enthusiastic, so that we gradually went into the business of having scientific visitors". Thus, the remaining forests stayed relatively untouched until in 1968, when the Organization for Tropical Studies bought "Finca La Selva".

Upon purchase of La Selva, OTS abandoned Holdridge's plantations which subsequently have undergone considerable vegetative succession. One of the cacao plantation areas has since been maintained as the "Holdridge Arboretum" that supports over 100 tree species (650 individuals) (Hartshorn 1983). Another area has been set aside as "Arboretum II", and a series of 4 ha plots are dispersed on the original property. A series of trails make visitor access possible. Aside from the abandoned plantations which cover less than 10 percent of the property, the original farm remains in primary forest (OTS 1987).

**Annex A:** In 1970, OTS purchased a section of a ranch on its eastern border known as "Hacienda Santiago". Much of this parcel, now called Annex A, was in pasture from the late 1940s to 1967. Santiago is thought to have acquired the land from Acción Nacional de Trabajo in the late 1940s. He cleared a small area for a hut near the Puerto Viejo river and raised a few animals (pers. comm. A. Benevides). Andrés Benevides Dobles, a lawyer from the Heredia vicinity, bought the Santiago farm in 1952 and owned it until 1969. At the time of his arrival, the land had a small pasture with some laurel trees left standing for a future source of wood. Benevides proceeded to create pastures throughout the 1950s and a few in the 1960s. Axes were used to clear the forests. Fire was used only when weather permitted in order to burn the stacks of forest debris that were otherwise left to decompose. Grasses planted include Para *(Panicum purpurascens)*, Guinea (*Panicum maximum*) and *Estrella africanus*. (pers. comm. A. Benevides). Benevides regularly ran about 750 head of cattle on his farm, which was approximately 1 head per manzana (1 manzana = ¯.7 ha) In 1970, Benevides sold the half (330 ha) of his farm which lies just outside of Annex A to a Cuban who resided in Miami, Florida.

The land that Benevides kept included Annex A. The pastures, now comprising part of Annex A were abandoned in approximately 1967, though Benevides continued to run cattle in other regions of his ranch (pers. comm. G. Hartshorn). OTS decided to purchase Annex A in 1970, primarily for 2 reasons; 1) La Selva needed a buffer zone between the primary and older second growth forests and the pastures of Hacienda Santiago; and 2) it was thought to be important to have natural boundaries (the Quebrada Sábalo-Esquina) (pers. comm. G. Hartshorn). The first of 5 half hectare strips, each cut 1 year apart, was cleared in 1971. These strips, known collectively as the "Successional Strips", are maintained via annual clearing of 1 plot to provide a scope of 0-5 year successional conditions. The remaining area of Annex A consists of secondary forest and swamp in the northern half and primary forest in the southern half.

**Annex B:**  Included in the "Hacienda Santiago" land was the area now known as Annex B. This annex is a 400 meter wide strip located on the southern border of La Selva and consists of primary forest. In 1973, concern grew over a logging road that ran along the Esquina creek to the edge of the primary forest on the present Annex B. After inquiry, it was discovered that plans were set to log the entire area. OTS responded quickly and bought the parcel in 1973, an action that many people feel made possible the future establishment of the corridor connecting La Selva to Braulio Carrillo National Park (pers. comm. G. Hartshorn).

**The Western Annex:**  In 1981, La Selva doubled its area with the purchase of a 600 ha ranch on its western border, now referred to as the Western Annex. The region to the north of Rio Sarapiquí in the Western Annex is called "La Guaria Annex" and that to the south, "Sarapiquí Annex". This property was first deeded to Acción Nacional de Trabajo in the 1940s, though no land manipulations occurred until the land was under its second owner, Juan Heison. Heison, who was a consulate from Columbia, bought the land in 1952 (pers. comm. E. Vargas). He began selectively logging Spanish cedar, of which huge individuals are said to have lined the riverbanks (pers. comm. S. Gonzolez and H. Morena). Logging of these large trees was accomplished with up to 4 men using axes, standing on a system of wooden supports that were placed around the tree (pers. comm. Cascantes). On the banks of the Sarapiquí, 2 ha of bananas, root crops such as yucca (*Manihot escalenta*), and malanga (*Colocasia*) were planted by workers for their own use. In 1956, Heison sold his farm to Domingo García Villalobo. García had the farm for only a short while (2 to 3 years) during which time he kept a small herd of cattle and continued to log the remaining Spanish cedar and other quality woods (pers. comm. Cascantes). The northern two-thirds of the La Guaria Annex were converted to pasture between 1956 and 1958, as was a portion of the Sarapiquí Annex. The southern one-third of La Guaria was planted in banana and yucca, most likely for worker consumption (E. Vargas).

In 1958, the Vargas family from Coronado, Heredia province, bought the farm and over the next 23 years made a series of land use changes. The Vargas' original intention was to run a cattle ranch on their new property, but initially they took advantage of the wood resources found on their land.  Though many of the largest and finest quality trees had already been taken out by the previous owners (especially in the La Guaria Annex), there remained a wide variety of trees suitable for timber.  During the first few years logging was focussed on La Guaria Annex, rather than the Sarapiqí Annex, because of access made easier by a road that traversed La Guaria. (pers. comm. E. Paniagua).  More subsistence crops were planted, some in areas that were already cleared, and others in areas where it was necessary to cut down the forest (pers. comm. E. Vargas).  The Vargas family were among the area's large landholders who participated in the mechanized rice era.

The northern two-thirds of the La Guaria Annex that was in pasture or "charrales" (bushy secondary growth devoid of trees) was plowed by a bulldozer and leveled. Memories differ on the role of fire in clearing areas for mechanized rice.  While 1 worker said fire was never used (G. Cascantes), another insisted that the fields were burned with a hot fire during the dry season (E. Paniagua).  Efren Vargas, 1 of the 5 sons that was in charge of running the ranch, claims that rice fields were burned only after harvest, and only when the weather permitted.  He further states that burning in general was not practiced as much as it is done currently in the region because" there was not as much summer as there is now" and the vegetation was often too wet to burn.  Two years (1960-1962) of cultivation with 4 harvests was the extent of the rice era on the Vargas ranch (pers. comm. E. Vargas).

An increase in cattle production followed the abandonment of mechanized rice and more forest was cut and pastures planted on the property.  The Vargas family received a loan from the "Banco de Costa Rica" to enable them to cut down 200 manzanas (142 ha) of primary forest in the northern section of the Sarapiquí Annex to create pasture.  Efren Vargas was frustrated with Costa Rica's forest management practices when, ironically, years later his request for a loan to help *reforest* the area with laurel was turned down (pers. comm. E. Vargas).  In addition to beef cattle there were also about 30 dairy cattle and a shed in La Guaria Annex where cheese was sold locally (pers. comm. Cascantes and E. Vargas).

The Vargas' designated the southern half of the Sarapiquí Annex as a forest-logging area shortly after they bought the land.  This decision was based upon a land use study that was conducted by neighboring La Selva (at that time owned by Dr. Holdridge).  In approximately 1960, this study indicated that the region had no agricultural potential, but did possess forestry potential.  A few years later, ITCO (Instituto de Tierra y Colonización, today IDA), accused the Vargas family of having idle land because they were only using it as a forest reserve and not cultivating the land.  This resulted in heavy taxation for the Vargas'.  By the 1970s problems arose from landless people entering and squatting in the forest reserve.

This time however, ITCO accepted the Vargas' argument that the forest was indeed being used for timber (pers. comm. E. Vargas). Efren Vargas claims that a part of his farm was put into pasture to counter invasions of squatters who would have entered had he not cut the forest.

After 23 years of ownership, the Vargas family sold their ranch to OTS in 1981. Since then the Western Annex has been used for several experiments requiring large-scale site manipulation and abandoned pasture conditions.

**Las Vegas Annex:** Located in the northeastern corner of La Selva where the Rio Puerto Viejo and Rio Sarapiquí unite is a piece of property known as the Las Vegas Annex. This land originally belonged to Acción Nacional de Trabajo and was later sold to 1 of Holdridge's employees, Guiermo Chaverria. Señor Chaverria planted bananas and harvested timber on the land (pers. comm. S. Gonzalez). By 1955, the land was purchased by an American named Robert Hunter, (co-founder of the Tropical Science Center) who claims Las Vegas was a mixture of bananas and secondary forest with some very large trees remaining far from the rivers (pers. comm. R. Hunter).

Plantations of cacao were established by Hunter with financing from Hershey Company. In addition, what was once the world's largest peach palm plantation, totaling 14 ha, was established at Las Vegas. With the assistance of Robert Baker, a former hog farmer from Kansas and currently a resident of San José, a hog operation was set up on the property, and there were 50 to 100 hogs at Las Vegas (pers. comm. R. Hunter and R. Baker). Root crops and pejibaye were planted for the hogs. Due to an infection of the cacao pod disease, *Monilia*, Hunter's cacao production dropped drastically, leading to the donation of the Las Vegas Annex from Hershey Company, to the Organization for Tropical Studies in 1986 (pers. comm. R. Hunter). OTS abandoned the plantations upon purchase and they are undergoing succession.

*Peje Annex:* The Peje Annex (105 ha) is located to the east of the Sarapiquí Annex. This land was originally owned by a small farmer who planted only subsistence crops and therefore cut very little of the forest (pers. comm. R. Orlich). The farm was sold in 1950 to Romano Orlich, a man from the Central Valley who had a great influence on both the mechanized rice trend and on Costa Rica's cattle enterprise. The Peje Annex made up a part of his expansive ranch but, due to difficult access, was not included in mechanized rice production, prevalent on other sections of his ranch.

Most of the annex is thought to have been cleared in the 1960s and early 70s. Only a few species of trees were logged from the area, including Gavilan (*Pentaclethra macroloba*), laurel, and the remaining Spanish cedars. The Peje Annex was burned 3 times; for initial clearing then twice to clear secondary growth (pers. comm. J. Brenes). Approximately 1,000 head of cattle were kept on the ranch, and two-week pasture rotations were practiced (pers. comm. R. Orlich). Part of Orlich's ranch, including the Peje Annex, was sold in 1972 to a United States citizen, who continued to raise cattle for 5 more years before he sold the ranch to Gerardo Loelder. Loelder kept 1,000 head of cattle on the land, of which 150 were dairy cattle, until he sold it to OTS in 1987 (pers. comm. Loelder). This is the last annex purchased to date by OTS and is now used by researchers studying aspects of forest regeneration on abandoned pastures.

## The Importance of Environmental History to Biological Research at La Selva

There are many ways in which the environmental history of Sarapiquí and of La Selva may affect present day research sites and thus the conclusions derived from research at the biological station. The environmental history and social context of each research site play an important role in determining actual site conditions. La Selva's primary forest is often considered as untouched, unaltered, "virgin", tropical wet forest and it from this premise that much research is conducted. As the studies of Sanford (1986) and Quintanilla (1990) indicate, this "untouched" state simply cannot be assumed for these forests.

Both soils and vegetation have been anthropologically altered throughout the region. Researchers need to be aware of the proximity of pre-Columbian sites to their research sites and the potential impact that these civilizations may have had on the area. Is there charcoal within the soil, or are there introduced species on their site? By considering these questions the researcher can obtain a better idea of just how "virgin" the primary forest is and therefore make a more accurate assessment of the actual state of the site.

The majority of site descriptions written for ecological research publications do not include historical factors (Hamburg and Sanford 1986) and more often than not this is due to the fact that the researchers themselves are not familiar with these factors. How then can site descriptions be accurate portrayals of the condition of the site?

Among the direct biological factors that historical events or land use practices can directly affect is species composition. In the Puerto Viejo region for example, logging which was targeted primarily on Spanish cedar has had an immense impact on the population of this species. According to historical accounts, enormous Spanish cedars once lined the river floodplains. Presently, this species is extremely rare in the region. The yellow fever epidemic in the 1950s is another example of how a historical event virtually decimated a local population (in this case howler monkeys, as well as large reductions in spider monkeys and white-faced capuchin). To study primates at La Selva and to make valid conclusions on present day primate conditions one must be aware of this critical past event. The reduction of large animals such as the jaguar and the white-lipped pecary and its effect upon the ecosystem need be considered as well.

Understanding the original source of species also requires historical knowledge. Plant species commonly considered native to the region may actually have been introduced by pre-Columbian people or early colonizers. Species introductions have had large repercussions in many situations (such as the introduction of rats and pigs to islands). Cattle are probably the source of the largest impact within this study area; the forest is cut to create pastures, and new species are attracted to this new habitat and to the cattle themselves. Species invading a cattle pasture, including vampire bats (*Desmodus rotundus*), cattle egrets (*Bubulcus ibis*), chiggers (*Eutrombicula*), various grass seeds and several species that are not directly connected to the cattle prefer edge conditions, such as the birds, Oropendolas (*Zarhynchus wagleri*).

Knowledge of when a small piece of forest was separated from a large tract is essential in understanding sources of seeds, species interactions, and succession within the remnant forest.

The duration and effect of pollutants that historically entered the research area must also be considered in current research. For example, the effect of the pesticides used on Rio Frio's banana plantations and other local agriculture should be of concern not only to stream ecologists but to all researchers at La Selva.

Soil conditions are also influenced by environmental history. The parent material, (whether it be volcanic, alluvial, etc.), is often considered, while more recent events are frequently ignored. When was the forest cut? Was fire applied for clearing purposes? ... how often? Were trees felled by heavy machinery or by axe? Was the debris left to decompose or was it burned? The answers to these questions are essential in order to understand site conditions. The method and duration of land use also greatly impacts soil conditions. Soil compaction by cattle or heavy machinery is common. The type of crop plays an important role. The nutrient status of areas which were under mechanized rice cultivation, a nutrient consumptive activity, compared poorly to those which were used to produce native root crops. Soil affects the entire structure of ecosystems, and no ecological research can by properly evaluated without this base knowledge.

Historical factors can help to discriminate between those species that are native to the area and those that have been introduced. This must be taken 1 step further by incorporating the possible effects of these species upon each other (eg. the possible impact of the introduced vampire bat, which is an edge species, upon the less aggressive native bats, or the influence of the Orolpendolas upon other canopy birds). The study of a species in isolation is not realistic, because it is only by considering the species within its environmental context that valid conclusions can be drawn.

If one does incorporate environmental history into their research, then more criteria is available to determine whether or not findings can be more readily extrapolated to other regions. This is particularly relevant in the case of La Selva Biological Station. La Selva is basically a small tract of forest surrounded by agricultural land, except on the southern border, where a land corridor connects it with a larger tract of forest (Braulio Carrillo National Park). Despite this, a great percentage of research on tropical wet forest is conducted at this station and the findings derived from La Selva are often extrapolated and said to represent tropical wet forests, or in the extreme case, "rain forests" in general. Can conclusions from a station such as La Selva (with its unique environmental history and context), truly be extrapolated to huge tracts of the Amazon Basin? These differences need to be emphasized when scientists (and politicians) extrapolate site-specific conclusions beyond the borders of the original study.

By looking at the dynamics of historical land use trends, a better prediction of the future can be made. What governmental policies were made and why, and do the same conditions exist today that brought about these policies? For instance, by following the development of land reform in Costa Rica, a better understanding of where this movement is heading may be formulated. Present and future land use trends must in turn be considered when developing station policies and interpreting research.

Research personnel need to know how to build a good rapport with the local community. By being aware of forest and land use history, the present day environmental and social context in which the research station is located, and future possibilities, steps can be taken to establish this rapport. This positive relationship with the local community is crucial to the research station's survival in a world where competition for land is becoming increasingly intense.

Figure 3. The Braulio Carrillo Complex

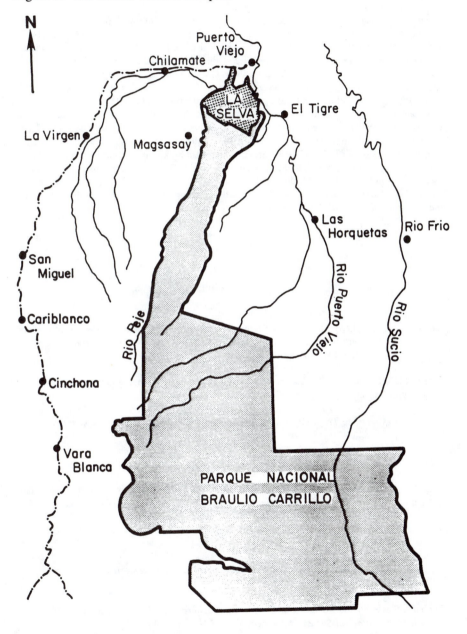

## Acknowledgments

I would like to thank my major advisor, Robert L. Sanford, who informed me of this research project, edited my work, secured funds and gave his encouragement. I am also greatly indebted to both Steven Hamburg for writing the proposal with assistance from Dr. Sanford and to the Andrew W. Mellon Foundation for funding this research. Other advisors and editors who helped me at Colorado State University include Kathern Galvin, George Wallace, and Greg Aplet. The diagrams were made presentable by Michelle Nelson. The research from which this article is based was a true community effort, therefore, my thanks go out to all of the people of Puerto Viejo, La Guaria, Chilamate, and Magsasay and to many individuals in the Central Valley, Costa Rica, all of whom received me warmly and shared memories which oftentimes contained previously unrecorded information. Many long-time La Selva researchers assisted me greatly. Gary Hartshorn, Efren Vargas, Rebecca Butterfield, and others provided pertinent background literature and shared it willingly. A special thanks goes to the local employees of La Selva, all of whom contributed to my research, and to Dr. Leslie Holdridge and his wife Lydia for the hours spent reminiscing with me. My family and friends have been a source of limitless support and encouragement throughout the research process, for which I am very grateful.

## References

Budowski, Gerardo 1976. "Poblacion y Recursos Naturales en Costa Rica: un caso de confusion con peligrosas consecuencias." In: Budowski, G. *La Conservación Como Instrumento Para el Desarrollo*, EUED, San José, Costa Rica.

Butterfield, Rebecca P. 1990. " La Selva in a Regional Context: Changing Land Use Patterns in Sarapiquí," manuscript in preparation.

CCT (Centro Cientifico Tropical) 1982. *Costa Rica, Perfil Ambiental Estudio de Campo*, CCT and USAID.

Clark, David, 1990. "NSF Proposal La Selva Station Improvement"

CODESA (Corporación Costarricense de Desarrollo), 1983. "Diagnostico Socio-Economico del Canton de Sarapiquí, Provincia de Heredia". Puerto Viejo de Sarapiquí, Costa Rica.

Fernández Guardia, Ricardo, 1985. *Costa Rica en el Siglo XIX: Antológia de Viajeros*, EDUCA, San José, Costa Rica.

Hall, Carolyn, 1985. *Costa Rica: A Geographical Interpretation in Historical Perspective*, Westview Press, Boulder, Colorado.

Hamburg, Steven P. and Robert L. Sanford, 1986. "Disturbance, Homo sapiens and Ecology". *Bulletin of Ecological Society of America*, Vol. 67 No. 2:169-171.

Hartshorn, G. 1983. "Plants" pp.136-140 in: Janzen, *Costa Rican Natural History*, The University of Chicago Press, Chicago.

Holdridge. L.; Grenke, W. C.; Hatheway, W. H.; Liang, T.; and Tosi, J.A., Jr. 1971. *Forest environments in tropical life zones: A pilot study.* Pergamon Press.

Keogh, R.M. 1984. "Changes in the Forest Cover of Costa Rica Through History", *Turrialba*, Vol. 34, No. 3:325-331.

OTS (Organization for Tropical Studies) 1987. "La Selva Biological Station" San Pedro, Costa Rica.

Petriceks, J. 1956. "Plan de ordenación del bosque de la Finca `La Selva'". M. Agr. Thesis, Instituto Interamericano de Ciencias Agrícolas, Turrialba, Costa Rica.

Quintanilla, Ifigenia J. 1990. "Occupaciones Precolumbinas en el Bosque Tropical LLuvioso; Evaluación Arqueologica de la Estación Biologica La Selva. Report.

Sader, Steven A. and Armond T. Joyce, 1988. "Deforestation Rates and Trends in Costa Rica, 1940 to 1983", *Biotropica* 20(1):11-19.

Sanford, Jr., R.L. 1986. "Evidence of fire as a recurrent rain forest disturbance at La Selva Biological Station, Costa Rica" IV International Congress of Ecology Bulletin: 298.

Sludd, Paul 1960. "The Birds of Finca "La Selva", Costa Rica: A Tropical Wet Forest Locality", *Bulletin of the American Museum of Natural History*, Vol. 121: article 2, New York.

Sylvander, Robert B., 1981. *Los Bosques del País y su Distribución por Provincias*, Universidad Estatal a Distancia, San José, Costa Rica.

Tosi, Joseph. A. Jr. 1974 "Los Recursos Forestales de Costa Rica". Primer Congreso Nacional sobre Conservación de Recursos Naturales Renovables. Universidad de Costa Rica, 1974. pp. 89-107.

Vargas Mendez, Efren, 1974. " Bases Para la Programación del Desarrollo Agropecuario del Canton de Sarapiquí. Herida, Costa Rica", Facultad de Agronomia, Universidad de Costa Rica.

Deforestation and Changing Land-Use Patterns in Costa Rica

Mary Pamela Lehmann
University of Florida

**Deforestation: An Economic Overview in Historical Context**

According to Hartshorn (1988), "Economic incentives and strong foreign influences over the national economy [have been] part of the complex factors behind the apparent irrational destruction of forests [in Costa Rica]." Before the 20th century, the majority of Costa Rica's deforestation was associated with agricultural settlement and took place in the Meseta Central (central valley), where homesteaders had originally settled and where the coffee boom was centered. Portions of Guanacaste province (in the country's northwest) and the Atlantic lowlands were also areas of original settlement and hence forest conversion, but not on the scale of the Meseta Central.

Deforestation accelerated in the 20th century when, according to Hall (1985), migration away from the Meseta Central began to increase markedly with the burgeoning population and the emergence of other export-agricultural activities alongside subsistence production. The Atlantic lowlands became dominated by the commercial production of bananas, and the Pacific northwest region (provinces of Alajuela and Guanacaste) was transformed mainly from forest to the production of subsistence crops and cattle ranching (Bonilla, 1985). The construction of the Pan-American Highway during World War II was a further cause of deforestation, as it provided access to otherwise isolated areas of the country, facilitating both settlement and marketing of goods (Williams, 1986). Thus deforestation before World War II was a function of both population and agricultural growth.

**Export Policy Change**

World War I and the Great Depression created a lull in the market demand for bananas, and the closing of European coffee markets during World War II caused the Costa Rican economy to suffer (Villarreal, 1983). Though the economy recovered significantly following the war, economic policy eventually shifted toward export diversification in agriculture and import substitution in industry by the 1950-60s. This change became manifest in the production of "non-traditional"

58

agricultural products (principally cotton, sugar, and beef) and the formation of the Central American Common Market (Bulmer-Thomas, 1987).

## Growth of the Cattle Industry

The cattle industry was a major thrust of Costa Rica's agro-export diversification beginning in the 1950s. In 1954, large scale exports of live cattle began, but were soon replaced by exports of chilled beef (Hall, 1985). The amount of beef exported, as well as export prices, rose substantially during the 1960s and 1970s, until beef became the third largest agro-export earner (IMF, 1989).

The Costa Rican government pumped federal and international credits into the cattle industry in order to support this new sector of economic development (according to Leonard (1987) nearly one-half of all agricultural credit went to cattle in the early 1970s). A 1974 Central Bank document entitled "Development and Incentive Project for the Costa Rican Beef Industry" attests to the fact that these credits "were important for the tremendous growth of the [beef] industry throughout the early 1970s."

Thus agro-export "diversification" was accomplished with beef alone during this period, and Costa Rican beef joined coffee and bananas as yet another traditional export in an export-led economy.

## Forest-to-Pasture Conversion

The development of the cattle industry had a notable effect on overall forest loss. It was responsible for much more deforestation in the pre-1980 period than all other economic activities including commercial logging, which at the time was not a strong economic force (Hall, 1985; Leonard, 1987; Ramirez & Maldonado, 1988). Cattle are the most land-extensive of Costa Rica's major export commodities, and by 1973 more than one-third of the country's 5,135,900 hectares was in pasture. Parsons (1976) refers to the transformation of forest to grassland from the 1960s to mid-70s as a "grassland revolution".

## Response to Forest-to-Pasture Conversion

The first signs of alarm at the rapidly growing conversion of tropical forests in Costa Rica appeared in the mid-1970s. During the Oduber Administration (1974-78) "a more systematic and rational strategy was adopted for the country's forests"

(Myers, 1980). These initiatives included financial support and physical expansion of the National Parks Service and important additions to the Forestry Law in 1973, all of which contrast markedly with the "cattle" policy under the same administration.

Oduber also assigned 5 new areas to the National Parks Service and incorporated 8 additional forest reserves. In addition, 6 existing protected zones were enlarged, and the first wildlife refuge was created. In all, the Oduber administration's contribution to the country's protected areas totaled 42.2 percent of all current parks and reserves (Rodriguez & Vargas, 1988).

## Agro-Export and Environmental Policies of the 1980s

Despite efforts to protect remaining forest resources during the mid-1970s, at the end of that decade, Costa Rica still had many symptoms of the jungleburger syndrome. Norman Myers sums up the Costa Rican case as follows:

> Costa Rica represents an extreme illustration of the "hamburger connection". In 1950, cattle-raising areas covered only one-eighth of the country: they now account for over one-third. In 1960, the country's cattle herds totalled slightly over 900,000 head: by 1978 they had surpassed 2 million. During the 1960s and 1970s, beef production more than tripled. But during the same period, local consumption of beef actually declined to a mere 12.6 kilograms per head per year. ...Almost all the extra output was dispatched to foreign markets. (Myers, 1981, 6)

But the decade of the 1980s was characterized by important policy transformations, changing the "hamburger connection" outlook.

## The Debt Crisis and Agro-Export Policy Reform

The second oil shock and world recession of 1979-82 threw Costa Rica into a severe foreign debt crisis. As a result, the country accrued one of the highest per capita foreign debts in the world (Umana, 1987). The debt crisis caused Costa Rica to embark on what Bulmer-Thomas (1988) refers to as the "new model of development".

The new model basically emphasized the export of non-traditional goods to the rest of the world, effected through price policy, fiscal policy, monetary policy, and institutional reform. Perhaps the most important of these was institutional reform, under which the Ministerio de Exportaciones (Ministry of Exports, or MINEX) was formed in 1983 to oversee non-traditional export promotion (Bulmer-Thomas, 1988). The Caribbean Basin Initiative (CBI) of the United States and the European Economic Community's (EEC) Cooperation Agreement with Central America were helpful in opening markets for the new Costa Rican exports.

In 1987, the Arias administration instituted an "Agriculture for Change" policy (Latin American Monitor, 1988), providing subsidies as an incentive for production of the new exports; these included ornamental plants, flowers, pineapples, and melons, to name a few. Concommitantly, the beef sector suffered. In 1983, cattle ranchers applied for a state of emergency status in order to receive government aid. The government responded in 1985, and monetary incentives totaling nearly US$100 million were supplied to farm debtors, including cattle ranchers. Despite the aid, the new non-traditional exports provided Costa Rica with half of its export earnings, filling the gap that was left by the decline in prices for coffee, bananas, and beef during periods of the 1980s (U.S. Department of Commerce, 1988).

**Debt and the Environment**

Sader and Joyce (1988) estimate that the rate of primary forest clearance in Costa Rica reached more than 7 percent annually by 1983 for primary forest. Concerns over Costa Rica's loss of forest ecosystems, coupled with the debt crisis, led to 2 new approaches to combatting both problems by the mid-1980s: these were "debt-for-nature" swaps and "eco-tourism".

Debt-for-nature swaps were first proposed in 1984 and allow a country to retire a portion of its foreign debt portfolio for local currency bonds, the interest of which is funneled into nature conservation-related projects. By 1987, the first debt-for-nature swap was arranged for Costa Rica under the Arias administration, and soon thereafter additional nature swaps were arranged. Costa Rica managed to retire US$40 million of its US$3.38 billion foreign debt in this manner by 1989 (Tripoli, 1989).

Eco-tourism is another new market which was given priority by the government in the last half of the 1980s to help spark economic recovery (Mendoza, 1988). Eco-tourism caters to tourists who are most interested in visiting wildland areas. This specialized market has been a means for Costa Rica to expand its tourist industry. Likewise, eco-tourism is dependent upon continued efforts to protect forest ecosystems through such mechanisms as debt-for-nature swaps (Coone, 1989).

Tourism replaced beef as the number 3 export earner as early as 1980, and has since consistently outweighed beef. The number of tourists increased by 20 percent between 1987 and 1988, "probably the sharpest increase in the world," according to the president of the Costa Rican Chamber of Tourism (Gugliotta, 1989). Estimates for the eco-tourist portion of the 1988 tourists range from 6 to 36 percent, depending upon one's eco-tourist definition (Budowski, 1989).

Finally, the case of Guanacaste National Park merits special attention. It is perhaps the first example of pasture-to-forest conversion, marking a reversal of the forest conversion trend. The park was established in the 1980s on lands that were purchased from former cattle ranchers, and scientists are attempting to biologically restore this deforested land to its previous life zone status as dry tropical forest (Allen, 1988). The park has a rather symbiotic relationship with nature swaps and eco-tourism, as it will benefit from the input of nature swaps and the interest generated by eco-tourism, while the swaps and tourism function as a result of such innovative parks as Guanacaste.

**The Jungleburger Re-Examined**

The jungleburger connection seems to have had an undeniable relation to pre-1980 conditions in Costa Rica. Changing events throughout the 1980s, as previously described, require a re-examination of the pervasiveness of the connection.

It must be noted that the cattle cycle is characterized by dynamic factors such as herd inventories, withholding of slaughter, and grain prices, which make it difficult to eliminate the jungleburger concept in the long run for Costa Rica, despite the trends of the 1980s. At present however, the rainforest - North American hamburger connection is questionable in a literal sense for 2 reasons:

- overall beef exports from Costa Rica declined in the 1980s as compared to the 1970s

- domestic consumption of beef in Costa Rica rose during the same period of comparison

Though this may put the original jungleburger theory into question for now, it does not rule out the possibility of a domestic jungleburger. Certainly the intensification of transportation networks, population growth and public demand have allowed for a higher consumption of beef within Costa Rica and may easily offset the decline in beef exports. Parsons foreshadows this outcome in his 1988 article "The Scourge of Cows", stating that "even if Central America were to end its [beef] exports, feeding the population growth would soon require all the land now used for maintaining livestock." And though Costa Rica's current population is only approximately 3 million, the possibility that domestic demand could meet current supplies can already be supported by data (see Simpson & Farris, 1982). Central Bank figures from 1985 showed the highest level of domestic beef consumption ever, at a time when beef production also reached a high, yet only 39 percent of the commodity was exported. By 1989, 76 percent of all beef produced was consumed internally, while the remaining 24 percent of total production was exported to foreign markets (Amador, 1990, personal communication).

Nevertheless, forest-to-pasture conversion as the primary indicator of Costa Rica's forest loss is questionable as applied to the past decade. It seems to counter Costa Rica's "agriculture for change" policy, launch of eco-tourism, debt-for-nature swaps, and precedent-setting pasture-to-forest conversion. In addition, expansion of other agricultural land uses, road transportation networks, and commercial logging have stepped in to "compete" with cattle ranching in the conversion of Costa Rican forests.

**Alternate Deforestation Explanations**

**Agriculture**

In a comparison of the 1973 and 1984 Agricultural Censuses of Costa Rica, percentage changes in different land uses on the national and province levels facilitate an evaluation of agricultural trends (see Figure 1). Land use categories include annual and permanent crops, pasture, primary and secondary forest (forested portions exploited for timber and non-timber products on an otherwise agricultural farm), and "other" (portions of agricultural farms occupied by buildings, roads, waterways, or any other land use not explained by the other land use categories).

At the national level, these changes reflect minimal growth for pastureland, with a 6 percent positive change during the period 1973-1984, whereas annual crops show the highest positive change at 36 percent, followed by permanent crops at 14 percent for the same period. The sharpest decreases in land-use were in primary and secondary forest, at 31 and 17 percent, respectively. These figures seem to indicate that conversion to pasture has not been the driving force behind deforestation in this 1973-1984 period, since pasture growth was minimal. However, they do indicate that the cultivation of annual crops has played an increasing role in forest conversion.

Figure 1.  Percentage Change of Land Use in Costa Rica, 1973-1984

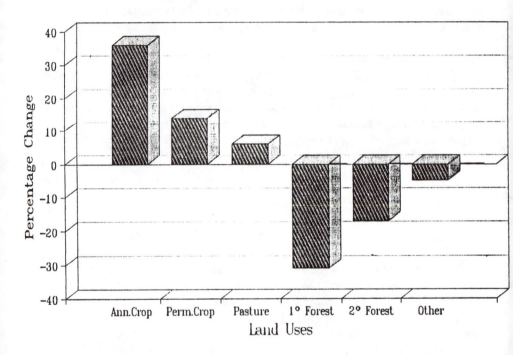

Source:  DGEC.  Compiled from the *1973 and 1984 Agricultural Census*.  San José:  1976, 1987.

A more profound look at land-use trends on the province level indicates that, except for the provinces of Heredia and Cartago, there was a negative change in usage of primary forest coverage on agricultural landholdings of between 31 and 37 percent. Annual crops, however, reflected positive change of between 74 and 177 percent in all provinces except for Puntarenas, Guanacaste, and San José. Permanent crops increased most in the province of Guanacaste with a 256 percent change. This is also the province where pasture experienced the greatest decrease at 19 percent. Though pasture overall did not experience much growth, it did boom in the north and Atlantic provinces of Heredia and Limón with positive percentage changes of 56 and 70, respectively.

**Road Construction**

Figure 2, depicting road construction projects carried out by the Ministry of Public Transportation (MOPT, 1990) for the years 1980-89, indicates that the majority of road-building activity occurred in the northern and Atlantic sectors (Alajuela and Limón province) and in the south (Puntarenas province). These are frontier areas which are experiencing a trend of expanding settlement for crop cultivation. Likewise, this activity indicates a relationship between extensions of road systems and deforestation activity. This trend has often been cited about areas elsewhere in the tropics (Browder, 1989; Bunker, 1985; Leonard, 1987; Schminck, 1985).

**Commercial Logging**

Information from the General Forestry Directorate (DGF) regarding logging activity indicates that, during the period of 1975 to 1985, the number of total timber extraction permits peaked in 1981. In this year alone, 1,270 permits encompassing 24,206 hectares were issued. The number of permits then declined steadily to reach 553 issued in 1984, but rose again in 1985 to 664 permits. The last figure is still more than double the number of permits allotted in 1975. But despite this, the permits for 1975 encompassed almost as many hectares as in the peak year (1981), while in 1985 twice as many permits issued only encompassed 16,340 hectares. This information would indicate that the area of land legally deforested in the first half of the 1980s has sharply declined as compared to the area deforested in the latter half of the 1970s. On the other hand, the DGF (1990) claims that 60 percent of ongoing deforestation is illegal; this amount would not be indicated by the extraction figures cited above.

The 1986-87 Forest Industry Census shows Alajuela province as having the greatest logged area (103,357 hectares between 1975 and 1985, with Limón next at 43,815 hectares, followed by Puntarenas at 27,399 hectares). Also, the northern

Figure 2.  New Road Construction Projects, 1980-1989

Source:  MOPT.  General Planning Directorate.  Based on information from the
Construction and Finance Directorates.  San José:  1989.

Atlantic, and southern Pacific zones are primary areas where *madera en rollo* is extracted. And despite limits of primary wood, there is still much waste in the forest at extraction sites and saw mills (DGF, 1988).

## Summary of Land-Use Trends

A perusal of Figure 3 (DGF, 1989) aids in summarizing some of the above findings. According to views 1-4, the greatest overall decreases in forest coverage occurred up until the 1980s, whereas relatively little change occurred between 1983-1987 (views 5-6). Exploitation of partially forested agricultural lands seems already to have peaked. Views 4-6 also aptly depict the only regions where forest use took place according to the agricultural census, namely Alajuela, Limón, and Puntarenas. Additionally, the agricultural censuses indicated a tremendous growth in annual crops and/or pasture in the above-mentioned provinces, as well as logging activity indicated by the forest industry census and road activity indicated by construction projects.

Though these maps are based upon land-satellite photographs, none of the photographs for the years depicted after 1961 were taken in a single year, meaning that maps depicting forest coverage after 1961 are not entirely accurate with respect to the year indicated. For instance, view 6 (marked 1987) is based upon a compilation of photographs from different sections of the country taken in 1979, 1982 and 1987, and thus does not reflect actual forest coverage status for 1987 (Garita, 1989).

## Rates of Deforestation

Predictions regarding the deforestation rate in Costa Rica are contradictory, which is not surprising given the uncertainty conveyed in the aforementioned satellite mapping. The rate of 50,000 hectares per year is often quoted in reference to deforestation activity, though some citations have the country's deforestation rate slowing for the first time in 1989 (Tripoli, 1989). A DGF official noted that, since most of the remaining forested land is under park, refuge, or reserve status, and practically no deforestation is occurring in these areas, the forest conversion rate would have to be decreasing. However, the World Resource Institute's (WRI) 1990 report cites Costa Rica as having the highest deforestation rate in the world. Data used for the WRI report, however, were based on research conducted between 1977 and 1983 (Dudenhoeffer, 1990). Thus, it would seem that WRI's claim may have held true for 1985, but perhaps not for 1990.

Figure 3.  Dense Forest Coverage (80-100%) in Costa Rica, 1940-1987

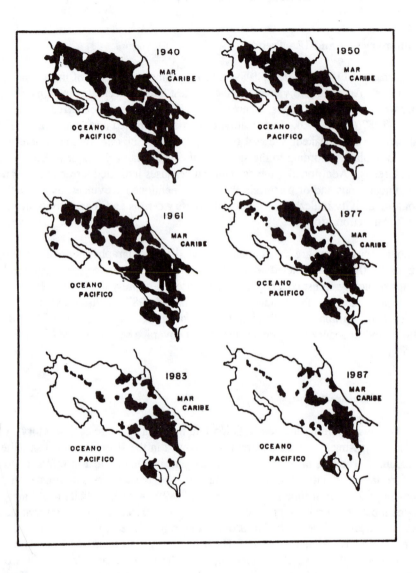

Source:  DGF.  Cartography Unit.  San José:  1989.

The following points from the DGF (1988) merit consideration in any discussion of the current status of deforestation:

1. There has been a rise in sawmill output, from 58.2 percent in 1983 to 82.60 percent in 1986.

2. A diminution in the number of tree species used in production was noted from 115 in 1980 to 102 in 1987, which may indicate that the 13 species no longer used may be difficult to find.

3. If current output remains constant, the few remaining timber reserves will disappear within a few years unless fixed extraction levels are set, and reforestation and natural forest management are implemented.

In the meantime, the Costa Rican government is boosting efforts at reforestation by means of fiscal incentives under the Forestry Law, including tax breaks and subsidies. A total of 43,498 hectares were reforested between 1964 and 1990, of which 42,898 hectares (98 percent) were reforested in the years 1980-1990 alone. Figure 4 re-emphasizes the nearly exponential growth in reforested areas per year in the last few years. Although reforestation has not outpaced deforestation, it is nevertheless possible that (a) timber is not being supplied solely from primary forest and/or (b) the reforestation rate may some day offset the deforestation rate. Indeed, as the director of timber extraction at the DGF reflected:

> In the moment that plantations begin to produce, it is likely that they will have an effect upon the deforestation rate; although these species will never be able to replace the quality of primary species... (Valerio, 1990, personal communication)

Under the new Forestry Law passed in July 1990, incentives for natural forest management are provided, and should be implemented in 1991 according to Decree number 20187 (La Gaceta, 1991) to stimulate maintenance of remaining privately owned forests. Also, under the present Calderón administration, the reforestation goal for the next 4 years is set at between 80,000 and 100,000 hectares. These policies fall in line with the recommendations mentioned above under point (3) by the Forest Industry Census of 1987.

Figure 4.  Reforestation in Costa Rica, 1964-1990

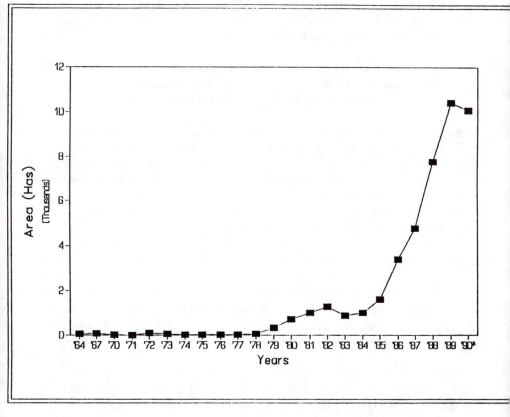

Source:  DGF. *Boletín Estadístico #3*. San José;  1989; Valerio. *Análisis del Programa de Incentivos Forestales en Costa Rica*. San José:  1990.  Preliminary data for 1990.

## Recommendations and Conclusions

The first step in lending proof to the argument that expansions in agriculture, road systems, and commercial logging are just as pervasive as cattle ranching in the conversion of forests would be in generating more accurate correlations of deforested areas to specific land uses via reconnaissance missions on a regional basis. Sader and Joyce (1988) point out the persistence of uncertainty regarding rates and trends of tropical forest clearing. Regarding their map digitization efforts for Costa Rica, they propose continued monitoring of forest clearing using satellite change detection techniques and quantitative estimates of change. This would allow one to assert with more reliability the nature of forest cover loss, and then more accurately to correlate loss to specific economic activity.

Also important are the efforts being made by the DGF to complete forest inventories on the national level, a tremendous task which has been underway for several years now (Cyrus, 1990, personal communication). This would allow the government to better plan its forest policies and extraction quotas based on a more precise appraisal of Costa Rica's forest resources.

Costa Rica's continued deforestation, at whatever rate, is not a result of only 1 specific land use, such as the often-cited conversion to pasture. There are many land-use activities, ultimately tied into socioeconomic forces and policy issues, which have had a strong bearing on conversion of forests in the last decade. Systematic investigation, including a recognition of social needs, followed by thorough analysis of current policies, would aid Costa Rica in its task of making rational decisions regarding economic development and environmental conservation.

## Acknowledgments

I thank the Organization for Tropical Studies for their support of some of the research reported here. I also acknowledge the *Latinamericanist* for editorial comments on an earlier draft appearing in their May 1991 edition.

# References

Aguilar, Irene and Manuel Solis. *La Elite Ganadera en Costa Rica*. San José: Editorial de la Universidad de Costa Rica, 1988.

Allen, William H. "Biocultural Restoration of a Tropical Forest." *Bioscience*, 38:3 (1988), 156-161.

Banco Central de Costa Rica. *Anuario Estadístico Fascículo del Sector Externo 1982-1986*. San José: Litografia e Imprenta Lehmann, S.A., 1988.

___. *Pregrama de Credito Agropecuario BIRF/SBN: Proyecto de Desarrollo y Fomento de la Ganaderia Bovina de Carne en Costa Rica*. San José: BCCR Departamento de Credito de Desarrollo, 1974.

Bonilla Duran, Alexander. *Situación Ambiental de Costa Rica*. San José: Ministerio de Cultura, Juventud y Deportes, Instituto del Libro, 1985.

Bonilla Castro, Marco. Estación Experimental Jiminez Nuñez de Libéria, telephone interview by author, 04 December 1990, San José. Hand-written.

Brajer, Victor. *An Analysis of Inflation in the Small, Open Economy of Costa Rica*. Research Paper Series No. 18. Albuquerque: University of New Mexico, 1986.

Brockett, Charles D. *Land, Power and Poverty*. Winchester: Allen & Unwin, Inc., 1988.

Browder, John O. "Development Alternatives for Tropical Rain Forests." Chap. in *Environment and the Poor: Development Strategies for a Common Agenda*. eds. H. Jeffrey Leonard *et al.*

___. "The Social Costs of Rain Forest Destruction: A Critique and Economic Analysis of the 'Hamburger Debate'." *Interciencia*, 13:3 (1988), 115-116.

Budowski, Tamara. "Ecotourism." *Tecnitur International Magazine* (1989) 12-14.

Bulmer-Thomas, Victor. *The Political Economy of Central America Since 1920*. Cambridge: Cambridge University Press, 1987.

___. *Studies in the Economics of Central America*. New York: St. Martin's Press, Inc., 1988.

Bunker, Stephen G. *Underdeveloping the Amazon*. Chicago: University of Chicago Press, 1988.

Caufield, Catherine. *In the Rainforest*. New York: Alfred A. Knopf, Inc., 1985.

Coone, Tim. "'Eco-tourism' takes off for Costa Rica." *Information Services Latin America*. 39:5 (1989) 26.

Corporación de Fomento Ganadero. "Consumo percápita de carne en Costa Rica." Unpublished statistics. San José.

"Costa Rica: Farmers Question Agricultural Policy." *Latin American Monitor*, 8 (August 1988), 558-559.

"Costa Rica: Nueva Prorroga del BM para el PAE II." *Inforpress Centroamericana*, July 1989, 5-6.

Cyrus, Edwin. Cartography Unit, Dirección General Forestal, personal interview by author, San José. Hand-written. 04 December 1990, San José.

Dirección General de Estadística y Censos. *Censo Agropecuario 1973*. San José: Imprenta Nacional, 1976.

___. *Censo Agropecuario 1984*. San José: Imprenta Nacional, 1987.

Dirección General Forestal in Solórzano, Jorge. "Bosques a merced de la deforestación." *La Nación*, 29 October 1990.

___. *Boletín Estadístico No. 3*. [San José]: Dirección General Forestal, MIRENEM, September 1989.

___. *Censo de la Industria Forestal 1986-1987*. [San José]: Dirección General Forestal, MIRENEM, 1988.

___. *Mapa de Cobertura Boscosa de Costa Rica*. by Damaris Garita Cruz [San José]: Dirección General Forestal, MIRENEM, June 1989.

Dudenhoeffer, David. "Out-of-Date Data Spark Doubts." *The Tico Times* 17 August 1990, 2.

Dudley, Nigel. *The Death of Trees*. London: Pluto Press, 1985.

Economic Research Service. *World Agricultural Trends and Indicators, 1970-88*. Statistical Bulletin # 781, 1989.

Economist Intelligence Unit. *Nicaragua, Costa Rica, Panama: Country Profile 1989-90*. London: The Economist, 1989.

Edelman, Marc and Joanne Kenan, editors. *The Costa Rica Reader*. New York: Grove Weidenfeld, 1989.

Fraser, Bruce W. "Costa Rica Woos Foreign Investors." *Information Services Latin America*, April 1989, 88-89.

Gugliotta, Guy. "Costa Rica Stability Attracts Investors." *Information Services Latin America*, 38:6 (1989), 90.

Guppy, Nicholas. "Tropical Deforestation: A Global View." *Foreign Affairs*, Spring 1984, 928-965.

Hall, Carolyn. *Costa Rica, a Geographical Interpretation in Historical Perspective*. Boulder: Westview Press, 1985.

Hamilton, Nora *et al.*, editors. *Crisis in Central America*. Boulder: Westview Press, 1988.

Instituto Costarricense de Turismo. *Costa Rica Anuario Estadístico de Turismo 1989*. San José: A.G. Covao. 1990.

International Monetary Fund. *International Financial Statistics Yearbook 1989*. Washington, D.C.: IMF Publications Services, 1989.

Jacobstein, Helen Lemlich. *The Process of Economic Development in Costa Rica 1948-1970: Some Political Factors*. Ph.D. diss., University of Miami, 1972.

Leonard, Hugh Jeffrey. *Natural Resources and Economic Development in Central America*. Washington D.C.: International Institute for Environment and Development, 1987.

Mendoza, Rolando. "Ecoturismo en Costa Rica." *Biocenosis*, 3:1 (July-Sept. 1988), 14-16.

Ministerio de Obras Públicas de Transporte. Mapa de Proyectos Ejecutados, 1980-1988. Dirección General de Planificación, based on information from the Direcciónes de Construcción y Financiera. San José: 1989.

Myers, Norman. *Conversion of Tropical Moist Forests*. Washington, D.C.: Office of Publications, National Academy of Sciences, 1980.

___. "The Hamburger Connection: How Central America's Forests Became North America's Hamburgers." *Ambio*, 10:1 (1981), 3-8.

Nations, James D. and D. Komer. "Rainforests and the Hamburger Society." *The Ecologist*, 17:4/5 (1987), 161-167.

Nations, James D. and Roland B. Nigh. "Cattle, Cash, Food, and Forest." *Culture & Agriculture*, 6 (1978), 1-5.

Nelson, Harold D. *Costa Rica, a Country Study*. Washington, D.C.: U.S. Government Printing Office, 1983.

Parsons, James J. "Forest to Pasture: Development or Destruction?" *Revista de Biología Tropical*, 24:1 (1975), 121-138.

___. "The Scourge of Cows." *Whole Earth Review*, (Spring 1988), 40-47.

Paus, Eva, ed. *Struggle Against Dependence: Nontraditional Export Growth in Central America and the Caribbean*. Boulder: Westview Press, 1988.

Porras, Anabelle and Beatriz Villarreal. *Deforestación en Costa Rica*. San José: Editorial Costa Rica, 1985.

Ramirez Solera, Alonso and Tirso Maldonado Ulloa, eds. *Desarrollo Socioeconómico y el Ambiente Natural de Costa Rica*. San José: Editorial Heliconia, 1988.

Repetto, Robert. "Deforestation in the Tropics." *Scientific American*, 262:4 (1990), 36-42.

Rodriguez C., Silvia and Emilio Vargas M. *El Recurso Forestal en Costa Rica: Políticas Públicas y Sociedad 1970-1984*. Heredia: Departamento de Publicaciones de la Universidad Nacional, 1988.

Sader, Steven A. and Arnold T. Joyce. "Deforestation Rates and Trends in Costa Rica, 1940 to 1983." *Biotropica*, 20:1 (1988), 11-19.

Sanderson, Steven E. *The Transformation of Mexican Agriculture*. Princeton: Princeton University Press, 1986.

Shane, Douglas R. *Hoofprints on the Forest*. Philadelphia: Institute for the Study of Human Issues, 1986.

Schminck, Marianne. "Sao Felix do Xingu: A Caboclo Community in Transition." In *The Amazon Caboclo: Historical and Contemporary Perspectives*. Publication No. 32, Studies in Third World Societies. College of William and Mary, 1985.

Simpson, James R. and Donald E. Farris. *The World's Beef Business*. Ames: Iowa State University Press, 1982.

Soley Monge, Alberto. *Administración de Explotaciones Ganaderas en Costa Rica*. San José: Editorial Costa Rica, 1978.

Solórzano, Jorge. "Bosques a merced de la deforestación." *La Nación*, 29 October 1990.

Stammer, Larry B. "Costa Rica Will Preserve Forests in Exchange for Reduction of Debt." *Information Services Latin America*, 38:1 (1989), 96.

Tripoli, Steve. "Costa Rica Halts Assault on its Fragile Tropical Forests." *Information Services Latin America*, 38:1 (1989), 97.

Umaña, Alvaro. "Costa Rica Swaps Debt for Trees." *Wall Street Journal*, 6 March 1987, 31.

Unnamed official, Ministerio de Agricultura y Ganaderia, personal interview by author, 27 June 1990, Bagaces. Hand-written. MAG office, Bagaces.

U.S. Department of Commerce. *Foreign Economic Trends and their Implications for the United States*. Washington, D.C.: U.S. Department of Commerce, 1988.

Valerio, Ricardo. Análisis del programa de Incentivos Forestales en Costa Rica. Unpublished Manuscript. October 1990, San José.

___. Director of Timber Extraction, DGF, personal interview by author, 06 December 1990, San Jose. Hand-written.

Vasquez, Alexis. "Agricultura." Chap. in *Memoria 1er Congreso Estrategia de Conservacion para el Desarrollo Sostenible de Costa Rica - Octubre 1988*. San José: República de Costa Rica, 1989.

Villarreal, Beatriz. *El Precarismo Rural en Costa Rica 1960-1980*. San José: Editorial Papiro, 1983.

Whitefield, Mimi. "Costa Rica Starts to Branch Out." *Information Services Latin America*, 39:3 (1989), 35.

Williams, Robert G. *Export Agriculture and the Crisis in Central America*. Chapel Hill: University of North Carolina Press, 1986.

World Resources Institute. Part I: The Plan. In *Tropical Forests: A Call for Action*. Washington, D.C.: World Resources Institute, 1983.

Perspectives on the Educational Programs and Policies
Underlying Natural Resources Development in the Canton of Coto Brus:
A Case Study of a Rural Costa Rican Community

Robert D. Leier
Pennsylvania State University

This case study is an inventory, description, and collection of perceptions on the educational programs and policies underlying natural resources development in the cantón of Coto Brus, Costa Rica.

Participants were asked to reflect upon development programs and policies experienced through local, national, and international institutions, associations, and development agencies within the community as related to changing environmental, social, and economic factors. The objective was to reveal local perspectives on the forms, problems, and possible solutions of natural resource development in Coto Brus and its relationship to the quality of life, degree of contentment and hopes for the future of the community. Data are intended for use by development planners and implementers to link social, agricultural, and environmental activities within the development and implementation of integrated natural resources educational programs and policies.

## Physical Aspects of Coto Brus

Coto Brus is located in the province of Puntarenas in the South Pacific Zone (Brunca Region) of Costa Rica. Politically, Coto Brus is bordered to the north by the Cantón of Talamanca and the Republic of Panamá. The Cantón of Buenas Aires borders it to the west, and to the southwest is the Cantón of Golfito. The eastern border is shared with the Republic of Panamá. Coto Brus has an area of 935 km$^2$ and it is divided into 5 districts. (IFAM, 1981).

## Overview of the Qualitative Research Methodology

The ethnographic approach utilized in this study, modeled after the qualitative work of Spradley (1979, 1980), Lincoln and Guba (1985), Werner (1987), and Patton (1990), involved residing in a selected Costa Rican community (Coto Brus) for an extended period of time (in this case 5 months, from November 1990 through March 1991) where development education programs and policies have occurred. This allows the researcher, according to Lofland, "to get close enough to the people and situation being studied to personally understand in depth the details of what goes on". (Patton, 1990).

This approach considers the researcher as an integral component of the research instrument where prejudices and biases are not denied, but are recognized as being pertinent to the study. The researcher's background and experience working with Central American Natural Resources issues since 1977 and interest as a socio-ecologist (defined as someone who collects information about the natural resources conditions within a region or a community as related to its social and economic factors) are significant elements to the "perspectives" that were brought to the study.

A benefit in undertaking an ethnographic and participatory approach to the study of development programs and policies is that it provides a "voice" for local and indigenous people to express their experiences and knowledge. It also gives them an opportunity to participate in the problem solving and decision making processes of issues which directly affect them. Another important advantage is that local perspectives are revealed which otherwise may not be available to development planners and implementers who develop and implement programs and policies to rural communities such as Coto Brus.

Three stages of research:

1. The selection of the community was determined by observations of and discussions with members from potential communities in Costa Rica. Criteria used in the selection process were: the community's exposure to development education programs and policies, its agriculturally based economy, and its diverse ethnic and class composition.

2. The "ethnohistory", patterned after the work of Werner (1987), describes and identifies the uniqueness of the community as related to natural resources development education programs and policies. It provides a concise "view" of the community through revision of local documents and endemic perspectives. It describes issues and policies that have been important to the community, development education programs that have occurred, and identification of program participants. It additionally provides the background for the selection of the participants in stage 3.

3. The "ethnographies" are constructed from the participants' perceptions of the policies that underlie programs which they have been aware of or have

experienced. The data will be validated by triangulation of methods characteristic to naturalistic inquiry and participant observation. According to Patton, "participant observation simultaneously combines document analysis, interviewing of respondents and informants, direct participation and observation, and introspection". The ethnographies of the community members reinforce the data gathered through the "ethnohistory" of the community.

## Cultural and Class Aspects of Coto Brus

Coto Brus contains various cultural and class groups that generally interact in a constructive manner. The following list summarizes these groups and their relationship to each other and the development of the region.

1. The Guaymí, an indigenous community with an estimated  population of 500-1,000 in the Coto Brus Indigenous Reserve, have been present in Coto Brus since the early 1900s. Their impact in the area has been minimal. Many live a marginalized existence on the reserve and generally do not integrate with other groups.

2. Italians immigrated to Coto Brus in the early 1950s and established the colony of San Vito. Their colonization brought in external resources and infrastructure which expanded coffee production commercially, which persists today as the economic base of the region. They have integrated within the Costa Rican community. (Weizman, 1982).

3. Small populations of Costa Ricans were present in the region before the Italians arrived.  Within the past 4 decades there has been a steady migration of rural Costa Ricans into Coto Brus from the Central Valley and Guanacaste, and most recently from Golfito. They work on coffee plantations, extract wood and clear lands for homesteading.  They have competed with the Guaymí for agricultural lands. (IDA, 1988)

4. Seasonal Costa Rican migrant workers (identified more by their class than any ethnic similarity), who follow the harvest of coffee, bananas, and pineapple throughout the country, reside in Coto Brus during its coffee harvest season. Panamanian migrant workers cross the border without documentation to pick coffee in the county.

5. Central American political refugees, mainly from El Salvador and Nicaragua, in recent years have entered the area in search of a safe haven

from war and to start a new life. They have routinely been employed as day laborers and coffee pickers, or utilized to perform jobs others do not care to do.

6. The Mennonite community arrived in the region a few years ago and have not integrated into the community.

7. North Americans and Europeans have settled in the area over the past 30 years. Many are pensioners and others have started small agricultural export projects.

**General History of the Colonization of Coto Brus**

Coto Brus is relatively a recently colonized area and does not share the centuries of colonization which is characteristic of other areas of Costa Rica, especially the Central Valley. To understand the factors influencing natural resources development in Coto Brus, it is essential to outline its brief history of colonization.

According to (Drolet, 1988), early human occupation of the area dates back to 1,000 B.C. where there is evidence of a long period of agricultural colonization. The subsistence based economy was characterized by hunting, fishing, and gathering in the forest zones close to the settlements, along with the cultivation of basic grains (beans and corn) mixed with tubers and fruit crops. W. Haberlan (1959) also affirms "dense populations of pre-Columbian settlements" in the area during this period. (Lange, 1980).

In the years 1500-1520, Manuel Arguello calculated a population of 1300 Brunca Indigenous, dominated by the Coctos tribe, occupying the area. (IFAM, 1982)

Don Peralán de Rivera in 1571 is credited to be the first non-aborigine to enter the Coto Brus region. (La Republica, 1988) and as Spanish colonists filtered into the area from the Central Valley, the indigenous populations slowly disappeared.

One of the major intrusions into the area was accelerated by the "Mule Road", a trail that stretched from Cartago to Chiriquí, Panama in 1601. (IFAM, 1982)

In the early 1900s, several Latin families inhabited the area and the Coto War (Frontier conflict between Costa Rica and Panamá in 1921) provoked an interest in the Coto Brus area. This began the migration into the region by entrepreneurs to look for "riches". (IFAM, 1982)

In the 1940s, the Pan-American highway was programed to pass through the region of Coto Brus. This provoked the construction of feeder roads and subsistence farmers began colonizing the area from both the Costa Rican and Panamanian sides. (IFAM, 1982) Some began experimenting with coffee in the region by 1949, but not on a commercial basis.

In 1950, the population of the Coto Brus region was estimated to be 2,000 (La Republica, 1988), but this was soon to change. Beginning in 1951, the government of Costa Rica and the Italian Society of Agricultural Colonization (SICA) made an agreement for the acquisition and subsequent development of a large portion of the Coto Brus area. Some of this newly "acquired" land was already being occupied by the Indigenous Guaymi and Los Coctos, as well as colonial Costa Ricans at the time. (IDA, 1988). The agreement of colonization featured a minimum of 300 colonists, 20 percent being Costa Ricans, to be established in the zone within 10 years. They were expected to cultivate cereals, vegetables, fruits, raise cattle and introduce coffee and sugar cane to the area. They were to experiment with raising grapes, wheat, and sheep. Other stipulations for colonization included installing a sawmill, an electric plant, a school, a church, a medical center, and a market place to buy and sell products. The Costa Rican government's contribution to the new colony was to provide state lands for the construction of access roads to markets. (Bodas de Plata, 1975).

Until 1954, SICA was purchasing the agricultural products of the colony, providing a market for the colonists. When this arrangement ceased, the colonists were forced to seek alternative agricultural products and markets. What resulted was the formation of a monocultural environment of coffee that was dependent on the established coffee market in San José. With the new Pan-American highway scheduled to pass through Coto Brus, it would have been a perfect arrangement, but the highway never reached Coto Brus. (IFAM, 1982).

**Natural Resources Development Issues in Coto Brus**

The following natural resource development issues have been identified by the research participants in Coto Brus: land-use, agricultural extension (technology transfer), water availability and quality, watershed protection, soil erosion, deforestation and reforestation, park and wildlife management, eco-tourism, environmental education, road construction and improvements, and artisan development using local natural resources.

### A. Land Use and Land Distribution, The Coffee Culture

**Active Agricultural Use**

Agricultural use in Coto Brus has been based on the opening of forested lands for the planting of coffee. When coffee prices were low, coffee trees were cut or forests were leveled to make way for cattle to augment income.

In the 1973 Census, reports quoted by Laurent, (Laurent, 197?), Coto Brus consisted of 1,964 farms with 26,636 hectares (28 percent) of the approximately 93,552 hectares in the county of Coto Brus in active agricultural use.

According to the MIDEPLAN 1984 census report, cultivated land had increased to 41,163 hectares, (44 percent of the total county). This was classified as:

pastures - 17,775 hectares (19 percent)

forests and woody areas - 7,484 hectares (8 percent)

arable lands - 6,549 hectares (7 percent)

permanent crops - 5,613 hectares (6 percent)

recently overgrown areas - 3,742 hectares (4 percent)

others - 936 hectares (less than 1 percent)

To illustrate the intense push to open areas for the expansion of agriculture, it can be noted that in less than 12 years, from 1973-1984, approximately 14,500 previously uncultivated hectares (16 percent of the county) were producing export cash crops of coffee and, to a lesser degree, cattle.

**Protected Areas**

The concept of protecting areas in Coto Brus for non-agricultural activities has been a recent phenomenon brought about more by a function of development planning than by the residents themselves. The Ministry of National Planning and Economic Policy, (MIDEPLAN), in their 1984 census report, designated more than half (56 percent) of the Coto Brus area as being protected. These areas include:

Las Tablas Protected Area, watershed of Coto Brus (21 percent).

Robert & Catherine Wilson Botanical Gardens and others (10 percent)

La Amistad International Park (10 percent).

Guaymí Indigenous Reservation (8 percent). The reservation is designated as protected in that technically only indigenous Guaymi can utilize its resources. The Guaymi cultivate small plots of land within the reserve, but on a low impact basis. Reservation chiefs estimate that today approximately 90 percent of their land is in forest.

Frontier belt with Panama (7 percent). This area is designated as protected only in that it provides a buffer between Costa Rica and Panama, where certain activities, such as granting land titles, are controlled. (Censo 1984).

**Distribution of Land**

As can be noted from the chart of land distribution in Coto Brus (below), almost half the area used for agriculture is owned by a very small percentage of the population. In general, the larger farms are characterized by cattle, coffee, and wood exploitation. The majority of the farms with less than 10 hectares are operated by family units and mainly grow coffee and basic grains and are forever dependent on slight changes in market economics.

| Size of Farm in Hectares | Percent of Total Farms | Percent of Total Land |
|---|---|---|
| Less than 10 H. | 52.1 | 11.4 |
| 10 to 50 H. | 39.9 | 40.2 |
| More than 50 H. | 08.0 | 48.4 |

Source: *Description Ecologica Población de Cafe en Coto Brus*. San Jose, Costa Rica, Sep de 1984.

According to local agricultural extension agents and farmers, a well managed farm of 5-6 hectares with initially good soils can support a typical rural Tican family of 6-8 members. As soil quality diminishes over time and use, it takes more land and an increase in external resources to produce the same quantity as before. Land in the Coto Brus area has steadily lost productivity due to erosion from high yearly rainfall coupled with lack of conservation measures and steep terrain. Little or nothing has been suggested to remedy soil loss, though farmers are aware of its occurrence. Traditionally the solution to poor productivity has been for farmers to abandon exhausted lands and move on to more potentially productive areas. What is problematic with this system presently is that unclaimed new areas are no longer available to farmers. Instead of moving to new lands within the county, farmers are attempting to sell their small farms and move out of the county.

For years, Coto Brus was considered a region by many where farmers from other parts of Costa Rica could acquire the "Costa Rican Dream" of land ownership and autonomy if they were willing to work hard. The soils were fertile since much of the land had never been cropped. This influx of human resources has resulted in Coto Brus becoming one of the principal coffee producing counties in Costa Rica, with approximately 81 percent of all farmers participating in coffee production (MAG, 1988). It claims to be the only county in Costa Rica which has 7 coffee warehouses and 3 coffee cooperatives (Coto Brus, 1991. Only 40 years ago, coffee was hardly known in the region.

In recent years, prices for Costa Rican coffee have been relatively low compared to the increase in production costs. Some research participants contribute low profits to "unfair" special considerations given to South American Andean countries (Peru, Colombia, and Bolivia) to produce coffee rather than cocaine. Regardless of the origin, farmers with 2 or 3 hectares of coffee are finding it increasingly more difficult to support their families on agricultural activities alone and are migrating from the area in search of employment on a permanent or semi-permanent basis. The alternatives for farmers include:

1. to borrow money from family or friends and "if God is willing, coffee prices will rise."

2. to migrate to areas within Costa Rica such as Buenos Aires - to pick pineapple, or Limon - to pick bananas.

3. to travel to the U.S. and work as menial laborers with plans to
   eventually return to Costa Rica and live off their savings until
   economic conditions are better. (MAG, 1990)

In 1984, it was estimated that 22 percent of farmers worked outside their
farms in order to sustain their families. In recent years this figure has
increased dramatically (estimated 80 percent by participants). The small
farmer is more likely to have family members with off-farm employment
than medium or large farmers. Off-farm employment in many cases is not
available and this has fueled the current economic crisis to where an
estimated 5,000 to 8,000 people (20 percent of the population) have migrated
from Coto Brus this year.

With fertilizer and maintenance costs on the increase and the price of coffee
low, recent policy has been to encourage agricultural diversification with the
planting of macadamia. (Comité 1990). The main problem with macadamia
production for small and marginal farmers is its long start-up period before
financial return, (approximately 4-7 years). Most farmers need a return no
more than a year from initial investment. A macadamia processor is essential
for processing and AID has promised to loan money through the cooperative
"Apromas" to acquire one for the community, but nothing has materialized.
Only 400-500 hectares have been planted with macadamia in Coto Brus. The
macadamia trees are usually planted among coffee trees and the farmers
anticipate to weed out the crop that performs the poorest. There is much
hesitation by farmers to gamble with an unproven crop such as macadamia
and believe it will have the same marketing and price problems as coffee
presently has. (IDA, 1988). Coffee has become too much part of the identity
of the farmer to easily have him relinquish it to another cash crop.

## B. Deforestation or Reforestation, Trees for the Taking

According to Stiles, "The rate of deforestation in Costa Rica is one of the
highest in the world: more than half the country's forests have disappeared
since 1940, and the remainder is being lost at a rate of about 3 percent of the
country's land area each year." (Stiles, 1989). This scenario is also true for
the Coto Brus region.

Laurent, in 1973, estimated that 500 hectares yearly were being deforested in
Coto Brus. Much of this lumber was being directed to the 8 sawmills in Coto
Brus for internal use and the remaining unknown quantity leaving the county
for external markets. Stiles states that the area "between the Golfo Dulce and

and the Panamanian border, has been completely shorn of forests in the last
10 or 15 years. Within a few years, nearly all Costa Rica's remaining forest
will presumably lie within the system of parks and equivalent reserves".
(Stiles, 1989).

"Reforestation" programs recently have been created and promoted in the
region as an alternative to the mono-culture of coffee. There appears to be
little concern and no pretense, among the participants, to reforest with mixed
tree species or to replant forests composed of native species. The aim is to
plant proven fast growing trees for maximum harvest. In some instances,
mixed stands of mature native forest have been leveled and marketed in order
that the owner could take advantage of the program and "reforest" his land.

The programs provide the land owners payment and production credit for
maintaining parcels of land in trees for a period of at least 15 years or when
it is decided the trees are marketable. In 1990, approximately 28 farmers
participated planting from 1 to 50 hectares totaling, 249 hectares. From
1991-1992, 300 hectares are scheduled to be planted each year. This is a
relatively small percentage of land being replanted compared to the quantity
of wood used each year in the Coto Brus region. Forestry engineers in the
area have indicated that within 5 years, even with these reforestation
programs, wood will need to be imported into the county. (MAG, 1989)

The reasons that farmers participate in the program has been reported by
research participants to be: they are ecologically minded; they have extra
land which is not productive; they want to make money from reforestation
programs. Only individuals that have land available for long term investment
can realistically participate in this program. Many farmers are excluded,
especially the small farmer who can not keep his land occupied without
monetary returns for long periods of time.

The species available for reforestation in the region are walnut (*Juglans
alanchona*); laurel, (*Cordia allidora*); eucalyptus (*Eucalypto deglupta*); and
amarillón (*Terminalia amazonia*). In the past, exotics, mostly eucalyptus,
have constituted over half the trees planted. Trees have usually been planted
in agriculturally exhausted soils and in a mono-cultural fashion.

Trees appear to be an inexhaustible resource to most people in Coto Brus.
There is little consciousness (40 percent reported by some participants) of the
need for reforestation, conservation of existing forests, or the relationship
between trees and water availability and quality. Wood sources are still available and convenient for many, though being depleted rapidly. In the past 40

years, trees were seen as hindrances to agricultural and economic development. As lands were cleared for agriculture, trees were disposed of by cutting and burning or more commonly, in wet environments such as Coto Brus, leaving them to rot. When asked about firewood sources for family needs, farmers usually indicated some area in the distance as their wood source. Small farmers saw no need to manage woodlots on their land. There is a tendency for all available land to be planted in coffee.

Numerous research participants have reported a relationship between deforestation and climate change in the region. Original settlers claim that when they arrived in the 1950s, it rained consistently and there was no dry season. Residents, from 20 years ago, remember only 8 days to 2 weeks of dry season in January and/or February. Now the dry season can begin in December and go through April. Most people have found this climatic change favorable. They noted how depressing it was to have "cold" and rain everyday of the year. Less rain has also aided their attempts to grow coffee on a larger scale because of less moisture to promote fungal diseases. Adversely, rains are more concentrated throughout the rainy season and violent downpours are washing away crops, soils and the expensive chemicals, which therefore need to be applied more frequently. Participants have also indicated that their wells dry up in March, limiting water availability for agricultural or domestic use.

## C. Eco-tourism, The Last Resort

Eco-tourism has been promoted as the easiest alternative to "a developing country that has exhausted its agricultural frontier and is struggling to cope with a heavy burden of external debt." (CI, 1990). "On the part of the Costa Ricans, eco-tourism provides a partial solution to economic development that will address large external debts, short-term needs (generated primarily by the need to earn foreign exchange) and will be in conflict with long-term conservation goals in Costa Rica." (Stiles, 1989)

Coto Brus has generally been overlooked by the Costa Rican tourist promotion literature as possessing attractions for eco-tourists. The exception is the Robert & Catherine Wilson Botanical Gardens administered by the Organization of Tropical Studies. To alter this image, organizations such as the local environmental protection association (APRENABRUS), Conservation International (CI), Organization of American States (OAS), Ministry of National Planning and Economic Policy (MIDEPLAN), Ministry of Natural Resources, Energy and Mines (MIRENEM), and others have suggested additional areas for eco-tourism development in Coto Brus. They include: a section of the La Amistad International Park, the Las Tablas Protected Area and the Guaymi Indigenous Reserve. According to a CI, "these areas are designated for sustainable development and traditional use to provide long-term social and economic benefits to the people" (CI, 1990). What is not clearly understood is how these protected areas will be effected ecologically when eco-tourism planners "develop them or the socio-ecological question of how native populations interpret "traditional use".

Conflicts between eco-tourism development, agricultural development and conservation in Coto Brus will need to be addressed. The continuous progression of forest clearing, the construction of temporary access roads, incompatible agricultural activities, and the development of tourist attractions near protected areas may create environmental degradation. It will be a formidable challenge to protect these fragile areas and at the same time provide for the demands of tourism and the infrastructural needs of the local people.

An example is The Las Tablas Protected Area which has been designated as the watershed for much of Coto Brus. AYA-IDB projects are currently constructing the infrastructure to provide potable water from Las Tablas to this area. (AYA, 1988). Eco-tourism development planners also project constructing paved roads and eco-tourism facilities around the Las Tablas Protected area to take advantage of the pristine environment that still exists. (Wyeth & Horrell, 1988). Providing accessibility to this area may intensify the rate of its destruction. For years, the "opening up" of frontier areas has meant uncontrolled exploitation and eventual environmental degradation even under the best management plans. How will Las Tablas endure illegal logging of trees, invasion of peripheral lands by squatters, expansion of agriculture and livestock activities, and hunting or poaching of wild animals within the protected area? Uncontrolled as well as regulated use of this area by humans will inevitably jeopardize the quality of potable water and the protected environment.

A few research participants have expectations that eco-tourism will rescue them from low coffee prices and the lack of agricultural diversification. Others view eco-tourism as "another scheme by the government to take lands away from the poor farmers and provide a way for the rich to get richer". They do not see eco-tourism benefiting the rural campesino but the rich businessman from San José.

The Guaymí Indigenous Reserve is another proposed eco-tourism attraction. With the recent interest in indigenous cultures, eco-tourism planners anticipate a heavy influx of eco-tourism dollars from the promotion and development of the 20 or more indigenous reserves within Costa Rica (Solano Martínez, 1989). This has resulted in Costa Ricans now proposing to exploit the "treasured" indigenous populations they denied the existence of just a few years ago.

As with most of the indigenous reserves adjacent or near the Talamanca Region, the Guaymi Reserve, which has been in "traditional use", has one of Costa Rica's lowest socio-economic standards, lacks adequate health care, transportation, and education facilities. This has resulted in one of the highest mortality, malnutrition, and illiteracy rates in Costa Rica. (Tenorio, 1987). These problems are slowly if at all being addressed by the Costa Rican development community.

This year Coto Brus discovered "their" Guaymi population and invited them for the first time ever to participate in the annual county celebrations. (Coto Brus, 1991). Non-indigenous in Coto Brus appear to overlook their prejudices, that the Guaymis are "unintelligent, dirty, and unmotivated", if it means that Guaymi people and lands can be exploited for eco-tourism. With the possibility of enticing foreigners with "eco-dollars" to experience the Guaymis in their "natural setting" as a part of eco-tourism, attention has been brought to the traditional qualities of their indigenousness such as native dress and crafts. Meanwhile, minimal efforts are being made to alleviate the socio-economic and political problems of the Guaymis in the Coto Brus reserve. The Guaymis still have not found ways to protect their lands from the onslaught of non-indigenous squatters and poachers. The Guaymis do not know where there reserve boundaries exist nor have little power to force intruders off their land. As they become surrounded by eco-tourism "development", this once isolated and forgotten indigenous community will face challenging future self-determination decisions.

**Conclusions**

Participants expressed that development education programs and policies which addressed the above natural resources development issues had little effect on their way of life. In general, government programs and policies originating in San José to improve the lives of rural people were unknown to the rural community members. If known, they offered little that was tangible for the majority of poor farmers. Because of physical, economic, and class barriers, many found it impossible to participate in development programs or to take advantage of policies. Before the 1980s, most people believed development organizations and programs were designed to help them. Now they contend that development programs are measures by the government to implement policies that support the "development" of the controlling rich, not the rural poor. The recent policies and resulting programs of government controlled land distribution, reforestation, housing improvement, food assistance, and the official hour change are examples of how the government has excluded the poor from the decision making process of their own development.

The majority of people in Coto Brus, especially the poor, do not see themselves having a positive impact on natural resource development issues. Generally, the reaction has been to ignore the issues and continue along the same line of development in hope that someone else or an "act of God" would be the answer. Or, participants perceive that people in power dictate, through development agencies, the answers to their problems and then proceed to encourage them to participate in the programs. The best strategy for many people from Coto Brus has been to associate with programs or policies that are offered with the hope that "this time it will be different", or that "something is better than nothing".

Policies and programs underlying natural resource development in Coto Brus have been directed to the exploitation of the region without interest in sustaining its natural environment. Since Coto Brus is one of the last frontiers in Costa Rica to be exploited and pressures for the opening of new lands continue, more marginalized areas are coming under cultivation, becoming quickly exhausted, and subsequently less able to sustain agriculture each year. Coto Brus is another victim of Costa Rica's recent economic policy to increase export income and improve its balance of payments by further exploiting the environment through promotion of monoculture export cash crops, uncontrolled logging of remaining forests, and promotion of eco-tourism.

As evident in this case study, development programs and policies that continue to support export mono-crop economies (regardless if the crop is coffee, macadamia or trees) and frontier exploitation mentalities (get as much as fast as possible), such as exist in Coto Brus, will not address the natural resource development issues plaguing the community. This is resulting in a continuous migration from the area. There is no data available which indicates that people will cease to exploit the remaining natural resources base in Coto Brus until it is exhausted. Restoration of a sustainable resource base to support a growing population, with increasing material demands, appears doubtful without major changes in development attitudes.

# References

Bodas de Plata 1952-1977. (1977). *Memoria, Veinticinco Años de la Fundación de San Vito*. Coto Brus, Costa Rica: s.e.

*Censo Agropecuario 1984*. MIDEPLAN No. 22 de Junio 1987.

Comité Agropecuario de Coto Brus. (1990). *Diagnostico de la Situation Agricola en Coto Brus*. Coto Brus.

Conservation International (CI), Organization of American States (OAS), Ministry of National Planning and Economic Policy (MIDEPLAN) and Ministry of Natural Resources, Energy and Mines (MIRENEM). (1990?). *A Strategy for the Institutional Development of La Amistad Biosphere Reserve, A Summary*. River Press.

Coto Brus 25 Años de Fundación. (1991). *Comunidades La Revista de Costa Rica* no. 19 - 1991, San Jose, Costa Rica: Grafel S.A.

Drolet, R.P. (1988). "The Emergence and Intensification of Complex Societies in Pacific Southern Costa Rica. in Lange, F.W. (ed.) (1988). *Costa Rican Art and Archeology, Essays in Honor of Frederick R. Mayer*. Colorado: Regents of the Univ. of Colo.

IDA document files. 1988. *Reseña Historica de Coto Brus*.

Instituto Costarricense de Acueductos y Alcantarillados (AYA). (1988). *Acueducto Regional de Coto Brus*. San José, Costa Rica: Unidad Ejecutora AYA-BID.

Instituto de Fomento y Asesoria Municipal (IFAM). (1981). *Cantones de Costa Rica*. San José, Costa Rica: IFAM.

Instituto de Fomento y Asesoria Municipal (IFAM). (1982). *Programa de desarrollo municipal fronterizo Costa Rica Panamá*. San José, Costa Rica: IFAM.

Lange, F.W. (1980). "Cultural Geography of Pre-Columbian Lower Central America". in Lange, F.W. and D.Z. Stone (eds.) (1984). *The Archeology of Lower Central America*. Albuquerque: University of New Mexico Press. pp. 33-60.

Laurent Sanabria, J.C. (197?) *Informe de la Sub Comisión de: Titulación, Explotación y Conservación de la Tierra*.

La Republica, viernes 11 de Marzo de 1988. "Synopsis histórica del Cantón de Coto Brus".

Lincoln, Y.S. and E.G. Guba. (1985). *Naturalistic Inquiry*. Newbury Park, CA: Sage Publications.

Ministerio de Agricultura y Ganaderia (MAG). (1988). "Dirección Regional Pacifico Sur, Programas - Objectivos - Metas, Cantón de Coto Brus". San José, Costa Rica: MAG.

Ministerio de Agricultura y Ganaderia (MAG). (1990) Interview, Ing. Alvaro Chavez on December 18, 1990 at the San Vito office.

Ministerio de Salud, Costarricense de Seguro Social, Centro Integrado de Salud, Hospital San Vito. (1990). "Diagnostico Situacional 1990". Por: Consejo Técnico Basico y Consejo Técnico Local. San Vito, Costa Rica, Diciembre, 1990.

Patton, M.Q. (1990). *Qualitative Evaluation and Research Methods*. CA: Sage Publications.

Solano Martínez, Carmen María. (1989). "Alternativas Para El Desarrollo Turístico Del Cantón de Coto Brus". Universidad International de las Americas Turismo. San José, Costa Rica.

Spradley, J.P. (1979). *The Ethnographic Interview*. New York: Holt, Rinehart and Winston.

Spradley, J.P. (1980). *Participant Observation*. New York: Holt, Rinehart and Winston.

Stiles, F.G., A.F. Skutch and D. Gardner. (1989). *A Guide to the Birds of Costa Rica*. Ithaca New York: Cornell University Press.

Tenorio Alfaro, L.A. (1987). *Situatión Actual de Los Grupos Indigenous de Costa Rica*. San José, Costa Rica: Comisión Nacional de Asuntos Indigenous (CONAI).

Weizman, H.G. (1982). *Emigrantes a la Conquista de la Selva*: Estudio de un Caso de Colonización en Costa Rica, San Vito de Java. Ginebra, Suiza: Comité Intergubernamental para las Migraciones.

Werner, O. and G.M. Schoepfle. (1987). *Systematic Fieldwork, Volumes 1 & 2*. Newbury Park, CA: Sage Publications.

Wyeth, J. and D. Horrell. (1988). "Proyecto de Desarrollo Rural Integrado Binacional en la Zona Fronteriza Costa Rica y Panama". Informe Final, European Economic Community. San José, Costa Rica: MIDEPLAN September 1988.

History of Protected Areas and Their Management
in Central America

James R. Barborak
University for Peace

Even though the practice of reserving certain lands for resource extraction is as old as civilization, the modern history of protected areas began in the late 1800s with the development of the first national parks in North America, New Zealand, Australia, and several other nations. By the 1930s national parks were first established in some parts of Latin America, such as Argentina, Chile, and Mexico (Miller, 1980). However, the development of both strictly protected "park-like" protected areas managed primarily for production of environmental services, and "reserve-like" areas managed to produce a mix of environmental goods and services is a relatively recent phenomena in the 7 nations of the Central America region (Barborak, 1987; Hartshorn, 1983; MacFarland and Morales, 1981).

Although the protected areas movement in the region is quite recent, it has grown and changed dramatically in just a few decades. This paper reviews the development of the regional protected areas system to date, with the particular objective of identifying factors which have been associated with successful efforts plan, establish, and effectively manage parks and reserves in Central America. Identification of such factors can help to improve management of "paper parks" and non-managed and threatened protected areas, and can also help guide planners involved in attempts to fill in the gaps in existing park and reserve systems in the rapidly shrinking areas where this is still possible.

**Early History of the Regional Protected Areas System**

As long ago as the 1920s, a small number of reserves were established in Central America. In 1928, the colonial government of Belize, then British Honduras, established the Half Moon Caye Crown Reserve as a bird sanctuary, which was upgraded to national monument status through post-independence park legislation in 1982. The Lancetilla Biological Reserve, next to the famous Wilson Popenoe Botanical Garden in Tela, Honduras, was being effectively protected as long ago as 1925 as a watershed reserve for Tela by the United Fruit Company, even though it did not receive official governmental designation as a protected area until 1990. Likewise, the island of Barro Colorado, formed in 1914 with the filling of Lake Gatun in the Panama Canal, was already being visited by scientists by 1916, and had received U.S. government designation as a reserve by 1923. Under the U.S.-Panama Canal Treaty of the late 1970s, it became a national monument under Panamanian legislation.

These early pioneer parks and reserves, while significant, were all established through the direct intervention of colonial powers of U.S. private companies. This trend continued in the 1940s with the protection of Cerro Uyuca near Tegucigalpa, Honduras, as a private forest and watershed reserve for the Panamerican Agricultural School of nearby El Zamorano. Like the other areas already mentioned, this area received official governmental protection, long after its initial establishment, in 1986.

**Early Consolidation of the Protected Areas System: 1950-1970**

In the 1950s and 1960s, a series of protected areas were established in the region, and for the first time most of these efforts were on national land and designated for management by national governments--true national parks and reserves in the global sense. Such reserves included the San Juancito, Cusuco, Guanaja, and Gulf of Fonseca Forest Reserves in Honduras, Santa Rosalia, Tikal, and Volcan de Pacaya National Parks in Guatemala, several forest reserves in Belize, and oak forest national park bordering the newly opened land Panamerican highway in the Talamanca Range of southern Costa Rica, and a radius of 2 km of land surrounding all volcanic craters in Costa Rica. Unlike the previous colonial and private protection efforts, most of these legal initiatives were not accompanied by active field management of these sites, giving rise to the "paper park and reserve syndrome" in the region. During this period, widespread environmental consciousness did not exist in the region, institutional structures for managing wildlands were non-existent, and training and technical and financial assistance programs for local conservation efforts were still absent.

Some of the early parks and reserves created in this period have since been degazetted, such as the Panamerican highway park in Costa Rica. Others, such as the San Juancito and Cusuco Forest Reserves in Honduras, have been upgraded in protection status to national parks in the 1980s and only recently have begun to have active field management. Some, such as Poas and Barva volcanoes of Costa Rica, are now managed in a relatively effective manner as part of larger parks created through subsequent legislation. Poas receives upwards of 100,000 visitors a year. Other volcanoes in Costa Rica, such as Turrialba, Tenorio, and Miravalles, were later declared part of larger forest reserves and have never received effective field management, even though still on the books as national parks.

**The Boom Years: 1970-Present**

A tremendous surge in interest in conservation occurred globally in the 1970s and 1980s, and this was also the case in Central America. A growing and increasing vocal network of nongovernmental conservation groups arose in the region. By the late 1970s, a major increase had occurred in technical assistance, training programs, and financial support to regional conservation efforts by bilateral and multilateral agencies and private conservation groups, such as the World Wildlife Fund, CATIE, the World Conservation Union, the Rockefeller Brothers Fund, the Caribbean Conservation Corporation, and the U.S. Peace Corps. All of this occurred during a period when rampant deforestation and habitat degradation, which had been occurring at an increasing pace since after World War II, made it obvious that time was rapidly running out to create both national parks and extractive reserves throughout the region.

This increase in human and financial resources, and national and international concern for the plight of Central American wildlands and associated cultural resources, led to a veritable boom in the creation of new parks and reserves. Most of the nearly 500 such protected areas which now exist in the region, covering over 10 percent of its land territory and a third of remaining forests, have been created in that period (IUCN, 1982; 1990). In Belize, over one-third of the country is now in some type of protected status, mostly as multiple-use forest reserves, although a number of forest reserves are now begin upgraded to park and sanctuary status. In Costa Rica, over one-fourth of the country has protected status, including over 10 percent in a national park system considered to be one of the best preserved and managed in the neotropics. Panama has an even larger park system, though not as well managed, covering over 12 percent of the indigenous reserves for Kuna and Embera peoples. Guatemala also has approximately 15 percent of its territory in parks and reserves, and passed an omnibus protected areas bill in 1989 creating 43 new reserves. The same year, it created the largest single protected area in the region, the 1.1 million hectare Maya Biosphere Reserve, incorporating Tikal National Park.

In Honduras, an omnibus cloud forest park and reserve bill in 1987 greatly increased the size of the protected area network of that country, which is now approaching 8 percent of its territory. In Nicaragua, the largest Central American nation, with the largest remaining pristine forests, the first large strictly protected area was just established in 1990, with the creation of the 290,000 ha. Rio Indio-Maiz Biological Reserve, part of a binational system of border parks and reserves long proposed along the Costa Rica-Nicaragua border. Even El Salvador, a country already mostly deforested by the turn of the century, has just completed a park systems plan. Primarily using lands acquired by the government agrarian reform program, the National Parks and Wildlife Department there hopes to protect over 50 small remnant forests and wetlands.

## Rationale for the Protected Areas Boom in the Region

In spite of widespread economic problems in the region and governmental budget cutting, which often hit "non-essential" programs like natural resource management disproportionately, the greatly increased support from the international community has led to ever growing total levels of investment in protected area land acquisition, personnel and training budgets, and in many cases, park-specific endowments. Innovative funding approaches such as debt-for-nature swaps and international children's campaigns, as well as much greater revenue for many parks due to a boom in ecotourism, particularly in Belize, Guatemala, and Costa Rica, have helped offset government austerity measures. Bilateral agencies from the U.S., Canada, the Scandinavian countries, and the Netherlands have greatly increased their support to conservation efforts in the region. U.N. programs such as FAO, UNESCO, and UNEP have also pitched in. WWF and IUCN have more recently been joined by other private conservation groups as major donors to conservation programs in the region, such as Wildlife Conservation International, the Nature Conservancy, Conservation International, the Rainforest Alliance, Nepenthes, and several U.S. foundations.

International scientific interest in Central American wildlands has also increased greatly over the past several decades. The La Selva research station, part of the Central Cordillera Biosphere Reserve of Costa Rica, and the Barro Colorado Island National Monument of Panama are 2 of the most important research sites and training grounds in the world for tropical ecologists, and the number of research projects and field courses given in the region on such topics is growing rapidly, both through local and foreign universities. A lot of this interest is due to the combination of the relative accessibility of Central America, and the fact that it is a biological hotspot: though covering less than one-half of 1 percent of the earth's land surface, it contains about 8 percent of the world's plants and possibly 10 percent of its vertebrates. An important number of species present, particularly migratory birds, are shared species with North America. Declining populations of many such forest, shore, and wetland bird species has greatly increased interest in the U.S. and Canada in Central American conservation efforts. It is hoped that such bird population declines can be arrested in part through expanded conservation efforts to save wintering habitat for these birds.

Another reason for increased action to establish and manage protected areas in Central America is due to the growing awareness of the economic importance of wildlands to watershed protection, promotion of nature-based tourism, protection of indigenous cultures and archaeological sites, and to the economic value of long term sustained harvest of wild species and derived products (Ledec and Goodland, 1988). Central America has very limited oil resources and most energy is generated through hydroelectric projects. Many areas with good soils are in dry regions, requiring irrigation for year round cultivation. And most urban centers, whose population is growing very rapidly, are near the seasonally dry Pacific coast. All of these factors make long term protection of upper watersheds, often covered with cloud forests where high rainfall and fog drip make them veritable water factories, vital to continued economic development in the region. Likewise, nature-based tourism is seen as one of the few viable alternatives requiring limited capital investment and capable of producing sizable flows of foreign exchange, considerable employment in depressed rural economies, and income to help to manage and protect parks and reserves. Coastal mangroves and wetlands are increasingly recognized to play a vital role in ensuring sustainable production levels for coastal fisheries (Barborak, 1987).

The region's ever more savvy and vocal tribal groups, still representing 5 percent of the total population and a much higher figure in Guatemala, have successfully lobbied for creation of indigenous reserves, particularly in Panama and Costa Rica. One such project, in the Kuna Yala Comarca in Panama, is managed by a highly trained corps of Kuna rangers and professionals. A similar, Mosquito Indian-run wildlife refuge is now being planned in eastern Nicaragua. The Talamanca Indian reserves of Costa Rica form part of a biosphere reserve and are important forested buffer zones nearly surrounding the Costa Rican portion of La Amistad International Park.

In spite of considerable progress, particularly in the past decade, much remains to be done to really consolidate protected areas management in the region. A decreasing, though still important number of parks and reserves lack minimum adequate field protection and management. In many, large areas of private lands must be acquired and just compensation paid. Scientific research, public education, and recreation and tourism programs are non-existent or minimal at best at far too many reserves. And the threat of unplanned roads, squatter invasions, mining operations, mass tourism, and degazetting hangs over some such areas (Machlis and Tichnell, 1985). Most of the so-called multiple use or extractive reserves lag far behind strictly protected parks and preserves in regards to management effectiveness, and conservation efforts to date have largely ignored the need to create both strict and controlled harvest marine reserves, in a region with the most diverse and spectacular coral reefs in the Western Hemisphere (OAS, 1988).

Illegal drugs continue to be produced and shipped through many parks and reserves. And many wildlands, particularly in El Salvador and Guatemala, are still unmanaged due to continued guerrilla and counter-insurgency operations in these countries. While Costa Rica and more recently Guatemala, Panama and Belize have gained considerable international support for their conservation efforts, far too little has been directed to the protection and management of equally important large wildlands in Honduras and Nicaragua, nor to help the Salvadorans save a tiny remnant of their natural endowment. However, in the midst of a generally dismal panorama regarding environmental deterioration in Central America during the past several decades, the tremendous progress made in planning, establishing and management of a protected areas system in Central America is noteworthy.

Experience to date with protected area in Central America has shown that legal creation and production of formal plans for reserves is not enough to guarantee their long term survival. Those reserves with the longest management history, most visitor use and least threats share common characteristics. These include the following:

- Full clear land title to all of the reserve

- On site, stable, well trained and paid permanent protection and management staff assigned from the moment of reserve creation

- Diverse and strong constituencies including local communities, local, national and international conservation groups; tourists and the tourism industry; and national and international political, business, and scientific leaders;

- Stable (though not necessarily massive) funding, usually derived from several complementary sources including regular government budgets, entrance and use fees, and private, bilateral, and multilateral aid

- Ongoing efforts to involve local communities in protected area management directly as staff members, through employment in research and tourism projects, and through buffer zone management projects aimed at stabilizing and improving land use and tenure and social welfare in communities surrounding protected area

- Ongoing research and operational planning efforts to improve information on protected area resources and constantly fine tune management based on improved knowledge and changing socioeconomic conditions

- The presence of a strong, relatively autonomous management entity, which if governmental is aided by parallel private groups with greater flexibility to raise funds and influence policy

- Outreach, recreation and tourism programs designed to show utilization of the park or reserve to fend off complaints that parks and reserves are off-limits and underutilized areas

While no panacea nor recipe, a careful review of the current status of management for all Central American parks and reserves shows that those area possessing all or most of the above factors are the "success stories" and those lacking many or most of these factors are the "distress stories" and paper parks.


**An Outlook for the Future**

While the era of initial establishment of large parks and reserves is rapidly drawing to a close in Central America, much remains to be done to consolidate the regional protected areas system (Barborak, 1987; MacFarland and Morales, 1981). Many existing area are too small or lack necessary altitudinal diversity to protect genetically viable populations of many threatened species they contain. Several conservation projects in Costa Rica, such as the Guancaste National Park project and the La Selva corridor project, have dealt with this problem on a park-specific level by increasing the size of parks and incorporating elevational transects into them. A new, joint project of the U.S. Agency for International Development, Wildlife Conservation International and the Caribbean Conservation Corporation, called Paseo Pantera, is even working to try to link existing and planned parks and reserves in a nearly continuous macrocorridor from southern Mexico to Colombia. The project goal is to maintain the 3 million year old biological land bridge through Central America, responsible for the high biological diversity of the region and a large share of that far to the north and south.

In addition to filling in final gaps in the regional system, upgrading the conservation status of some protected areas from forest reserve to park status, and fine tuning and linking existing parks and reserves, much greater attention is now being placed on landscape ecology and biosphere reserve approaches to natural areas management. These include more attention to buffer zones and extractive reserves ringing existing strictly protected areas than has been the case to date. More work can be expected on restoration ecology as well, as the area of degraded and abandoned pastureland in the region grows. Novel institutional approaches to managing wildlands, including total or more likely partial privatization of park management, a growth in privately owned small ecotourism reserves, and regional and municipal parks and reserves, can be expected as well.

Finally, much greater use of parks and reserves for recreation and tourism can be expected, as the population of Central America grows, the urban middle class expands, nature-oriented recreational pursuits increase in appeal, and international ecotourism expands in the region. With peace breaking out throughout the region, and improved road links through Mexico and Central America being planned, it is simply a matter of time before the tremendous cultural and natural attractions of Central America are firmly within the range of the local, regional, and North American and European traveler, and visits to currently remote parks become much more common. Thanks to the foresight of an initially small and rag tag band of local conservationists and foreign do-gooders dating back to the early part of this century, these natural cathedrals will still stand as living monuments preserving a sizable remnant of the rich and diverse natural endowment and cultural heritage of this tiny region.

## References

Barborak, J. R. 1987. Wildlands Conservation in Central America: Current Status and Trends. Paper presented at the IV World Wilderness Congress, Estes Park, Colorado. Unp. manuscript.

Hartshorn, G. 1983. Wildlands Conservation in Central America. Pp. 423-444 in Sutton, S. L. *et al.*, eds. *Tropical Rain Forest: Ecology and Management*. British Ecological Society, Special Publication Number 2. Blackwell Scientific Publications, Oxford.

IUCN. 1982. *IUCN Director of Neotropical Protected Areas*. IUCN, Gland, Switzerland.

IUCN-The World Conservation Union. 1990. *1990 United Nations List of National Parks and Protected Areas*. IUCN, Gland, Switzerland and Cambridge, U.K.

Ledec, G. and R. Goodland. 1988. *Wildlands: Their Protection and Management in Economic Development*. The World Bank, Washington, D.C.

Machlis, G. E. and D. L. Tichnell. 1985. *The State of the World's Parks*. Westview Press, Boulder, CO.

MacFarland, C. and R. Morales. 1981. *Planificacion y Manejo de los Recursos Silvestres en America Central: Estrategia para una Decada Critica*. CATIE, Turrialba, C.R.

Miller, K. R. 1980. *Planificacion de Parques Nacionales para el Ecodesarrollo.* FEPMA, Madrid.

Organization of American States. 1988. *Inventory of Caribbean Marine and Coastal Protected Areas.* OAS/USNPS, Washington, D.C.

Maya Urbanism and Ecological Change

Elizabeth Graham
York University

David M. Pendergast
Royal Ontario Museum

**Introduction**

In the first section of this paper we discuss our current archaeological research and its bearing on the ecological approach to analysis of past and present forest resources. In the concluding portion we consider possible Precolumbian scenarios of Maya urban-forest relationships that may have meaning for us today.

The archaeological research in which we are involved in Belize is not directly related to the issues in forest and environmental history that are the focus of this volume. Nevertheless, the results of work we have recently begun at the site of Marco Gonzalez on Ambergris Caye in Belize (Graham and Pendergast 1989; Pendergast and Graham 1990) bear on forest and environmental matters in 2 ways.

The first is practical and direct, in that the outcome of interdisciplinary studies-- some of which have already been initiated, whereas others are in the planning stages--will provide data that we hope will be of use in contemporary reforestation efforts.

The second way in which our work relates to forest and environmental matters is conceptual and far more indirect. We hope that the results of botanical, soil, and nutrient cycling studies in Belize at Marco Gonzalez will have impact on biogeographical, ecological, and other scientific studies of the so-called "natural" tropical environment. One of our goals is to impress upon natural scientists that intensive urban activity--that is, construction, processing, and refuse deposition on a grand scale--has been a force, along with geomorphology, hydrology, and other factors, in the development of the soil sequence, soil morphology, and vegetational patterns observable today in forested zones of the Maya lowlands.

## Archaeological Research

Our perspective on modern tropical forests and on ancient Maya forest use is based on excavation of a number of different kinds of Maya sites over a protracted period in Belize. Some of these sites are on the coast, but most are inland. Some settlements are urban centers with monumental temples; others are regional towns with far less impressive architecture. Some sites are even located in swamps, and others on beach sands. Most recently we each have completed the direction of archaeological excavations at a sixteenth-century site and, with an ethnohistorian studying contact-period documentary evidence, have cooperated in an attempt to elucidate the Conquest experience in Belize (Graham, Pendergast, and Jones 1989). The areas we excavate are by and large in bush--covered with relatively mature forest growth, with the second-growth vegetation that results from clearing for cultivation, or with a mixture of naturally occurring trees and cultivated plants. This association between archaeological sites and the vegetation of varying nature that covers them is the core around which our environmental studies are being built.

## Present and Future Research

Examination of the relationship of the archaeological research at Marco Gonzalez to the themes of this volume serves as a vehicle for describing some of the work in more detail.

In the practical sense, the results should bear on the potential for reforestation in areas where forest has been cleared, or where human activity has been intensive. This is because one focus of the work at Marco Gonzalez is to qualify, and if possible to quantify, the processes of sedimentation, nutrient cycling, and soil development that have taken place during and since the time of human occupation.

The site of Marco Gonzalez--like a number of similar sites along the Belize coast-- is distinguished from the mangrove swamp that surrounds it by the occurrence of a community of broadleaf forest species and by the presence of soils that often permit the cultivation of root crops, or sometimes fruit trees. There are naturally occurring phenomena, such as a limestone ridge (Dunn 1990) and an underground freshwater lens, that undoubtedly contribute to the distinctiveness of the Marco Gonzalez zone, but its general characteristics appear to mirror those at archaeological sites throughout Belize's coast. The association between coastal "islands" of distinctive vegetation and archaeological sites is so consistent that it constitutes a strong indication that debris from human occupation must have been a critical factor in the development of such zones (Graham 1989).

At the present time we are designing research strategies in an effort to document exactly how these zones developed, and to what extent human occupation debris can be said to have been responsible for their distinctive character. If we are correct in suggesting that debris from human activity--construction remains, processing refuse, and garbage--was a critical factor in the development process, then one of the outcomes of the research should be detailed data on the sources of critical nutrients that can affect soil development, on the time required for morphological changes, and on the critical mass necessary to effect long-term changes. What ought to make this sort of archaeological "waste management" study different from most modern efforts is the time depth involved--approximately 1,500 to 2,000 years. This is a span outside the range of consideration for in most modern organic waste management experiments.

In addition to the potential practical results of the archaeological work, the research at Marco Gonzalez, and indeed the study of pre-industrial tropical urbanism in general, can open avenues of inquiry that have heretofore remained underexplored. One such avenue crosses the science/humanities boundary. Despite the florescence of the field of ecology, a sharp dichotomy remains between natural science research, in which humans and their habits are seen as more of a nuisance than a systemic variable, and cultural/historical studies, in which most levels of analysis are generally inappropriate to the consideration of humans as part of cyclical biological systems. We are not critical of the existence of this dichotomy; it is intrinsic to the sciences/humanities interface. The concept of ecology bridges the dichotomy, and many of the papers in this volume are written from an ecological point of view. The existence of an ecological viewpoint does not mean, however, that it will be easy to fit humans into an ecological framework with illuminating results.

Archaeological soil-settlement studies provide an example of the problems inherent in attempts to apply an ecological perspective to relationships between human beings and their environment. Although most archaeologists consider themselves ecologically inclined, archaeological settlement pattern studies assume, rather than prove, a one-way relationship between soils and settlement. The procedure is as follows: Having used modern soil studies and soil maps to derive information on soil drainage and fertility, archaeologists then chart the remains of ancient settle-ment and lo and behold, they discover a correlation between good agricultural soils and ancient settlement. The conclusion then drawn is that soil character determined settlement pattern and the nature of agricultural exploitation (see Sanders 1977). Such studies give little thought to the fact that 2,000 years or so of occupation, much of it urban, may have been a factor in shaping modern soil morphology and vegetation, and that there may be considerably more to the soil-settlement association than the one-way causality normally proposed.

The suggestion we make here is not an original line of inquiry. A number of scholars have examined the relationship of soils, vegetation, and ancient land use and wrestled with the complexity of that interrelationship (e.g., Lambert and Arnason 1982, 1978). However, we know of no studies, except those begun at Marco Gonzalez (Emery 1991), that have taken a systematic look at the urban society that was set in the Mesoamerican tropical lowlands and considered the effect such settlement may have had on the character of modern mainland forests as well as the structure and populations of the offshore barrier reef. If such systematic examination of a highly complex interrelationship ultimately verifies the enduring environmental effects of ancient urbanism, the results should assist modern long-term planning and decision-making in areas of concern to environmentalists.

In focusing on the weaknesses of settlement-soil correlations in the Maya lowlands, we do not mean to ignore the extensive research that has been carried out on the possible relationship between ancient land-clearing and the maintenance of savannas (Bourliere 1983, Harris 1980). The persistence or spread of savanna vegetation, where it can be proved to be affected by human activity, is seen by many (but not necessarily by savanna biogeographers) as a process of degradation from broadleaf forest to pine, from high to lower species diversity, from good soil to poor. Savanna research contrasts sharply with studies of forest environments, in which far less attention has been given to the role of human activity in vegetational succession (some exceptions are Gomez Pompa 1987, Gomez Pompa et al. 1987, Balée, this volume). Even in situations where humans can be directly implicated in processes of degradation, there are very few studies--an exception that comes immediately to mind is Elinor Melville's (this volume)--in which the complexity of the interrelationship between human activity and environmental degradation has been charted in any detail.

Part of the problem in forests is the complex nature of species associations and the nature of their relationships to soils and hydrology. It is for this reason that we decided to undertake our studies on the coast, in a zone in which we have hopes of being able to "factor out" natural plant-soil associations.

Another part of the problem may be persistent misconceptions about the nature of urbanism in the humid tropics. That is, we tend to carry around with us a concept of urbanism that is rooted in the Western European experience. The source of our ideas is the urban tradition of the Middle East, and the end product is the world in which we grew up. We are not the first to propose that there are urban traditions besides that of the Middle East (Sanders and Webster 1989) but comparative studies of urban traditions, although they are scholarly and informative, are also filled with assumptions about what it is to be urban as regards use of space, assessments of density, and attitudes towards the function of vegetation within cities.

Suffice to say that our reconstructions of Maya cities may be based as much on our view of how we think cities should look as much as they are based on archaeological remains. It is not that data are manufactured or consciously misrepresented, but rather that what we expect affects what we seek and what we find. One apparent result of such expectations is that the focus is universally on major rather than ancillary structures. Kitchens and processing areas, for example, almost always extensive but separated from residential structures in the tropics, are rarely excavated, let alone included in site surveys. Without excavating ancillary activity areas and structures, how can we pretend to know how residential units functioned or how many people were supported? Perhaps more important, how justified are we in mapping only mounds or platforms when the ancillary structures that can provide critical information on building use and nature of activities are left to be mapped as "open spaces" between buildings?

The foregoing observations provide, we believe, enough information to support the case for an ecological approach to urbanism in the humid tropics as a fertile area for future research. Our experience in the Maya lowlands suggests in addition that that a focus on ecology can be complemented by speculations regarding relationships between the ancient Maya and their environment, and whether they might have some value in structuring present-day attitudes toward forest use.

**Precolumbian Scenarios**

As is true of us, the fundamental element for the ancient Maya as regards forest management--to use a modern term--was their understanding of their role in the world around them. As far as we can tell from the placement of offerings and caches, the nature of the effigies represented, and from ethnohistoric documents, the Maya did not hold the view that they had been granted dominion over nature but rather were constrained to deal with their environment in a number of ritually prescribed ways. Respect for the environment and atonement for environmental damage were given formal expression in religious rituals of sorts that are still carried out in many parts of the Maya area when forest is felled for slash-and-burn agriculture. It is very unlikely, however, that the Maya world view extended to the sort of conservationist attitude that is characteristic of the latter part of our century.

The Maya attitudes towards their forest surroundings were manifested directly in the process of field clearing. Judging by practices extant today, forest clearing in ancient times was enmeshed in ritual from its outset to its completion. Ritual activity was not, however, a safeguard against excessive forest destruction, because (as far as we can ascertain) Maya religion offered no more in the way of specific proscriptions in this regard than do the Christian and other modern belief systems. As a result, significant localized soil degradation as well as changes in biomass composition are very likely to have been produced by Maya forest clearing.

In addition to establishing a context of ritual for land clearing, Maya belief, as exemplified in ethnographic data, proscribed the felling of certain trees and thus effected a measure of forest conservation. The principal species granted immunity from destruction was the ceiba (*Ceiba pentandra*), a tree of great significance in Maya cosmogony. The copal (*Protium copal*), source of the aromatic resin highly prized by the Maya for ritual use, was surely also preserved. Mature ceibas, and to a slightly lesser extent copals, cast enough shadow to render a considerable area unsuitable for agriculture; hence it is probable that tracts of forest in the vicinity of large specimens of these 2 species were spared simply because development of the land for planting would have been unrewarding.

Numerous other species with recognized values as living trees were very probably also spared whenever circumstances permitted. Included in this group were very likely ramon (*Brosimum alicastrum*), cohune (*Orbignya cohune*), sapote (*Achras sapota*), and wild rubber (*Castilla elastica*), and probably wild cacao (*Theobroma cacao*), huano or bay leaf palm (*Sabal mauritiiformis*), soapseed (*Sapindus saponaria* [source of both washing material and fish poison]) and several fruit trees such as mamey (*Mammea americana*) and guava (*Psidium guajava*). Practical considerations may, furthermore, have protected some other species not valued for their products or their wood but simply too difficult to fell. Outstanding among such species is the deep-forest axemaster (*Krugiodendron ferreum*), with the water's-edge bullet tree (*Bucida buceras*) close behind.

What kept the problems that derive from forest clearing within manageable bounds for the ancient Maya was, first of all, a scale of activity that was small by modern standards. A second important factor was highly intensive utilization of riverine, lacustrine, and swampy soils for irrigation-supported agriculture. Finally, there was the fact that the Maya used a number of essential forest products that could only remain available through maintenance of forest communities.

As attitudes towards forest preservation have changed in what was once the land of the Maya and as modern machine technology has provided the means to implement new attitudes on a dramatically grand scale, human communities in the region have been able to close their eyes to their place in the tropical ecosystem. They have also been encouraged by the availability of an array of chemical fertilizers to disregard the soils factors that motivated ancient Maya irrigation agricultural practice. The Precolumbian Maya appear to have been far more cognizant of the limitations inherent in most tropical soils, and the resultant need for highly focused husbanding of the best soils, than are the modern inhabitants of their territory. With the developments that distinguish modern from ancient agricultural practice has come, furthermore, the casting aside of most products whose availability is dependent on forest maintenance in favor of those that are produced through forest exploitation. The increasing concern with such issues in tropical forest destruction must focus on the present, but at the same time it can profit from examination of the ancient past. In the development of effective and feasible systems of forest management it appears that both ancient Maya beliefs and the modern investigation of the urban centres where those beliefs held sway are very likely to contain lessons of value for today's world.

# References

Bourlier, F. (editor) 1983 *Ecosystems of the World, Vol 13, Tropical Savannas.* Elsevier, Amsterdam.

Dunn, Richard Kirk 1990 Holocene Paleocoastal Reconstruction of Southern Ambergris Cay, Belize, and Archaeological Geology of the Marco Gonzalez Maya Site. M.S. thesis, Department of Geology, Wichita State University.

Emery, Kitty 1991 Marine Resource Availability and Use at the Marco Gonzalez Archaeological Site: Present-Day Evidence from the Hol Chan Marine Reserve. Preliminary Report prepared for E. Graham and D. Pendergast, on file, Department of Anthropology, York University, and Department of New World Archaeology, Royal Ontario Museum.

Graham, Elizabeth 1989 Brief Synthesis of Coastal Site Data from Colson Point, Placencia, and Marco Gonzalez, Belize. In *Coastal Maya Trade*, edited by Heather McKillop and Paul F. Healy. Trent University Occasional Papers in Anthropology, No. 8. Peterborough, Ontario, Canada.

Graham, Elizabeth and David M. Pendergast 1989 Excavations at the Marco Gonzalez Site, Ambergris Cay, Belize, 1986. *Journal of Field Archaeology* 16:1-16.

Graham, Elizabeth, David M. Pendergast and Grant D. Jones 1989 On the Fringes of Conquest: Maya-Spanish Contact in Colonial Belize. *Science* 246:1254-1259.

Harris, David (editor) 1980 Human Ecology in Savanna Environments. Academic Press.

Lambert, John D. H., and J. Thor Arnason 1978 Distribution of vegetation on Maya ruins and its relationship to ancient land-use at Lamanai, Belize. *Turrialba* 28(1):33-41.

Lambert, John D. H., and J. Thor Arnason 1982 Ramon and Maya Ruins: An Ecological, not an Economic, Relation. *Science* 216:298-299.

Pendergast, David M. and Elizabeth Graham 1990 An Island Paradise (??): Marco Gonzalez 1990. *Archaeological Newsletter*, Royal Ontario Museum, Series II, no. 41.

Sanders, William T. 1977 Environmental Heterogeneity and the Evolution of Lowland Maya Civilization. In *The Origins of Maya Civilization*, edited by R.E.W. Adams, pp. 287-298. University of New Mexico Press, Albuquerque.

Sanders, William T. and David Webster 1988 The Mesoamerican Urban Tradition. *American Anthropologist* 90(3):521- 546.

# La Importancia Ecologica Y Economica De Las Tecnologias Tradicionales En La Agri- Y Silvicultura En Areas De Bosque Tropical Humedo En Mexico

Rhena Hoffmann
Instituto Latinoamericano
Universidad de Rostock
República Federal de Alemania

La naturaleza cumple una amplia diversidad de funciones dentro de los procesos de producción de bienes materiales: proporciona recursos y medios de producción así como el *locus standi* para actividades económicas. Es, asimismo, el objeto que el hombre transforma mediante el trabajo y según sus necesidades. A largo plazo el hombre puede conseguir una productividad económica permanente sólo si respeta y alienta la productividad ecológica, es decir, las leyes y las necesidades de reproducción del medio ambiente.

Las selvas tropicales húmedas constituyen ecosistemas con un grado extremadamente alto de diversidad y productividad biológica, pero, al mismo tiempo, con una gran vulnerabilidad ante las actividades económicas humanas. El problema de la destrucción, deforestación y degradación de las selvas tropicales ha salido del marco regional o nacional en el momento en que los científicos, primordialmente del llamado "Primer Mundo", descubrieron la importancia de las selvas, situadas exclusivamente en países "subdesarrollados", para diversos procesos químicos, biológicos y climáticos a nivel global, de los cuales depende la continuación de la vida en el mundo.

Una de las reacciones más fuertes ante el peligro de catástrofes ecológicas ha sido la de cuestionar, criticar y hasta prohibir ciertas tecnologías aplicadas durante milenios por la población local de las zonas de bosque tropical húmedo, sin considerar el contexto económico y social. Por otra parte, también se da el fenómeno inverso, llegándose a idealizar el sistema agrosilvícola tradicional de los campesinos indígenas.

El concepto de *tecnología tradicional* se refiere, en el contexto de este estudio, tanto a las tecnologías desarrolladas y empleadas originalmente por las antiguas civilizaciones mesoamericanas, como a las evoluciones y transformaciones adoptadas o promovidas por las generaciones posteriores, con el objetivo de adecuarlas a las circunstancias cambiantes. Las tecnologías conocidas actualmente como "tradicionales" (o también llamadas "indígenas") no constituyen relictos arcaicos y atrasados, sino que son el resultado de su adecuación permanente por parte de los grupos étnicos al entorno económico, social y ecológico.

La tecnología en general puede definirse como la organización de un proceso productivo concreto y la aplicación de técnicas e instrumentos al mismo. Mediante éste, el hombre transforma la naturaleza para obtener bienes de una utilidad especial, deseados y determinados aún antes de iniciarse el proceso de producción.

Los objetivos de la producción, en interacción con factores como el desarrollo científico y técnico, los conocimientos empíricos y la visión sobre la naturaleza, determinan la adopción de una estrategia económica en función de la cual se desarrollan tecnologías y se decide por una específicamente. Dicha tecnología está sometida a múltiples modificaciones dentro del transcurso histórico, dependiendo de los cambios en el desarrollo de las fuerzas productivas y de las condiciones sociales, culturales, económicas y ambientales.

Los ecosistemas del bosque tropical húmedo en México están situados en el sur y sureste del país con una extensión de 8-9 millones de hectáreas (selvas altas y medianas). La zona geográfica del Trópico Húmedo, un área de alrededor de 20 millones de has., originariamente cubierta de selvas perennifolias medianas y altas, constituye hoy en día una región donde predominan la ganadería extensiva (con una superficie de 10 millones de has.) y terrenos agrícolas (3.8 millones de has.)(Comisión Nacional Forestal, México 1988).

Los ecosistemas selváticos más importantes se encuentran en la península de Yucatán (especialmente en Quintana Roo y Campeche), en el estado de Chiapas (Selva Lacandona) y en la Sierra Norte de Oaxaca (Chimalapas). Existen también remanentes de selva en San Luis Potosí y Veracruz (región Huasteca, los Tuxtlas y Uxpanapa) así como una parte muy reducida en Tabasco, el estado con el mayor grado de transformación en pastizales y tierras de cultivo.

Pese a la gran influencia que desde hace cinco centenarios ejercen numerosos factores externos sobre la reproducción económica y social de los grupos étnicos, y a pesar de la heterogeneidad cultural, social y lingüística de las 17 etnias que actualmente habitan zonas de bosque tropical húmedo en México, predominan múltiples características comunes entre ellas en lo que respecta a estrategias económicas, tecnologías y factores determinantes para la elaboración y aplicación de dichas tecnologías.

En la mayoría de los ejemplos seleccionados para este estudio como son: grupos *mayas* de la península de Yucatán, *chinantecos* y *mazatecos* de Oaxaca, *lacandones* de Chiapas, *huastecos* de Veracruz y San Luis Potosí, *totonacas* de Veracruz y *chontales* de Tabasco, las diferencias entre ellos consisten por lo general en aspectos muy particulares que están en función de características ambientales, desarrollo social y cultural y del grado de inserción del grupo en el mercado capitalista.

Durante la época prehispánica la tierra era propiedad de la *comunidad étnica*, la cual asignaba a cada familia una parte para cultivarla. La distribución de la tierra se efectuaba según la calidad del suelo y ubicación geográfica, procurando una cierta equidad entre todos los campesinos.

La *unidad familiar* constituía el núcleo de la comunidad y el elemento básico para la economía de subsistencia. Parte de la importancia de esta unidad familiar radicaba en la transmisión de conocimientos y de técnicas productivas generación trás generación.

El proceso productivo abarcaba a todos los miembros de la familia, correspondiendo a cada uno una tarea especial. Los trabajos que rebasaban el ámbito familiar siempre fueron realizados con la ayuda de la comunidad, así como también las labores en beneficio de todos los habitantes (construcción de caminos y pozos, etc.). Formas de trabajo comunal basadas en principos de la reciprocidad se han mantenido hasta hoy en día en varios grupos étnicos, la "fajina" entre los mayas de Quintana Roo (como trabajo colectivo obligatorio de diez días anuales generalmente, para obras de infraestructura en las milpas, construcción y mantenimiento de edificios, caminos y pozos) (Nahmad, S., 1988) y el "tequio" entre los mazatecos de Oaxaca (que comprende todos los trabajos realizados en forma colectiva para dotar a la comunidad de los servicios necesarios) (Boege, E., 1990).

La sobrevivencia de estas relaciones de ayuda mutua, de mucho culto a la naturaleza y de costumbres tradicionales muestra que, pese a las alteraciones e influencias durante el desarrollo histórico, esos grupos han conservado gran parte de sus antiguos valores. A pesar de ello las relaciones salariales y de competencia, forzadas por la economía de mercado nacional e internacional, entran cada vez más en las comunidades sustituyéndolos.

El objetivo principal del trabajo productivo de los grupos étnicos ha sido siempre la obtención de bienes tanto para la *subsistencia* como para la *acumulación de excedentes*, y es necesario reconocer que, aún en tiempos prehispánicos, esto era ya parte integral del proceso de reproducción. Los excedentes se destinaban entonces para: a) pago de tributos a la clase dominante en su propio grupo étnico;

b) entrega (también como tributo) a otras civilizaciones mesoamericanas más desarrolladas (salvo los mayas, zapotecos y mixes, los otros pueblos del Trópico Húmedo eran tributarios de los aztecas) o, posteriormente a los colonizadores españoles; c) intercambio y/o venta en mercados regionales.

La existencia de varios grupos étnicos en zonas de bosque tropical húmedo durante muchos siglos sólo podía ser garantizada mediante una adaptación tecnológica muy sofisticada a este medio ambiente sumamente vulnerable.

Refiriéndome ahora a las tecnologías originales, lo cual no significa que éstas no sigan siendo practicadas en la actualidad, aunque la mayoría ha experimentado sucesivas transformaciones y algunas se encuentran en decadencia o a punto de desaparecer.

La tecnología básica, realizada mediante las técnicas de *roza-tumba-quema*, es y ha sido el sistema de alternancia en el manejo de la tierra, que consiste en la rotación, en una misma parcela, de los ecosistemas de: bosque primario - tierra agroforestal (*milpa I*) - bosque secundario (*acahual*) - tierra agroforestal (*milpa II*). Este sistema incluye el *policultivo* y el aprovechamiento de diversas especies silvestres que van desde herbáceas hasta arbóreas. Procurando el uso del área una vez desmontada el mayor tiempo posible y debido en parte a las dificultades de desmontar nuevas tierras de la selva primaria con herramientas de piedra, se puede calificar esta producción como semi-intensiva. Sin embargo,condiciones ambientales que limitaron un uso más intensivo y el aumento de población obligaron también a los campesinos a extender campos nuevos sobre la selva, es decir, a adoptar una producción extensiva.

Además de las *milpas* (campos de cultivo de alimentos básicos, especialmente maíz) y de los *acahuales* (tierras en descanso, pero aprovechadas para la recolección de frutos y otros productos no maderables, y para el cultivo de algunas especies útiles junto a la recuperación de la vegetación secundaria natural), existían otros subsistemas productivos como *huertos familiares, pequeñas hortalizas* (p. ej. el *kanché* de los mayas, cultivado en una cama de madera elevada sobre troncos de árboles con especies aromáticas y saborizantes), el *monte alto manejado* (p. ej. el *te'lom* huasteco) o terrenos para cultivos especializados, por lo general comerciales (p. ej. los huertos donde los chontales de Tabasco sembraban cacao bajo la sombra de los árboles después de haber rozado el soto-bosque).

Los periodos de manejo y las plantas cultivadas o recolectadas en estos sistemas variaban según las condiciones climáticas, edafológicas y geomorfológicas en las diferentes regiones y ecosistemas particulares dependiendo también de las necesidades de cada grupo étnico y del número y concentración de sus habitantes.

En varias regiones del país los campesinos llegaron a establecer sistemas agrosilvícolas con una productividad, intensidad y estabilidad muy altas, basándose en tecnologías muy precisamente adaptadas al medio ambiente. Ejemplo de esto son las *terrazas* de los grupos mayas en la península de Yucatán y en Chiapas, y los *campos elevados* de los mayas y de las etnias de Veracruz y Tabasco así como el sistema llamado *marceño* de los chontales de Tabasco.

Las *terrazas* de laderas y diques les permitieron aprovechar los suelos fertiles de estos sistemas microambientales, evitar la erosión y controlar mejor las condiciones climáticas (p. ej. la alta, pero variable precipitación y los vientos), aumentar y/o mantener los rendimientos agrícolas a largo plazo mejorando o conservando la fertilidad y estructura del suelo por medio de su enriquecimiento con nutrientes de los sedimentos ricos en minerales arrastrados por el agua.

Los *campos elevados* fueron construidos en terrenos bajos y pantanosos a lo largo de ríos que inundaban temporalmente los llanos. Mediante la excavación de canales en los pantanos y el amontonamiento de las arcillas, crearon plataformas elevadas por encima de los niveles de inundación, que fueron fertilizadas periódicamente con los sedimentos orgánicos y minerales provenientes de los canales y ríos. El cercamiento de los campos por las aguas ayudaba a controlar las plagas y los competidores de los cultivos y facilitaba el transporte de los productos por vías fluviales (Dahlin, B.H., 1985).

Los *camellones* (también llamadas "chinampas tropicales") de los chontales en el actual estado de Tabasco, constituyeron otra forma de campos elevados. Los chontales además aprovechaban la estación de sequía (febrero a mayo) para sembrar en las fertiles tierras bajas inundables cuyos suelos se rejuvenecían periódicamente y así ganaban una cosecha más por año, a la cual llamaban *marceño*. Su agricultura se adaptaba perfectamente a las condiciones de húmedad e inundación fluvial, que abarcaba más del 60 percent de la superficie de Tabasco.

Es posible considerar que el desarrollo de tecnologías para realizar una producción más intensiva hubiera sido una respuesta al aumento de las necesidades materiales por el crecimiento de la población en tiempos prehispánicos.

La tecnología original así como las tecnologías tradicionales aplicadas actualmente por comunidades que se encuentran lejos y poco insertadas en los mercados nacionales, han incluído siempre mecanismos que favorecen una regeneración del bosque lo mas pronta y ampliamente posible, tratando de asegurar una adecuada relación entre los periodos de cultivo y los de descanso. Dejaron espacios arbóreos entre las milpas para evitar la erosión, mantener cierta humedad y aprovechar las semillas de estos árboles para alentar el crecimiento de la vegetación secundaria en el período de descanso.

Respetaron, protegieron y cultivaron algunas especies arbóreas en las milpas o huertos familiares para utilizarlas para la alimentación, construcción o con objetivos medicinales. Existe la teoría de que el fruto del árbol "ramón" (*Brosimum alicastrum*) servía a los antiguos mayas, huastecos y a otros grupos como substituto del maíz en tiempos de mala cosecha (actualmente como forraje para los animales), y que fue conservado en la milpa e incluso introducido en los huertos y acahuales. Se piensa que, como resultado de esta selección el "ramón" es hoy en día una especie muy abundante y, en algunas regiones, la dominante de la vegetación secundaria.

*El cultivo múltiple en las milpas y huertos familiares tradicionales cumplía entonces, y sigue cumpliendo, una función económica y ecológica muy importante.* Su importancia económica para las familias y comunidades rurales radica en la gran variedad de productos que obtienen de un terreno reducido, basándose la producción, originalmente, en insumos internos. Dado el caso de que estos productos estén disponibles primordialmente para el autoconsumo, y no sean extraídos excesivamente por vía del comercio desigual, de relaciones tributarias y/o deudas, se garantiza una nutrición relativamente balanceada incluyendo las vitaminas y los minerales necesarios. Además otorgan una cierta seguridad económica ante los riesgos del mercado, de plagas o enfermedades a los cuales se expondrían especializándose en monocultivos.

El policultivo es un sistema muy intensivo respecto al aprovechamiento del espacio vertical y horizontalmente, del tiempo (combinación de varios ciclos productivos simultáneamente y/o sucesivamente) y de la mano de obra (ocupación permanente durante el año). El policultivo es, por otra parte, de gran eficacia ecológica a causa de sus interacciones complejas entre factores bióticos y abióticos, mismas que pueden ser aprovechadas para elevar la productividad económica. La simbiosis entre leguminosas y microorganismos (*Rhizobium*), promueve la fijación biológica de nitrógeno funcionando como un proceso de fertilización natural, lo cual, comparado con la aplicación de fertilizantes nitrogenados ofrece las ventajas de conservar mejor la estructura del suelo y evitar la contaminación.

Investigaciones realizadas en el estado de Tabasco comprobaron que en policultivos tradicionales de maíz, frijol y calabaza, los rendimientos del maíz se incrementan en casi un 50 percent, comparados con sus monocultivos aunque, en los casos del frijol y la calabaza la situación se invierte siendo menor la producción en policultivos, comparados con sus monocultivos respectivos. Es necesario destacar que los rendimientos de los tres cultivos juntos son más grandes que los de cada uno por separado y para la misma magnitud de área.

La presencia de calabazas en la milpa ayuda a controlar las malezas y la erosión del suelo. Los campesinos aprovechan también algunas hierbas silvestres (*Chenopodium ambrosioides, Lagascea mollis*) para combatir nemátodos patógenos o malezas nocivas para los cultivos (Gliessman, St. R., 1990). El policultivo agroforestal de las milpas, y aún más, el de los huertos familiares, en combinación con plantas silvestres, se acerca más a la diversidad biológica y estratificación del ecosistema selvático natural que cualquier cultivo especializado.

El techo vegetal relativamente cerrado del policultivo, y una franja arbórea alrededor de la milpa protegen la tierra de las influencias climáticas perjudiciales para el suelo y los propios cultivos. El tiempo de barbecho es de suma importancia para la regeneración ecológica y, por consiguiente, para la reproducción de las condiciones productivas. Mediante el reciclaje de nutrientes y minerales el suelo recupera su fertilidad, y la vegetación secundaria ayuda a inhibir el crecimiento indiscriminado de poblaciones nocivas. Además, las familias indígenas obtienen del acahual y del monte alto una cantidad inmensa de productos. Por citar sólo un caso: los huastecos utilizan del te'lom 221 especies para fines medicinales, 65 con utilidad artesanal, 33 para la construcción y 81 para el consumo humano (Alcorn, J.B., 1983).

Las tecnologías tradicionales, fundamentadas originalmente en los tres pilares: diversidad, intensidad y racionabilidad, aseguraban de esta manera tanto la subsistencia económica de la comunidad como la recuperación natural del ecosistema, para poder utilizarlo como base de un proceso productivo sustentable a largo plazo.

Los efectos de la colonización española y, posteriormente las condiciones económicas capitalistas se tradujeron en modificaciones de las tecnologías originales hacia una producción extensiva y una especialización, requiriendo cada vez más insumos externos y limitando la diversidad de cultivos. La tendencia a desplazar estas tecnologías por otras "modernas" como el monocultivo y la ganadería extensiva ha provocado la deforestación y la degradación de los recursos selváticos.

El grado de impacto que tuvo la colonización sobre las tecnologías de los distintos grupos étnicos dependía del control ejercido sobre cada uno de ellos y de los intereses económicos que se perseguían en las diferentes regiones. Mientras que pocas comunidades se encontraban suficientemente lejos del alcance o la influencia españolas (los lacandones de Chiapas), la mayoría (especialmente en la costa del

Golfo de México y en la península de Yucatán) estuvo sujeta a tributo y a jornadas laborales para beneficio de los encomenderos. Las comunidades se veían obligadas a especializarse en algunos productos que eran de interés para los conquistadores (p. ej. los chontales de Tabasco en la producción de cacao) reduciendo el terreno disponible para sus cultivos básicos. La introducción de nuevos instrumentos de producción (herramientas de metal) posibilitó una agricultura más extensiva sobre la selva degradando, al mismo tiempo, la estructura de los suelos por desuso del arado. En las haciendas se implementaban nuevas tecnologías (inapropiadas para el Trópico Húmedo) como el monocultivo de nuevas especies vegetales (cítricos, arroz, caña de azucar, hule, café, etc.) y la ganadería bovina extensiva (aunque ésta no representaba un peso significativo por la falta de demanda). Junto con las nuevas especies vegetales y animales, fueron introducidas malezas, plagas y enfermedades desde Europa.

El saqueo de mano de obra de las comunidades a través de la encomienda y la muerte de miles de indígenas redujo drásticamente la fuerza y el tiempo de trabajo disponible para las labores en los sistemas agroforestales tradicionales. Los más intensivos, como los campos elevados y terrazas casi desaparecieron en aquella época. Los huertos familiares, acahuales y montes manejados, empezaron a perder su función dentro de la economía tradicional, tendencia que subsiste hasta hoy en día.

A partir de la independencia política (1821) se incrementó la inserción del país en el mercado mundial que involucró más intensamente los recursos del Trópico Húmedo. Se desarrollaron diferentes ciclos productivos en los cuales la extracción o el cultivo de una o algunas especies, determinó la vida económica de las regiones hasta que la producción entró en crisis por la declinación de la demanda o por la aparición de plagas. Ejemplo de esto son el auge de la explotación del chicozapote (*Achras zapote*) en Campeche, Quintana Roo y Chiapas durante las guerras mundiales y la producción de plátano en Tabasco durante las primeras décadas del presente siglo para el mercado norteamericano.

A mediados de los años 30 del presente siglo se inició, bajo el gobierno de Lázaro Cárdenas (1934-1940), una nueva estrategia económica forzada por los efectos de la depresión económica mundial (1929-1933), la industrialización mediante la sustitución de importaciones y bajo la actividad directiva del Estado.

La continuación de la Reforma Agraria, uno de los puntos claves de la nueva política económica tuvo múltiples consecuencias para la economía campesina tradicional. El reparto agrario se efectuó de acuerdo al concepto del ejido moderno, como unidad social de producción, con el objetivo de estimular la producción nacional de alimentos. El proceso de reestructuración de la agricultura mediante la implantación de ejidos cambió la organización social y la estructura económica tradicionales de la población indígena de grandes zonas en la península de Yucatán y en los estados de Veracruz y Tabasco.

El fraccionamiento de la tierra en pequeñas parcelas fijas restringe las posibilidades de practicar la técnica tradicional de "roza-tumba-quema" y de mantener los periodos de rotación indispensables para la recuperación del ecosistema. Hoy en día, en todas las zonas ejidales del Trópico Húmedo ocurre que, por causa del crecimiento demográfico, los ejidos ya no disponen de extensiones suficientes para mantener el equilibrio apropiado entre cultivo y descanso agotando más rápido la capacidad productiva del suelo del Trópico. La única innovación técnica para solucionar este problema ha sido el empleo de fertilizantes lo cual, en muchos casos y debido a su aplicación desmesurada, perjudica la producción y la propia salud de los campesinos.

En los ejidos el trabajo individual y la competencia vienen sustituyendo las formas colectivas de trabajo de los habitantes. Más aún, los bajos precios estatales de garantía para los productos básicos como son el maíz, frijol y azucar, provocarón el decrecimiento de la producción basada en tecnologías tradicionales. Existen numerosos ejemplos para la suplantación de cultivos tradicionales por productos que tienen mejores precios en el mercado como el arroz y el sorgo.

A partir de los años 50 se produce un creciente desarrollo de la infraestructura en el Trópico: la construcción del Ferrocarril del Sureste, de una red de carreteras y una serie de presas y plantas hidrológicas. El cambio del régimen hidrológico de varias cuencas (Papaloapan, Grijalva) limitó drásticamente el proceso de rejuvenecimiento natural del suelo provocando su creciente salinización. Esto imposibilitó o dificultó el sostenimiento de la agricultura tradicional, especialmente en algunas zonas de Veracruz y Tabasco. La instalación de las presas originaba el desplazamiento de la población local del terreno previsto para la construcción y su traslado a otras regiones acompañado por proyectos de "colonización" y programas de modernización (Plan Chontalpa 1966, Balacán-Tenosique 1972, Uxpanapa 1975). Uno de los objetivos principales de estos programas fue la integración más amplia del Trópico Húmedo en la producción agropecuaria nacional como fuente de recursos naturales de bajo costo. Los "colonizadores" disponían de créditos de la banca nacional e internacional para la deforestación con fines agropecuarios y la promoción de la ganadería bovina extensiva. Se contó, además, con el apoyo estatal en forma de títulos de inafectibilidad, tecnología y creación de las condiciones de infraestructura para facilitar el transporte de los productos a los centros de demanda, a saber, la capital del país y las ciudades más grandes en México.

Actualmente el Trópico Húmedo suministra cerca del 44 percent de la producción nacional de ganado vacuno, y sólo el estado de Tabasco aporta el 75 percent del consumo de la Ciudad de México (Uribe Iniesta, R., 1990). El sector privado ha extendido su influencia y su propiedad mediante el acaparamiento y rentismo ilegal de tierras ejidales o comunales llegando a poseer terrenos de hasta mil o más hectáreas de "pequeña propiedad" en forma de pastizales. El proceso de ganaderización y ampliación de la frontera agrícola sobre las zonas tropicales propició entre 1940 y 1980 el desmonte de alrededor de 9 millones de hectáreas de selvas altas y medianas (Comisión Nacional Forestal, 1988), disminuyendo, además, las tierras disponibles para la aplicación de las tecnologías tradicionales agroforestales.

La más drástica degradación ambiental la experimentó el estado de Tabasco que perdió, en poco más de cuatro décadas, cerca de un millón de hectáreas de selva que cubrían, hasta 1940, el 49 percent de la superficie del estado y que hoy en día cubren apenas un 8 percent y con sólo un 3 percent de selva primaria (Tudela, F., 1990). El caso de Tabasco, donde se sostiene durante más de 50 años una importante producción agropecuaria, muestra que mediante la implantación de diferentes especies de pastos y el uso fertilizantes químicos y plaguicidas, se puede alcanzar una estabilización del ecosistema agropecuario en el Trópico Húmedo, aunque a un nivel de diversidad biológica sumamente inferior al ecosistema selvático, con consecuencias económicas negativas para la mayoría de la población indígena local (p. ej. el desempleo por la poca demanda de mano de obra de la ganadería extensiva). Esta estabilidad ecológica es, además, muy relativa porque depende en gran medida de las condiciones ambientales específicas. La tendencia hacia el sobrepastoreo provoca la cresciente degradacion de los suelos reduciendo los rendimientos.

Aunque en muchos casos los créditos fueron otorgados exclusivamente para actividades pecuarias, el Estado y los organismos financieros internacionales impulsaron también algunos proyectos agrícolas para desarrollar la pequeña producción maicera y modernizar el sector rural tradicional mediante la introducción de elementos "modernos" (semillas mejoradas, fertilizantes químicos y pesticidas) en la tecnología tradicional. El Banco Rural otorgó en 1978 un crédito a los campesinos mazatecos de Oaxaca para la compra de estos insumos con la condición de que no sembraran múltiples cultivos en sus milpas, favoreciendo la producción de maíz. Debido a esto y al uso de fertilizantes, los rendimientos de maíz efectivamente se duplicaron, pero la mayor parte hubo de ser entregada al propio banco como pago del préstamo (Boege, E., 1990).

El crecimiento de la población y su concentración en algunas zonas a causa de los movimientos migratorios planificados (programas estatales de "colonización") y los movimientos espontáneos (campesinos mexicanos despojados de su tierra y refugiados guatemaltecos) aumenta considerablemente la presión demográfica sobre la tierra intensificando, cada vez más, el desequilibrio entre reproducción económica y ecológica agravando, de paso, los conflictos políticos y sociales.

## Conclusiones

Las tecnologías tradicionales aplicadas bajo condiciones de relativa autarquía y autodeterminación de las comunidades indígenas sobre cuestiones de la organización social y reproducción, eran de gran importancia económica y ecológica porque garantizaban no solo la reproducción de las condiciones económicas y de las relaciones sociales, sino también la regeneración ecológica de su ambiente.

Mientras que originalmente, la práctica de estas tecnologías significaba una adecuación relativamente equilibrada y favorable al medio ambiente selvático, ésta se ha transformado, en el transcurso de la historia, a consecuencia de las modificaciones introducidas y/o forzadas por intereses económicos y políticos como son, entre otros:

- aplicación de nuevos medios de producción que posibilitan una producción extensiva.
- introducción de insumos "modernos" (fertilizantes, plaguicidas) que hacen innecesarios ciertos métodos tradicionales como el aprovechamiento de algunas malezas para combatir plagas.
- especialización de la producción según las demandas del mercado.
- decreciente importancia y reducción -en cuanto a espacio geográfico- de algunos subsistemas tradicionales como los huertos, acahuales y montes manejados.
- desequilibrio entre periodos de cultivo y descanso, paralelamente a una proliferación de la "roza-tumba-quema".
- substitución de las relaciones de colectividad y ayuda mutua por la competencia capitalista y las relaciones salariales que imposibilitan el empleo de tecnologías que requieren del trabajo comunal (campos elevados).
- pérdida de conocimientos.
- saqueo temporal de mano de obra y su empleo como jornalera.
- confrontación de los campesinos con nuevos insumos y nuevas tecnologías sin poseer, en muchos casos, una capacitación suficiente para aplicarlos.
- desaparición de disciplinas como la arboricultura o la recolección de ciertas especies.

Aunque la superioridad ecológica de las tecnologías tradicionales indígenas en zonas de bosque tropical húmedo ha sido reconocida ampliamente por los científicos como un manejo integral, ecológicamente racional y económicamente sustentable y eficaz, quisiera hacer énfasis en el problema que se presenta al confrontarlas con las condiciones económicas y sociales actuales. Las tecnologías tradicionales tal y como eran en su orígen ya no existen, y todas las modificaciones que en ellas se han desarrollado provocan buena parte de la sobreexplotación de los recursos naturales.

A todo esto se agrega el mínimo interes económico en la agri- y silvicultura tradicional del Trópico Húmedo y su gradual desplazamiento por la producción de carne bovina o de algunos otros productos agrícolas comerciales.

La evaluación de la aplicabilidad de las tecnologías tradicionales en la actualidad, y su participación en la elaboración de proyectos concretos de producción debería incluir, por lo tanto, un reconocimiento de la situación económica y social de los campesinos indígenas y, en el marco global, de la posición que guardan los países del llamado Tercer Mundo en el sistema de la división internacional del trabajo.

La aplicación de tecnologías combinadas con elementos tradicionales y modernos, que incluye un grado más elevado de elaboración de las materias primas en su lugar de producción y la transformación de la demanda y estructura del mercado, puede ser un camino hacia un manejo agroforestal económicamente eficaz y ecológicamente más racional. Este proceso sólo puede irse desarrollando como resultado de los esfuerzos de los propios productores por un lado, y por el otro, de toda una serie de cambios en la esfera económica nacional e internacional.

## Bibliografia

Alcorn Janis, B., "El te'lom huasteco: presente, pasado y futuro de un sistema de silvicultura indígena", *Biótica* 8(3): 315-331, México, 1983, (315-331).

Barrera, A., Gómez-Pompa, A., Vazquez-Yañez, C., "El manejo de las selvas por los mayas, sus implicaciones silvícolas y agrícolas", *Biótica* 2(2), México, 1977, (47-61).

Boege, E., *Recursos naturales, técnica y cultura*, El proyecto de desarrollo dominante, la economía de subsistencia y el manejo de los ecosistemas por los mazatecos, UNAM, México, 1990, (113-145).

Comisión Nacional Forestal; *Programa de Acción Forestal Tropical en México. Propuesta para la conservación y el desarrollo de las Selvas del Sureste*, México, 1988, (135).

Dahlin Bruce, H. *Historia de la agricultura*, Tomo 2. Epoca prehispánica, siglo XVI, La geografía histórica de la antigua agricultura maya, México, 1985, (125-196).

Gliessman, S. R., "Aspectos Ecológicos de las Prácticas Agrícolas Tradicionales en Tabasco-México", *Biótica* 5(3), México, 1980, (93-101).

Gliessman, S. R., *Agroecology and Small Farm Development*, The Ecology Management of Traditional Farming Systems, USA, 1990, (13-17).

Nahmad, S., González, A. y Rees, M., *Tecnologías indígenas y medio ambiente*, México, 1988, (282).

Tudela, F., *Medio ambiente y desarrollo en México*, Recursos naturales y sociedad en el trópico húmedo tabasqueño, México, 1990, (148-189).

Uribe Iniesta, R., *Recursos naturales, técnica y cultura*, Diagnóstico regional sobre prácticas tradicionales y tecnologías alternativas en el estado de Tabasco, UNAM, Mexico, 1990, (265-279).

Tropical Forest Policy and Practice During the Mexican Porfiriato, 1876-1910

Herman W. Konrad
University of Calgary

The long-lasting dictatorship of Mexico's Porfirio Diaz coincided with important national and international ideological and economic transitions. Within Mexico, Diaz and his select group of advisors were determined to transform what had been a basically rural and preindustrial society, into a modern state capable of competing successfully in the international marketplace. Building upon the secular, reformist trend established by Benito Juarez, and eyeing the "gilded age" optimism in the more industrialized countries of North America and Europe, the Mexican cientificos were determined to transform a country which had experienced over half a century of social, political, and economic upheaval. Through the implementation of an "order and progress" strategy, the porfirian administration was determined to modernize and industrialize the economic structures of the nation, educate and rationalize societal institutions, and exploit to the maximum the rich natural resources of the countryside. Lacking domestic capital and technology, these would need to be attracted from foreign sources. What could be accomplished domestically, in the form of establishing conditions of political stability and access infrastructures (transport, communication) for economic development, would receive the highest priority by the federal government as the necessary preconditions for international participation. Then, in close collaboration, both domestic and foreign interests would be able to reap the benefits of modernization, economic expansion, and gaining for Mexico its place in industrial, capitalistic, modern society. And the necessity to dominate and control nature, and to extract from it the ingredients of wealth, was central to this overall development strategy (Gonzalaz Navorno 1957; Hart 1990).

Policies and practices related to forest resources, therefore, included both domestic "push" and foreign "pull" factors. Those related to tropical forests, the central concern of this paper, also included additional influences of significant historical depth and ideological content. For throughout Mexico's colonial and national experience, its tropical lowlands had never been successfully integrated as political and economic factors. Apart from the importance of gulf ports (Veracruz, Campeche) and marginally integrated provinces in the Yucatan Peninsula, the enterprise of New Spain was focused largely in the interiors. The disasters experienced by Cortez in his fateful 1524-26 expedition (Martinez 1990) left a negative imprint upon future enthusiasm for any large-scale imperial activities in the forested tropics.  Despite the relative stability of the Yucatan colonial settlements, they retained their economienda status signifying a special borderland status.  And on the peninsula most of its more heavily forested areas remained, officially at least, as "unhabited" territory (de Vos 1988a; Farriss 1984).  In the colonial period tropical Mexico, with the port of Veracruz providing constant reminders, became associated with disease (*vomito negro* [yellow fever]), danger (foreign pirates), demographic decline, and dismal economic prospects.  The tropical forests remained as refuge zones for runaway slaves (*palenques*), apostate *indios* escaping the vigilance of clerics, and indigenous populations resisting the benefits of Christianity and Civilization.  And despite the periodic *entradas* of priests and military expeditions, along with foolhardy private efforts to establish economic footholds, colonial successes in the tropical forests were short-term events, more frequently dramatic, futile failures (Camparado 1987; de Vos 1988a; Zamurray-Stone 1932).  Accessible principally by sea or riverine channels, lacking in profitable mineral deposits, the riches of the tropical forests remained beyond the reach of Spanish colonial endeavor.

During the first half century, after national independence, attitudes toward the tropical forested areas and activities therein underwent little change (de Vos 1988b).  The Mexican national endeavor was clearly focused in the highlands.  It was here, after the shock of territorial losses resulting from the conflicts with the northern neighbor (War of 1846-48), after the traumatic defense of national sovereignty (French Intervention), that institutional order and systematic attempts to restructure the colonial rules of land tenure and access to natural resources began to unfold.  These, however, and evident in the reforms of Benito Juarez, were first and foremost oriented towards central rather than peripheral Mexico.  And although the principles of privatization of communal pueblo and Indigenous lands, and the elimination of the clergy as spokesmen for their interests applied to the nation as a whole, the Juarez reforms had little initial impact upon tropical areas.  As well, throughout the Porfirian epoch, most of the regulatory legislation had a temperate-zone rather than tropical-zone context.  Reinforcing this temperate zone concern, at the same time, were the foreign influences from Europe and North America, equally devoid of significant tropical forest insights (Lamb 1966).  The compartmentalization of both attitudes and experiences, which separated tropical from temperate landscapes into negative and positive habitats, even after successful economic penetration of the tropical forests, would not be resolved without contradictory consequences.

**Pre-Porfirian Policy and Practice**

The process of tropical forest exploitation followed 3 successive stages, initially being confined to coastal maritime access and shoreline activities, followed by inland riverbank activities, and subsequently inland penetrations with railroad connections to either rivers or coastal ports (de Vos 1988b). It would be the English successes in dyewood and cabinetwood extractions, on both sides of the Yucatan Peninsula, that forced Mexican attention to its own resources. As early as the 1660s British dyewood (*palo de Campeche*) camps controlled the Yucatan coasts from Rio Lagarto to Belize City. Using Jamaica as a point of departure that British had followed escaped slaves to the Mosquito Coasts, establishing the basis for its Belize Colony (British Honduras) dedicated to dyewood and, soon thereafter, cabinet woods (mahogany and cedar) (Cligern 1967).

By the 1670s British woodcutters were firmly entrenched in Isla del Carmen (Bife Island on English maps of the day) with access to interior coasts via the Laguna de Terminos. This occupation lasted for half a century, until finally displaced by Spanish naval forces (Lanz 1905). Mahogany, for ship-building purposes for the Spanish navy, came initially from the Caribbean Islands, but as early as the 1620s shipyards in Havana were receiving timbers from the Mexican Gulf coast. Throughout the 17th and 18th century mahogany timbers from the mouth of the Coatzacoalcos River, the Isthmus of Tehuantepec, and other Gulf river entries were being harvested. During the colonial period and the first quarter century of the national period, however, there was no attempt to regulate these activities. The central concern was to defend against foreign territorial encroachments. Yet in England, the period between 1725 and 1825, became known as the "golden mahogany century" (Lamb 1966:11-12; de Vos 1988:32-33) due to mahogany preferences of English cabinetmakers.

Continuing British and European demands and declining timber stocks in British Caribbean Islands, after Mexico's independence, encouraged national initiatives to supply this market. As of 1822 Mexican individuals began to explore the interior of Tabasco and Chiapas, via the Usamacinta River system, giving rise to the riverbank era of tropical forest exploitation in the area that became known as la Selva Lacandona (de Vos 1988b:37; Gonzalez Pancheco 1983:53-54). Earlier missionary and military expeditions from Guatemala in the 18th century had already provided some knowledge of this area, from highland attempts to create missionary settlements and private estates. Building upon this knowledge federal and state of Chiapas authorities, in the 1820s, began projects of establishing overland road access to the tropical forests via Ocosingo, to Tenosique, thereby creating both river and land access to the Lacandon forests. By mid-century, therefore, although little timber from this area reached the Gulf coasts via the only practical Usamacinta River route, there had been clearly established the feasibility of so doing. And, equally important, the overland route from the Chiapas highlands would become the principal means of accessing the labor force required to develop a flourishing forest exploitation (de Vos 1988b:45-55).

Since the borders between Mexico and Guatemala were not finalized to the satis-
faction of both countries until the 1890s, initial regulatory initiatives, prior to the
1860s, consisted in granting permits to extract timbers with a levy or tax for each
tree harvested. Because the Lacandon area was in dispute there was both a
question and a conflict over which authorities could issue permits; those from
Flores, capital of the Guatemalan province of El Peten; the governments of either
Mexican states, located in Chiapas (San Cristobal Las Casas), or Tabasco (San
Juan Bautista); or federal officials in Mexico City (de Vos 1988b:51, 103-129).
These latter authorities issued regulations in 1861, which required permit holders
to permit federal inspectors to allow permits and their locations and with the
additional obligation of lumber interests to plant 10 mahogany or cedar seeds for
every tree cut (de Vos 1988b:57). This early sign of concern for reforestation,
however, was more of a rhetorical gesture than a serious attempt to preserve forest
resources.

More important were the Juarez reforms regarding land-tenure allowing private
access to public vacant lands (terrenos baldios) through the legal mechanism of
denunciation (*denuncias*). Initially aimed at communal village lands and clerical
controls over property, and the encouragement of private small-scale landholding,
the 1856 Law of Expropriation (*Ley Lerdo*) had the effect of clearly delimiting
private, communal, and national lands. Access and use thereof, henceforth, would
be federally regulated with specific aims in mind, the increase of agrarian productivity and a shift from subsistence to surplus production. The liberal planner of the
day hoped to introduce known agricultural successes from temperate zone experiences in Europe and North America, the industrious family farm model, into
Mexico. The idea was to provide conditions that would attract great waves of
foreign agriculturalists whereas the result was increased alienation of native held or
claimed lands, and government regulated access to vast territories classified as
national and vacant lands. These included virtually all of the forested tropical
lands, which could now be claimed by private interests (de Vos 1988b:57; Kroeber
1983:2-7; Whetten 1948:85-86). A 1863 law, that increased to 2,500 hectares the
amount of land which could be claimed via *denuncia* in *tierras baldios*, was taken
advantage of by numerous individuals wishing to extract wood from the Chiapas
forests (de Vos 1988b:68), as well as other tropical forests.

The impact of the Juarez legislation, which remained in place until the porfiriato,
had little direct or significant immediate impact upon the tropical forests. But
these liberal reforms set the tone and direction for future exploitation; *first*, by
establishing the principle of federal jurisdiction and legislative supremacy; *second*,
by defining the purpose of national lands in terms of increased economic productivity in order to further the modernization of Mexico; *third*, by delimiting traditional village, clerical and native landholding in order to encourage privatization;

and *fourth*, by establishing policies favoring foreign investment as an important role in national economic development. What happened in the porfiriato was largely an extension and amplification of policies already set out earlier.


## Porfirian Policy and Regulation

Porfirian policy toward tropical areas were defined by the liberal economic planners whose objectives were to modernize and industrialize the nation, and who saw the agrarian sector as a key role. Since most of the tropical forest areas were either officially uninhabited or in areas of traditional indigenous habitation, this vast unexploited zone represented an untapped economic potential. In the highlands, where increased population pressures and extensive deforestation represented a cause for concern, token steps toward preservation can be seen in a 1881 repetition of 1861 regulations. Expressing it concern over irrational exploitation of forests the Secretaria de Fomento recommended that all states "takes opportune steps to conserve forest riches", including an obligation to plant 10 seeds or other trees for each one cut (de Vos 1988b:74). Of far greater significance was the 1883 colonization law which opened the door for large-scale forestry activities.

This *Ley de Colonización* eliminated the 2,500 hectare restriction on the size of land areas that could be claimed by individuals utilizing the legal process of *denuncias*. To encourage surveys of national and uninhabited lands it also allowed private survey companies to retain up to one-third of land formally surveyed. The intent was three-fold; *first*, to encourage the establishment of new settlements to expand agricultural activities, *second*, to encourage private participation and investment in agrarian activities and, *third*, to obtain a national inventory of lands allowing for a more rational direction of economic development. Two results were not long in forthcoming. On the one hand, there was a very substantial increase of activity by lumber interests in tropical forest areas who could now expand their territorial domains beyond earlier restrictions. This was particularly the case by Chiapas and Tabasco family firms which soon dotted the riverbanks of the upper Usamacinta and its tributaries with lumbercamps (*monterias*). On the other hand, these same firms and newly formed survey companies quickly laid claim to vast territories on the basis of having made the required surveys, not all of which could be substantiated (de Vos 1988b:75-101; Gonzalez Pacheco 1983:57ff).

News of the political stability established by the Diaz government, his friendliness toward foreign investments, and better transportation facilities with increased rail lines and better port facilities also began to attract increased capital to the forests. London capitalists formed the Mexican Exploration Company, in 1892, to take advantage of a concession in Quintana Roo, north of the Rio Hondo. Merida businessmen, investing henequen-plantation gains supplemented by German capital, formed the Cuyo y Anexas company to exploit the northern part of Quintana Roo. Their competitor, Fausto Martinez, claimed a concession of over 600,000 hectares on the east coast of Quintana Roo. And French capital was finding its way into the Lacandon area (Konrad 1991; de Vos 1988b:100ff). The combination of national and foreign investments were making increasing inroads into previously avoided tropical forest zones.

It is not surprising, then, that the Diaz government felt that the time had come to regularize and update its legislation. The plan of contracting survey work to private companies in exchange for one-third of area surveyed, by 1892 and in less than 10 years, had seen over 50 million hectares surveyed and 16.8 million hectares transferred to private ownership (Romero 1898:124-25). An additional 5.8 million hectares of national lands were sold directly to private companies and colonization companies, while 7,496 parcels with individual title. The dual processes of delimitation and privatization of native communal lands and expansion of large landholdings by individuals and companies were to be hallmarks of the rest of the Diaz regime. To further encourage this process 2 sets of regulations were passed by the federal legislature in 1894. The first, the Law of March 26, 1894 (*Ley sobre ocupacion y enajenacion de terrenos baldios y nacionales*), set the legal basis for occupation and annexation of national lands. The second, passed on October 1, 1894, was an amplification of Article 18 and 19 of the earlier law, referring to lands for the purpose of exploitation of timber, resins, gums and other products (*Reglamento para la Explotacion de los Bosques y Terrenos Baldios y Nacionales*) and had special significance for tropical areas (de Vos 1988b:133, 262-66). This legislation was comprehensive and became the basis for all subsequent forest legislation, including post-porfiriato forest codes.

These federal laws were reflection of rational management principles concerned both with production and conservation, an enlightened and forward-looking concern that was attempting to balance present and future economic needs. Details of requirements for renters or concession-holders wishing to cut and export timber (mahogany and cedar) and/or extract gums and resins can be summarized under the following points:

1. Permits were required, issued by the Land Agency (*Agencia de Tierras*) of the Ministry of Development (*Secretaria de Fomento*);

2. All concession or permit holders were obligated to manage their operations in complete conformity with the stated regulations of the 1894 laws and other special edicts passed by the Ministry, so that:

the destruction of national forests will be avoided, ensuring, to the contrary, their repopulation by conserving necessary trees with fertile seeds, in order that the existing species in the forests have their reproduction guaranteed.

3. Specific restrictions were placed on size of trees that could be cut, amount of gum or resin that could be taken from designated areas, quotas that must be paid per tree trunk or unit of resin and gum extracted, areas within forest that could be converted to agriculture, and head of livestock that could be pastured;

4. All trees cut for export must be clearly marked by forest inspectors;

5. Federal employees had the right, at all times, to watch over all phases of forestry operations, including transport from forests and ports;

6. At their own expense, concession-holders were obligated to marked boundaries and produce a map of their holdings, to be provided to federal agents within 2 years of being granted their concessions;

7. No transfer to third parties of any exploitation rights were permitted without prior permission by federal authorities, and under no conditions were such rights to be transferred to a foreign state or government, nor admitted as a partner;

8. No use other than stipulated in contract was permitted;

9. If valid claims (*denuncias*) against land were presented by private individuals the concession holders were obligated to turn over such lands to the government within6months;

10. Within rented areas, and subject to previous notice to Ministry of Development, buildings for residence of workers and other constructions housing sawmills, supplies, and equipment might be made;

11. Students of national forestry schools must be admitted to operations when accompanied by a professor;

12. A guarantee of 2,000 pesos must be deposited with the national bank, to be forfeited if any conditions of contract violated;

13. All disputes ot be settled by federal courts, without any recourse to foreign rights;

14. All concession and/or contract holders would always be considered as Mexicans, even if any or all were foreign, and were subject to federal laws and courts, without right to any intervention or influence of a foreign or diplomatic nature;

15. All concessions and/or contracts would be null and void, when so declared by federal agents - after having heard defense of holders - if:

    i.   any regulations governing forests were violated,

    ii.  failure to pay quotas and fees,

    iii. any 6-month interruption of activities without due cause;

    iv.  failure to make and present map within stipulated period,

    v.   failure to build constructions for extractions activities,

    vi.  area or part thereof transferred to third party without consent,

    vii. transfer to, or partnership with, foreign government or state, or agent thereof (DO, 1894).

Contracts were for 2, 5, or 10 year periods, with options for renewal provided terms had been met. Stated conditions emphasized essentially 3 aspects; the federal right to regulate and oversee all aspects of forest activities; revenue accruing to federal coffers from quotas, fees, and other privileges; and conservation of forests through rational, controlled exploitation and their renewal. The concern about conservation of forest resources, among Mexican intellectuals and the Diaz *cientificos*, had already been recognized. A national Society of Friends of the Trees already existed in the 1890s, encouraging the planting of trees and dissemination of information about the negative effects of deforested areas upon climate, flood control, and rainfall. An article from the publication *Mexico Intelectual* titled "Nuestros Amigos los Arboles" became part of the official government record when reprinted in the *Diario Oficial* in September 1894. It reviewed the negative impact of deforestation in the United States and France and advocated restoration of affected lands in Mexico and the reforestation of already devastated forests (DOC, No. 1186, Sept. 21, 1894, pp. 3-4). The concern, at this point, was for the Mexican Highlands rather than tropical forest. This is clearly indicated in the writings of Matias Romero, when he wrote:

> The climatic conditions of Mexico are undergoing great changes on account of the destruction of the forests. The country had formerly a great deal of rain and much humidity in the atmosphere, being covered with thick forests; but with the difficulty of transporting coal already found, the population had had to depend entirely for their supply of fuel upon charcoal, and this has in the course of time denuded the mountains, changing very materially the climatic conditions of some regions of the country. But in the lowlands, being thinly inhabited, the case is different, and the country is still so thickly wooded that it is impossible to pass through it.... In this region abound forests of mahogany, cedar, rosewood, etc. (Romero 1898:37-38).

The emphasis in the tropical forests was to increase rather than restrict exploitation. Early in the Diaz period (1878) export of tropical woods (mahogany, cedar, rosewood, dyewoods) represented 5.07 percent of national exports (NARA/DM, May 21, 1879). At this time the per capita exports from Mexico were, as reported by the United States top official representative in Mexico, the lowest of "almost all other Spanish American countries", and with precious mineral exports representing almost 80 percent of exports. By 1896 tropical woods exports had increased significantly as had other exports, resulting in an actual decrease - down to 4 percent - of national exports as a whole (Romero 1898:161). By then, however, the dependence upon precious metals had declined to roughly 60 percent of the national exports, with agricultural exports having increased greatly. The legislation of 1894, rather than restricting tropical forest exports, was designed to provide a stimuli for greater activity. As pointed out by Matias Romero, then ambassador to the United States, the Mexican lowlands represented a "magnificent arboreal vegetation" with 114 "different species of building timber and cabinet woods...8 of gum trees...oil-bearing trees and plants, of which there are 17 varieties...59 classified species of medicinal plants", plus tropical fruits, sugar cane, tobacco, india-rubber, cotton, cacao, vanilla, chicle, yuca. He drew special attention to the building timber and cabinet woods which, "in the low, hot countries we have all the cabinet woods growing wild." (Robero 1898:43-55).

By the mid 1890s both foreign and national capitalists were taking such promised Mexican opportunities seriously, and continued to do so until interrupted by the Mexican Revolution. Leading newspapers in the United States and Europe were soon running full-page advertisements promoting tropical opportunities in Mexico (Harper 1910). By 1899 the French consul in Mexico was suggesting that the tropical timber business was the "best business in the world", a view also supported by the Belgium ambassador (de Vos 1988b:202).

The Diaz government, at the same time, continued its push to increase private initiatives. By 1896, 30,869,007 hectares of national lands had been transferred to private interests, both national and foreign, via payment for surveys, or through sales (Romero 1898:124-25) at modest prices. The cost per hectare, for acquiring national lands varied, from a low of 1.00 peso in arid zones (e.g. Coahuila, Durango, Sonor, Chihuahua), 1.80 pesos in Campeche and Yucatan, 2.00 pesos in Chiapas, 2.50 pesos in Tabasco, 2.75 pesos in Veracruz, and considerably higher prices in highland states near the capital (e.g. Puebla, 3.35 pesos; Morelos, 4.50 pesos) (Romero 125). In addition, rental contracts for extraction of tropical woods, resins, and gums to foreign and national companies, many of vast areas, were negotiated. The rate of tropical forest penetration and accompanying economic activities, by the late 1890s, increased dramatically and particularly throughout the Yucatan Peninsula, and intensified during the first decade of 1900.

Earlier modest activities in these tropical areas faced 2 main obstacles, the lack of access and export facilities and questionable effective control over territory by the Mexican federal government. Massive investments by the federal government and generous conditions and subsidies for private companies had, by 1900, resulted in an extensive network of railroads being built, plus telegraph lines and port facilities. This facilitated the introduction of industrial technology for commercial economic activities. Control over territory involved 2 aspects, the first being the resolution of boundary disputes with neighboring Guatemala and the British Colony of Belize, and the successful occupation by federal forces of territory held by the dissident Santa Cruz Maya (Lapointe 1983). The boundary questions were resolved in the 1890s and the Maya territory occupied in 1901, resulting, in 1902, in the creation of the Territory of Quintana Roo directly controlled by the federal government. With these problems resolved both commercial companies and the Diaz authorities were in a position to collaborate in extraction of forest resources (Konrad 1991).

The Territory of Quintana Roo, which included the bulk of what had previously been the tropical forests of the State of Yucatan and a significant part of those of Campeche, was quickly sold or rented to forest extraction interests. Already prior to 1902 concessions had been granted, involving over 2 million hectares and between 1902 and 1905 the national *Diario Oficial* published contracts with 6 Mexican, 3 Belizean and 2 Yucatecan parties involving 3,456,857 hectares (Konrad 1991). In effect, the entire territory was now under control of private interests for the purpose of natural resource exploitation.

In neighboring Campeche the transfer, although not as complete, was equally significant. Here the role of United States interests were most visible. Registered in their home country as tropical products (timber, fruits, chicle) and development companies they purchased 1,598,258 hectares of land, virtually all of it forested. Along with national private interests claim was laid to most of Campeche's tropical forests accessible via river or coastal entries. Tabascan-based companies acquired roughly 2 million hectares of forest lands in return for surveys or purchases, resulting in almost all of the Lacandon forest area being under either private national or foreign control (de Vos 1988b; Gonzalez Pacheco 1983).

The total amount of forested land that changed title or was rented has not been precisely calculated, but the intention of the Diaz government to extract for export the maximum potential certainly showed promising results. John Hart's claim that United States interests alone came into possession of 40 million hectares of Mexican land during the porfiriato, representing over 20 percent of the national area (1990:228-230) provides a more global view of the massive shift toward agrarian and forest-related economic activities being sponsored by federal authorities.

What these titles all shared were specific regulatory conditions on exploitation of forest products. And although these existed on paper any ability of enforcement was quite another manner, since title represented less of a control over production than a *carte blanc* stimuli to increase national exports and spur forward the "economic miracle" the Diaz *cientificos* believed they were engineering. In fact, it included not just extraction of exportable staples to hungry European and North American markets but also points of entry for new colonies, the introduction of modern agriculture and, hopefully, a rush of foreign immigrants to create agrarian methods and successes in agriculture, duplicating results in temperate areas of the more industrialized countries (Romero 1898).

The riches of the tropical forested areas were considered by these highland planners, and the foreign companies for that matter, as being virtually inexhaustible. Introduction and expansion of a capitalist economy within all regions of Mexico held a higher priority than conservation. This did not mean porfirian policy was unconcerned with conservation, rather the manner in which such concern was expressed and its focus. The *cientifico* view, expressed at the 1909 North American Conference on Conservation of Natural Resources (Washington, D.C.) by Mexico's president of its Forestry Commission, illuminates the larger picture of official federal concern. The objective of the tri-nation (Canada, United States, Mexico) conference was to examine methods that might be adopted to achieve the conservation of natural resources of the continent. Mexico's concerns were expressed as follows:

...the conservation of natural forests and other forest resources is for us, in Mexico, the most important element for our consideration, because this is the

most squandered or being exhausted; that we need most and upon which
depends, in large part, convenient advantage of exploitation and also the
conservation of other natural elements, such as local rainfall, streams or
water sources, the richness of our agricultural soils, wild animals, and even
the economic exploitation of our great mineral resources, which, without
sufficient sources of timber and water for motive power, becomes almost
impossible. (AGNG, S/S, Caja 830, Exp. 1, Feb. 17, 1909)

The Mexican delegate then went on to describe conditions in the one-third of the
country represented by the Central Highlands, where less than 10 percent of the
area was still forested.  Pointing out that economists of the day considered a 33
percent forest cover was necessary for maintaining "economic equilibrium", allowing for adequate ground cover for retention of precipitation preventing aridity,
floods in wet seasons and lack of water for urban populations in dry seasons, the
problems facing Mexico were serious.  Not only had the water levels of the Valley
of Mexico Lakes declined from 1 meter to a few centimeters in the past 20 years,
causing serious habitat degradation, but similar problems were  also in clear view
on the Pacific and Atlantic slopes of the Central Highlands, currently calculated of
retaining, respectively 25 percent and 30 percent of forest cover.  With the rapid
extension of communication infrastructures and increased access to remote areas,
he indicated an urgent need to take adequate steps to prevent abuse of forest
resources.  Ongoing abuses were identified as including careless burning of
pastures resulting in forest fires, slash-and-burn agriculture in mountainous regions
not suited for cereal production, and pastoral abuse in the form of overgrazing in
tropical regions.  The end result from such abuses was removal of forest riches,
their displacement by economically valueless scrub cover and increasing infertility
of soils (AGNG, S/S, Caja 830, Exp. 1).

The overriding highland, temperate zone, emphasis expressed in the Mexican
position accurately reflected dominating views.  The tropical lowlands, despite a
passing reference to over-grazing, were being targeted for more intensive intervention by commercial interests.  The remedial actions, claimed to be taken by the
Mexican Forestry Commission, included an 8-point program: "In order to guarantee the indefinite conservation of forested areas necessary for the nation, to create a
reserve of all national lands as national forests" (AGNG, S/S, Caja 830, Expl. 1).

Mexico's official plan of action, long on generalities but lacking in concrete details
on how such plans might be implemented, was clearly focused toward the
highlands, toward the implementation of modern forestry methods, and specifically
targeting indigenous slash-and-burn traditional agriculture as principle culprit.  In
general terms it was both a rhetorical restatement of some of the principles already

legislated in 1894, the regulatory jurisdiction of federal authorities, and incorporation of temperate zone forestry experiences.

The lack of concern over activities taking place in the tropical forests and in tropical zones in general is not that strange if one takes into consideration 2 overriding
contemporary considerations. On the one hand, despite the massive and large-scale
penetration of commercial activity into tropical forest areas, clearly seen in concessions granted, rental contracts, and properties transferred to private interests, actual
timber and forest resource extraction had not yet reached levels causing concern.
Federal and private sector interest was still primarily being directed to increased
activity and greater foreign exports from tropical zone agrarian activities. On the
other hand, there was both a national and international perception that the productivity of tropical zones was inexhaustible and that Mexico's extensive tropical
regions were not, in any way, at risk. Quite to the contrary, they were in dire
need of modern "development."

## Attitudes Toward Tropical Forests

The second half of the nineteenth century saw a steady stream of foreign visitors
coming to Mexico to see first-hand both the exotic peoples and country-side, and to
explore the potential of economic investment. Many also wrote about their observations and experiences, resulting in materials eagerly devoured by the North
American and European reading public. Agents of the Diaz government, as well,
made concerted efforts to spread the good word of countless and very profitable
economic opportunities in Mexico's least economically active areas (e.g. Romeno
1989). The attitudes and opinions in such sources provide insights about late 19th
and early 20th century attitudes toward the Mexican tropics. Such views included
not only habitat and flora and fauna, but also Mexican society in general and rural
landscapes, (Kaegrger 1986: Katz 1980). And frequently, in the case of many,
they revealed more about their own attitudes than they did about the nature of
Mexico.

The case of a *Chicago Times* correspondent, John F. Finerty, sent in 1879 to
accompany and report upon a 80 member American Industrial Deputation visit
interested in promoting trade with Mexico, illustrates a common trend. His view
of tropical regions was restricted to Veracruz, and from this port a journey to the
Mexican highlands. His emphasis was on the negatives, the horrors of *vomito
negro*, the dangers of snakes and scorpions, and the backwardness of Mexicans in
general. He was, however, somewhat overwhelmed by the sights and smells of
tropical regions the deputation passed, causing him to write that "No one has truly
lived who has not seen and enjoyed some, at least, of the glorious vistas bestowed

by the Almighty bounty on this beautious planet" (Timmons 1974:88). Regarding the native Mexicans, he considered them "incapable of advancement. Although the Mexican government has done its utmost to render them more progressive" (193). Col. Albert S. Evans (1880), who accompanied a United States diplomatic mission, literally gushed with enthusiasm. He argued that "With railways, a good and liberal system of revenue laws, and a few years of uninterrupted peace, Mexico could supply the United States, Canada, and much of Europe with all the sugar required, and control the market of the world" (466). And after passing by the tropical vegetation near Orizaba he proclaimed:

> To sum up in a word, the Republic of Mexico has within her limits the resources of wealth and comfort unbounded, and the day will come--I trust it may not be far distant--when she will be regarded, with reason, as the Paradise of the world. (467)

Such a mixture of views of limitless tropical bounty, if only harnessed by progressive government and modern technology, was widely shared and became the bedrock of porfiriato planning. Gilbert Haven, another United States traveller and book-writer of the 1870s, saw the central problem as one of "insecurity of property" (1875:57) and great potential of tropical economic activities in agriculture. He recognized the lumber potential but was not enthralled by such prospects:

> Many sorts of . . . hardwoods are here, awaiting the horrid steam saw-mill that shall eat them all up, and ship them to New York, and make this green, grand wilderness a desolation. (67-58)

Haven wished to inform his countrymen of the Mexico they knew little about but was not a promoter of tropical forest exploitation. As a widely travelled and observant writer he had already seen the impact of deforestation, and wished the tropical forests to remain intact:

> How sorry I am to be compelled to thank that some Yankee speculator in lumber from Bangor to Brainerd will read these lines, and will be up and off in the next steamer to Vera Cruz and the splendid woods . . . . Cortez did not sigh more for Mexican silver than these lumbermen will for these mahoganies and rosewoods, and other equally polishable delights. Black walnut will be of no account when the Mexican lumber reaches the Northern market. Give us good fill, dear ancient forests, of your green delights, for the Yankee wood-sawyer is coming, and you will soon be no more.(58)

Frederick A. Ober, travelling through Mexico in the 1880s, echoing the contemporary view of how climate affected inhabitants of tropical zones, claimed that they were wonderfully indolent, "They are not fond of hard work, nor have they any need of it. A few dozen bananas, a small field with manioc and maize, afford nourishment without much labor. . . . Many days are passed extended on the mat, playing the guitar, or staring up at the blue sky." (1887:677) Nowhere was this more clearly expressed than in the Yucatan peninsula and the forested areas occupied by the Santa Cruz Maya. A lead editorial published in Campeche's *Diario Oficial*, in 1891, expressed frustration that its "virgin forests" were controlled by the Maya, "possessors of a great part of our rich and exuberant territory." To exploit these riches was of great importance since "this would, in large part, stimulate the enrichment of the states [Campeche and Yucatan] which found themselves deprived of such powerful elements for their greater engrandizement. . . . Everything that tends to advance humanity" the editorial argued, "we desire . . . and that the savage hordes that exist in our peninsula . . . should also acquire the benefits of civilization." (DOC, March 13, 1891) The governor of the State of Yucatan, in his 1896 address to the state congress, repeated the same view, arguing that the pacification of the Maya was necessary for "the requirements of civilization" and in "agricultural and mercantile interest." (RP, Jan 1, 1896)

The Mexican federal position, directed to the United States businessmen by Matias Romero, coincided with many of the views of foreigners and Mexican capitalists. As he wrote in the late 1890s, Mexico had "immense elements of wealth . . . we shall be able to provide the United States with most of the tropical products . . . which they now import from several other countries" (1896:9). Stressing the need for foreign capital to develop natural resources he pointed out that "This will never be developed until those who have capital to invest are acquainted with the unparalleled opportunities for safe and profitable investment in Mexico" (126).

Henry H. Harper, a young United States investor, came to personally inspect the tropical economica potentials, having been lured by glowing reports of profits to be made by investing in vanilla, coffee, and rubber production (in 1896), later published a book, *Experiences, and Observation on Agriculture and Industrial Conditions* (1910). In it he explained:

> I can readily understand the tendency of writers to praise the beauty of Mexican scenery and to expatiate upon the wonderful possibilities in all agricultural pursuits. In passing rapidly from one section to another without seeing the multifarious difficulties encountered from seedtime to harvest, they get the highly exaggerated ideas from first impressions, which in Mexico are always misleading. The first time I beheld this country, clothed in the beauty of its tropical verdure, I wondered that everybody didn't go there to live, and now marvel that anyone should live there, except possibly for a few months in winter.(4).

Harper was, of course, correct in his evaluation of the superficial nature of traveller accounts, a criticism that was equally valid of highland federal *cientificos* bent
on developing the tropical regions and exploiting its forests.  For they, as well,
generally had only the most superficial knowledge of tropical forest dynamics of
longterm commercial impact of commercial agriculture on tropical regions.
Overall, the Porfirian attitude toward tropical forests was driven by economic
factors and rapid modernization goals.  Official views accepted, uncritically,
extremely optimistic views of potential and extremely negative views about native
inhabitants.  Borrowing its economic liberalism from abroad, the Porfirian
*cientificos* believed that human engineering--through education, the demonstration
effect of foreign immigrants--could gradually convert subsistence agriculture into
commercial industrialized methods of production.  Foreign experience, borrowed
without critical review, was integrated into official policy formation.  Disease and
backwardness came to be linked with the tropical forests themselves and its
inhabitants.

### The Regulations of Tropical Forest Activities

The optimism and *laissez faire* capitalism espoused by the Porfirio Diaz cabinets
considered the implementation of carefully crafted regulations as secondary to
more important economic goals.  Feedback from international diplomatic sources,
which praised the policy directions taken (NARA, DM 1821-1906), and the
increasing volume of foreign investments resulting in the expansion of industrial
infrastructures for modernizing economic production was, after all, concrete
evidence that the "peace and progress" strategies of the *cientificos* was on track.  It
is not surprising, then, that attention to the actual compliance to regulations or the
need to expend significant resources to supervise and enforce regulations had a low
profile.  Such was definitely the case in tropical zones and specifically in tropical
forest commercial activities.  This did not mean that a national structure for admin-
istrating regulations was not created, rather that such never received adequate
funding in order to even begin comprehensive enforcement of legislated regulations.

The Ministry of Development, Colonization and Industry was responsible for
agrarian matters and through it rental contracts, concessions, and property transfers
were implemented.  Within the ministry, its land (*tierras*) section had authority
over forestry matters, (Escuela Forestal de Guardas) under which there was a
Department of Forests.  Within this Dep there was a Central Forest Council, with
branches in each state called local councils (*Junta Local de Bosques del Estado*).
Such councils, located in state capitals, were not in a position to do much more

than make recommendations to political authorities about forestry matters. The Department of Forests was also responsible for the National Forestry School, located in Coyoacan in the Federal District. At the end of the Porfiriato this school had a director, 7 professors, 3 administrative assistants, and between 20 and 30 students. This is where the forestry personnel was trained and from which came the forest guards and inspectors who were to implement the regulations detailed in the 1894 forestry legislations (AGNF, Bosques, Caja 71, leg. 90, exp. 1522)

A report on the activities and facilities of the School of Forest Guards, in 1912, complained that the location and facilities of the school were entirely inappropriate, lacking in adequate classrooms, appropriate texts, and other necessities for such an establishment (AGNF, Bosques, Caja 29, Leg. 99. Exp. 1739). Annual financial statements concerning the forestry personnel of the Department of Forests, in 1910, indicate the following information about number of persons active, their location, salary and annual costs in 6 tropical zone states.

| State and Location of Administration | Number | Titles | Salary Pesos/Daily | Annual Cost |
|---|---|---|---|---|
| Campeche | 5 | subinspector | 1.98 | |
| Campeche | 9 | guardabosque | 0.99 | 6,865.65 |
| Chiapas | 4 | subinspector | 1.00-2.00 | |
| Tuxtla Gut. | 4 | guardabosque | 0.50-1.00 | 3,292.30 |
| Quintana Roo | 1 | subinspector | 3.00 | |
| Santa Cruz | 2 | guardabosque | 2.00 | 2,555.00 |
| Tabasco | 4 | subinspector | 1.32-1.66 | |
| S.J. Bautista | 7 | guardabosque | 0.66 | 3,737.60 |
| Veracruz | 1 | subinspector | 1.65 | |
| Minititlan | 2 | guardabosque | 0.83 | 1,208.15 |
| Yucatan | 2 | subinspector | 2.31 | |
| Merida | 7 | guardabosque | 1.65 | 5,902.05 |
| Total | 17 | subinspectores | | |
| | 31 | guardabosques | | 23,580.75 |

Source: AGNF, Bosques, Caja 71, Leg. 90, Exp. 1522.

Additional states with inspectors and forestguards were Chihuahua (1), Durango (8), Jalisco (1), Neuva Leon (1), and Zacatecas (2), resulting in a total of 60 active inspectors and forest guards for the nation. These national guards and inspectors were under the ministrative authority of a land agent (agente de tierras) located in each state capital. These land agents were functionaries of the tierras section of the Ministry of Development.

The data on national inspectors and guards and the minimal funds spent on the activities of the national school and for forestry regulatory control clearly indicates a symbolic rather than serious attempt to implement the exploitation and conservation regulations in federal legislation. With only 3 individuals for the entire territory of Quintana Roo, for example, it was impossible to exercise any sort of vigilance over forestry activities. Lumber companies in the tropical forests or those extracting chicle had free reign to conduct their activities at will. Occasionally there were expressed concerns over the lack of control, for example, in a Yucatan periodical article (RM, July 7, 1906) which argued that "the exploitation of chicle should be subject to more scrupulous vigilance. The devastation fo the zapote [trees] . . . suffering from deadly attack. The extraction of chicle has no order or vigilance, which of great prejudicial consequences, because it brings death to healthy trees. For this reason the federal government ought to pay attention to the problem, even in the most distant regions." Small wonder, as well, that territorial authorities complained that the inspectors were actually working for the forestry extraction concerns, aiding and abetting the evasion of quotas and taxes, and turning a blind eye to illegal smuggling of products into the neighboring British colony (Konrad 1991).

**The Impact of Porfirian Policy and Practice**

Our contemporary concerns with ecology, deforestation, environmental degradation, and the fragility of tropical forest ecosystems have provided an entirely different worldview than that of the late 19th and early 20th centuries. We have had, after all, the benefit of hindsight derived from a century of experiences. This needs to temper the temptation to merely critique past Mexican practices that sought to maximize exploitation of its tropical zones and, in particular, its tropical forest resources. Official Mexican opinion placed, we can say with hindsight, too great a confidence in the rationality of the investing foreigners, who, with their superior technological and scientific advances, would both know how to implement forward-looking extraction principles in their forestry endeavors. What was lacking, within both European and North American scientific circles, was even a basic understand of tropical forest dynamics. Thus they applied, with considerable vigor, the same techniques and strategies that had proved to be successful in temperate zone habitats.

The native populations of Mexico--with a much longer tradition of longterm successful residence in tropical forest habitats, and with a much better understanding of succession dynamics--were, unfortunately, considered part of the problem rather than having anything to contribute for the future. Such a fundamental misreading of the larger picture, mutually reinforced by the interactions between Mexicans and their trading partners, was to have impacts of great consequence. And the current state of affairs, wherein much of Mexico's tropical forest habitat has been irreversibly degraded, is a direct outgrowth of the policy and practice of the Porfiriato.

Yet most of this did not take place during the 1876-1910 period. By the end of the Diaz regime Mexico's tropical wood exports (cabinet woods, dyewoods) as a percentage of total non-mineral exports, actually declined (Gonzalez Pacheco 1983:136-138), even though the total volume increased. England and other European countries were still importing most of their tropical woods from other sources, and the United States was only beginning large-scale import from Mexican sources. In specific regions, Tabasco for example, tropical wood exports increased ten-fold in value, between 1889 and 1910 (de Vos 1988b:211). The coming of the Mexican Revolution and the insecurities it produced reduced tropical exports, and the advent of World War I cut off European markets. In the postwar period the United States became the dominant importer of Mexico's tropical forest products, with timber being the most important state source of revenue for Tabasco, while chicle became the most important source of revenue for Campeche and Quintant Roo.

The major impact of policy and practice, for neither changed significantly after the Mexican Revolution, was in the post-Diaz period. What was established during the porfiriato, however, were the basic conditions allowing for massive reshaping of Mexico's tropical forest areas and systematic extraction and export of its resources. For by opening these regions to foreign and domestic commercial activities, the wheels were set in motion, not only for exploitation of timber, resins, and gums, but also for an assault upon the flora and fauna as a whole. Ironically, the regulatory legislation passed during the Diaz period could have, if implemented, largely restrained what we now know to have been irrational exploitative practices. Overriding short-term economic objectives, the dominating paradigms of modernization imported from foreign sources, and failure to put any teeth into regulatory implementation, however, prevented this from happening. In retrospect, the rather romantic traveller and writer of books of the 1870s (Gilbert Haven) turned out to be a astute forecaster of events when he lamented what might

be in store: "Give us good fill, dear ancient forests, of your green delights, for the
Yankee wood-sawyer is coming, and you will be no more" (1875:58). Had he
substituted "modern, industrial society" for "the Yankee wood-sawyer", his
forecast would have been more precise.


## Primary Sources

| | |
|---|---|
| AGN | Archivo General de la Nacion, Mexico City |
| AGNF | Fomento |
| AGNFOP | Fomento y Obras Publicas |
| AGNG | Gobernacion |
| AGNG,S/SS | Sin Secciones |
| DO | Diario Oficial, Mexico City |
| DOC | Diario Oficial de Campeche, Campeche |
| ARA | National Archives and Records Administration, Washington DC |
| NARA/DM | Despaches from Mexico |
| RM | Revista de Merida, Merida |
| RP | Razon del Pueblo, Merida |


## References

Clegern, Wayne M., 1967. *British Honduras: Colonial Dead End, 1859-1900.* Baton Rouge:
Louisiana State University.

Comparato, Frank E., ed., 1987. *Fray Andrea de Avendano y Loyola: Relation of Two Trips to
Peten, made for the Convestion of the Heathen Ytzaex and Cehaches.* (Charles P., Bowditch
and Guillermo Rivera, trans.) Culver City, California: Labyrinthos.

de Vos, Jan, 1988a, *La paz de Dios y del Rey: la conquista de la Selva Lacandona (1525-1821).*
2nd ed. México: Fondo de Cultura Económica. 1988b, Oro Verde: la conquista de la Selva
Lacandona por los madereros Tabasquenos, 1822-1947. México: Fondo de Cultura Economica.

Evans, Albert S. 1889. *Our Sister Republic: A Gala Trip through Tropical Mexico in 1869-70.*
Hartford, Conn.: Columbian Book Co.

Gonzalez Pacheco, Cuauhtemoc, 1983. *Capital extranjero en la selva de Chiapas, 1863-1982.*
México: Instituto de Investigaciones Economicas.

Harper, Henery H., 1910. *A Journey in Southeastern México: Narrative of Experiences, Observations of Agricultural and Industrial Conditions*. Boston: De Vinne Press.

Hart, John M., 1990. *El México revolucionario: gestacion y proceso de la revolucion Mexicana*. México: Alianza Editorial Mexicana.

Haven, Gilbert, 1895. *Our Next-Door Neighbor: A Winter in México*. New York: Harper & Brothers.

Kaerger, Karl, 1986. *Agricultura y colonizacion en México en 1990*. (Pedro Lewin y Gudrun Dohrmann, trans.) México: Universidad Autonoma Chapingo.

Katz, Friedrich, 1980. *La servidumbre agraria en México en la epoca porfiriana*. México Ediciones Era.

Konrad, Herman W., 1987. "Capitalismo y el mano de obra en los bosques tropicales de México: el caso de la industria chiclera," *Historia Mexicana*, 43, 1: 465-505. 1991. "Capitalism on the Tropical Forest Frontier: Quintana Roo, 1880s to 1930," in J. T. Brannon and G. Joseph, ed., *Land, Labor, and Capital in Modern Yucatan: Essays in Regional History and Political Economy*. Tuscaloosa: University of Alabama Press.

Kroeber, Clifton B., 1983. *Man, Land, and Water: Mexico's Farmlands Irrigation Policies 1885-1911*. Berkeley and Los Angeles: University of California Press.

Lamb, F. Bruce, 1966. *Mahogany of Tropical America: Its Ecology and Management*. Ann Arbor: University of Michigan Press.

Lanz, Manuel A., 1905. *Compendio de historia de Campeche*. Campeche: El Fenix.

Lapointe, Marie, 1983. *Los mayas rebeldes de Yucatán*. Zamora, Mich.: El Colegio de Michoacan.

Martinez, José Luis, 1990. *Hernán Cortés*. México: UNAM y Fondo de Cultura Economica.

Ober, Frederick A., 1887. *Travels in Mexico and Life Among the Mexicans*. Rev. ed., Boston: Estes and Lauriat.

Revel-Mouroz, Jean, 1980. *Aprovechamiento y colonizacion del tropico humido Mexicano*. (Jose Barrales Valladares, trans) México: Fondo de Cultura Economica.

Romero, Matias, 1898. *Coffee and India-Rubber Culture in Mexico*. New York and London: C. P. Putnam's Sons.

Timmons, Wilbert H., ed., 1974. *John F. Fenerty Reports Porfirian Mexico, 1879*. El Paso: Texas Western Press.

Wells, Alan, 1985. *Yucatan's Gilded Age: Haciendas, Henequen, and International Harvester, 1860-1915*. Albuquerque: University of New Mexico Press.

Whetten, Nathan L., 1964. *Rural Mexico*. Chicago and London: University of Chicago Press.

The Long-Term Effects of the Introduction of Sheep
into Semi-Arid Sub-Tropical Regions

Elinor G. K. Melville
York University

In this paper I will use the sixteenth century history of the Valle del Mezquital, Mexico as a case study to examine the related problems of the ecological consequences of the introduction of alien species, and human action in landscape formation: specifically, the formation of eroded scrub-covered badlands characteristic of many semi-arid sub-tropical regions.

The Valle del Mezquital is the archetype of the poor and degraded regions of Mexico, and many have thought that its eroded and arid mesquite-dominated landscape is the reason for its poverty. The historian Miguel Orthón de Mendizábal, for example, thought that the exploitation of this region by cattle grazing in the seventeenth century and by pulque and rope production in the eighteenth and nineteenth centuries was a natural outcome of the poor resources of this region.[1]  Following Mendizábal, Sherburne F. Cook also argued that the southeastern quarter of the Valle del Mezquital was used for grazing because the Spaniards saw a region that was fit for little else.  He argued that deforestation and sheet erosion resulting from the pressure of the huge pre-conquest populations accentuated but did not change the essential poverty of the region.[2]

The sixteenth century history of the Valle del Mezquital challenges the assumption that this region is inherently resource-poor.  The Valle del Mezquital lies just to the north of the Valley of Mexico in the rain shadow of the Sierra Madre Oriental. Agriculture is plagued by frequent frosts and uncertain rainfall.  But at the time of the conquest, and for some time thereafter, huge quantities of grains (primarily maize) were produced for subsistence and exchange, and for tribute to be paid to the Aztec Empire.[3]  Harvests were secured by the widespread use of irrigation; and up to the last quarter of the sixteenth century, sufficient water was internally generated to maintain irrigation systems in practically all parts of the region.  The water supply was obtained by the construction of run-off dams and from springs.[4] The presence of springs and the extent of irrigation indicate that the catchment area was in good condition.  This supposition is confirmed by documentary evidence demonstrating that the hillsides--now bare rock--were forested with oak and pine up to the middle of the sixteenth century.  The documentation also indicates a greater vegetative diversity: willows and cedars marked springs and lined the bank of streams, and native cherries and mesquites grew in villages; it is significant that until the last quarter of the sixteenth century the mesquite grew as single large trees, not as chaparral.[5] It is true that the vegetation of the northeastern quarter of the region was dominated by maguey and nopal cactus,[6] but even here chiles and beans were grown in humid bottom lands. In fact, swamps were quite a common feature of the landscape of the Valle del Mezquital--something that is almost unbelievable today.[7]

By the end of the sixteenth century, however, this landscape had been replaced by that traditionally associated with this region: the forests had virtually disappeared and trees were restricted to mountain tops and quebradas.[8] The flat lands and piedmont were no longer covered with well-populated villages surrounded by fields of maize, beans, and squash; instead they were coated in a thick covering of mesquite, lechuguilla maguey, ocotillo, nopal cactus, yucca, and cardón.[9] The high steep hills had been eroded down to bed-rock, and the piedmont was scarred by sheet erosion and gullies and covered with the rubble of slope-wash debris. Springs were failing, and with them the water necessary to maintain irrigation on any but a reduced scale. Indian communities, decimated by disease, were hemmed in by sheep stations and isolated in a transformed and impoverished landscape.

I have demonstrated elsewhere a close correlation between this very rapid process of environmental degradation and the expansion of pastoralism in the Valle del Mezquital. I proposed that overgrazing by vast armies of sheep between 1560 and 1580 was a major variable in the transformation of this region.[10] But the simple addition of grazing animals to a new environment does not by itself explain environmental degradation. Studies carried out to ascertain the effects of introducing ungulates into New Zealand, for example, demonstrate that environmental degradation is not an inevitable result of the introduction of these species into new habitats.[11] These studies show that most changes in the composition of the vegetative cover caused by overgrazing are reversible if grazing pressure is removed.[12]

Why, then, did the introduction of sheep have such a devastating effect on the environment of the Valle del Mezquital? It is significant that the conclusions reached by the biologists are based on research carried out in populations of nondomesticated species, uncontaminated by human interference. But in the case of the Valle del Mezquital, the animals were introduced as part and parcel of the institution of pastoralism, and were thus closely affected by human decision-making. Pastoralists added flocks to this region during the period of rapid natural increase that follows the acclimatization of introduced species. By doing so they amplified the rate and degree of population growth and accelerated the speed and profundity of the associated environmental changes. The process of environmental change outran the capacity of the soils and vegetation of this semi-arid region to regenerate over the short period--especially since the pressure was maintained by the addition of yet more flocks--and the result was environmental degradation. In the first instance pastoralists added more flocks because they saw a very fertile and productive region; in the second instance because they perceived it as a region that was "fit only for sheep". In both cases pastoralists were responding to their perception of the nature of the natural resources of the region and to market forces. Unfortunately, the Spaniards did not understand the ecology of this region, and judged its potential on the productivity of the indigenous modes of production, not on its capacity for sustaining the addition of grazing animals. I suggest, therefore, that the ultimate cause of the degradation of the region was human decision-making based on imperfect information.

In order to demonstrate this hypothesis I will first compare the history of the Valle del Mezquital with the model of the introduction of ungulates into new regions, and then examine the attitudes and actions of the Spaniards.

The introduction of ungulates into a new environment results in an irruptive oscillation "where animals and vegetation (and thus the carrying capacity of the habitat) follow reciprocal trajectories".[13] This process has been broken down into 4 stages:

During stage 1, there is a progressive increase in numbers in response to the discrepancy between the carrying capacity of the habitat and the numbers of ungulates present. The death rate is low and as new generations reproduce, the population curve increases rapidly. Eventually preferred species in plant communities become over utilized. During stage 2, the population exceeds its carrying capacity as more extensive areas of vegetation are over utilized. Physical condition at the critical season drops noticeably; as juvenile death rate increases the population starts to level off in number. In stage 3, the population begins to decline, especially when an element of the environment such as climate becomes critical. Latterly, some plant species less susceptible to browsing or grazing pressure may begin to recover as the animal density declines to a level more nearly compatible with the current carrying capacity of the habitat. Stage 4 commences when a degree of stability has developed between numbers of ungulates and the new carrying capacity of the habitat. Population density at this stage remains lower than peak density, because plant communities have a lower cover of preferred species than when the population was first introduced.

Once this complete adjustment has taken place, the introduced population, ecologically, is the same as any established population.... Further irruptions of a dampened nature may occur only after creation of another discrepancy between carrying capacity and numbers present.[14]

The Spaniards first introduced cattle, horses, sheep, and goats into this region in the 1530s.[15] During the next 3 decades the number of individual flocks of sheep increased slowly: 34 flocks were introduced into the region during the 1530s, 41 in the 1540s, and 33 in the 1550s. At first the flock size also increased slowly. But during the third decade ganado mayor (cattle and horses) were expelled from this and other densely populated regions; ganado menor (primarily sheep, but also goats and pigs) were left. The sheep population responded to the removal of competition and began to increase rapidly: over the 1550s flocks grew fourfold from an average of 1,000 to an average of 3,900 head; and by the mid-1560s flocks of around 7,500 head were being corralled on individual estancias. The total number of sheep in the region increased from an estimated 75,000 head of sheep at the beginning of the 1550s to approximately 421,000 head at the end of the decade; and by 1565 there were 276 flocks and an estimated total of over 2 million head of sheep scattered unevenly over this large region of 10,000 km$^2$.[16] The increase in the number and density of sheep was associated with the conversion of croplands and hillsides to grasslands.[17]

The second stage lasted from the mid-1560s to the mid-1570s and was marked by huge numbers of animals, high grazing rates, and the failure of pasture. Stations of 780 hectares were now stocked with an average of 10,000, sometimes as many as 20,000 head.[18] I estimate that there may have been as many as 4 million head of sheep in this region in the early 1570s. (A quote that illustrates the density of the animals: "No hay ni se halla en este pueblo mas de ovexas y desto hay buen multiplico."[19]) Sheep were still grazed in common in the tierras baldias y realengas and in fallow fields, so the grazing rates were lower than might be expected from the stocking rates; but they were still very high: a peak grazing rate of 785/km$^2$ is estimated for the most intensively exploited sub-area. (While high, this rate does not come close to those calculated for other regions: Simpson, for example, reported a grazing rate of 2,857 sheep/km$^2$ in Tlaxcala in 1542.[20]) Not too surprisingly, many areas were left completely bare of vegetation under this sort of pressure; deforestation was extensive; and desert species resistant to grazing animals began to appear. But, despite evidence that the carrying capacity of the region had been overreached, another 123.5 stations were added bringing the regional total to 583.5 stations.

During Stage 3 (ca. 1575-1599) a process of profound and extensive environmental degradation became evident and flocks declined abruptly. Soils exhibited marked changes: erosion was particularly noticeable as the heavy summer rains formed swift flowing currents that washed down the hills, carried sheets of soil off the fertile piedmont areas, cut deep gullies through agricultural fields, and finally deposited debris on flat lands. Because these waters no longer soaked into the soil but ran rapidly over the top, the water tables dropped and springs failed.[21] And as the temperature of the denuded soils increased and moisture retention decreased, the desert species spread and increased in density; and by end of the 1580s many areas were covered by an impassable cover of mesquite-dominated scrub. Between 1580 and 1599 average flock size dropped from 10,000 head to 3,700 head; and the number of animals in the region at the end of the century was half the 1579 total. Despite the drop in productivity, however, yet another 462.5 new stations were taken up between 1580 and 1599, and many new flocks added to the region.

By the early seventeenth century the animal populations and the vegetation appear to have achieved some sort of mutual adaptation; and the process had moved into stage 4. As well, the bulk of the land had been moved into the Spanish system of land tenure and few new stations were opened up--and presumably few new flocks were developed.

The basic model of an irruptive oscillation has been modified to take into account the fact that ungulates introduced into large islands such as New Zealand can disperse from the point of liberation. It has been suggested that:

> the sequence of events...occurred successively in each new area occupied. At the newest dispersal front, and furthest from the point of liberation, density starts to increase (stage 1); further back in the range density is at its peak (stage 2); nearest the point of liberation the population attains relative stability at a lower density (stage 4). The sequence of stages following liberation could thus be observed either spatially or temporally, at one place over a period of time, or at one time over a range of distance.[22]

A temporal and spatial sequence of stages can also be discerned in the history of settlement of the Valle del Mezquital. The Valle del Mezquital is a region of approximately 10,000 km$^2$. All parts of this large region were not settled at the same time; nor were they subjected to the same intensity of exploitation during the sixteenth century. Pastoralists, primarily encomenderos in the early decades, shunned the poor areas and introduced their flocks into those areas which exhibited the best resources (water, soils, and pasture): i.e., the densely populated and highly productive agricultural areas on the northern border of the Valley of Mexico. As the density of flocks increased, pastoralists expanded their operations into the less attractive areas with fewer water resources, lower rainfall, and poor pasture. As the huge flocks flooded across the land environmental changes first noticed in the rich southern regions were repeated. And as the process extended into the more arid sub-areas, deterioration of the environment appeared after fewer years of intensive grazing than had been the case in the more humid southern areas, and this had the effect of producing a similar level of degradation all over the region at much the same time until, by the end of the sixteenth century, the region had been transformed into an homogeneous mesquite-dominated desert.

An indication of the Spaniards initial perception of the Valle del Mezquital is found in the geographic relations made ca. 1548. At this date they classified the Valle del Mezquital as a productive agricultural region with good possibilities for wheat production. Despite their perception of this region as a potential wheat producing region, however, the Spaniards left agriculture pretty much to the Indians until the late 1570s,[23] and began instead to add the flocks of sheep and herds of cattle, horses, and pigs so necessary to their culture and economy.

As the Spanish population of New Spain grew, and the number of Indians who ate meat, wore woolen clothes, and used tallow candles increased, the market for pastoral products, especially wool for the growing textile industry, grew accordingly.[24] A growing market proved irresistible, and Spaniards and Indians alike rushed to take advantage of the rich potential for pastoralism that the natural increase in numbers during the 1550s seemed to indicate. Between 1560 and 1565 alone, 168 new sheep stations were taken up by squatters and grantees.[25] The new flocks also increased rapidly, and during the following 15 years yet more flocks were added--despite evidence that the carrying capacity of even the most fertile southern areas had already been reached. By the mid-1570s intensive sheep grazing had displaced Indian agriculture and dominated regional production: a fertile agricultural region had become a rich grazing region--a place of endless increase.

But the period of intensive grazing was already at an end, and over the next 20 years the flocks dropped by nearly two-thirds and the process of environmental transformation had shifted into a process of profound environmental degradation. The period of intensive grazing, high reproduction rates, and high levels of production lasted a very short time: i.e., 5 to 7 years in any one place. And by the mid-1570s the pastoralists had begun to use fire in order to stimulate pasture regrowth, and those who could sent their flocks west to Michoacan in the long dry season.[26]

By the 1580s the Valle del Mezquital was no longer perceived as a fertile and productive region. It was an eroded and scrub covered badlands, a refuge for bandits and escaped slaves, a dangerous place where sheep were lost. It was now thought of--and written about--as being "fit only for sheep"; a place so poor that it could be used for little else. The fact that flock size was dropping rapidly only served to confirm the evaluation of the poverty of the region. And yet, still more stations were taken up, and more flocks added to the region: during the last 2 decades of the century more stations were taken up than in the previous 5 decades combined.

This seemingly irrational behavior can be explained as a response to a shift in the market for pastoral products, and to perception of the Valle del Mezquital as a sheep grazing region.

During the 1560s the market for pastoral products in New Spain was characterized by abundant supplies and low prices--the result of the spectacular increase in the introduced animals in all parts of New Spain.  Starting in the late sixties, however, flock size began to drop rapidly and the supply of pastoral products declined and prices rose.  A similar process of dropping supply and rising prices occurred in the market for agricultural products--brought about by the decline in indigenous production in the 1576-81 epidemic.  But the degradation of the Valle del Mezquital and the fact that it was perceived to be fit only for sheep, meant that agricultural exploitation was not a serious alternative; and sheep grazing, albeit now extensive rather than intensive, continued to dominate production.[27]

It is probable that not all the stations added to the region in the 1580s and 90s represented new flocks.  These decades were marked by the formation of latifundia as individual pastoralists moved to acquire large areas of land in order to monopolize pasture; and the accelerated process of land takeover may simply reflect this process.  But if just a half or a third of the new holdings were stocked with new flocks, they would have exerted increased pressure on an environment already under severe stress, and accentuated the process of degradation.

The activities of the sixteenth century pastoralists compromised the productive potential of this region for the next 350 years.  Following a period of cattle grazing in the seventeenth century, the region was exploited by latifundistas for the production of pulque using the domesticated maguey, and by peasants for rope production using the wild lechuguilla maguey.  The carrying capacity of the region has slowly decreased since the end of the sixteenth century, and only small mixed herds are now driven out daily into the hills by their peasant owners.  Where there is no water for irrigation, thorn-scrub grows in eroded and rock-hard sun-baked soils.  Where there is water, however, the amazingly fertile soils of the valley bottoms produce seemingly endless crops for the Mexico City market.

Sometime in the seventeenth century this region became known as "el Valle del Mezquital" and in the following centuries it achieved notoriety as the archetype of the degraded and poor regions of Mexico.  The perception of the Valle del Mezquital as inherently resource-poor has not only concealed the original status of the natural resources from later observers, and thus mystified its history, but has also discouraged any real attempt at improvement, and led to ever more exploitive modes of land use. And, starting in the 1940s the waste waters of Mexico City were channeled north to supply a very extensive irrigation system. But the success of this solution to the problem of what to do with Mexico City's waste has led, in a curious way, to the belief that the importation of these chemically-laden waters is improving the Valle del Mezquital. It is clear that if Mexico City's water was diverted from the region, it would lose the key to its productivity. But because this region is still thought of as inherently resource-poor, the idea that it could generate enough water internally to support irrigation is not seriously considered; and reforestation is not as far I know, planned. In the Valle del Mezquital, as in many other semiarid regions where sheep were introduced as part of the European diaspora, the evaluation of land as being fit only for sheep has become a self-fulfilling prophesy.

## Notes

1. Miguel Orthón de Mendizábal, *Obras Completas* 6 vols. (Mexico, 1946-47) vol. 6, p. 170.

2. Sherburne F. Cook, *The Historical Demography and Ecology of the Teotlalpan* (Berkeley, 1949) p. 57.

3. Cook, *Historical Demography*, pp. 33-41.

4. Elinor G.K. Melville, "The Pastoral Economy and Environmental Degradation in Highland Central Mexico, 1530-1600," Doctoral Dissertation, Department of Anthropology, University of Michigan, 1983, ch. 2, n. 24. The references are often very extensive and I will refer the reader to the dissertation when this is the case.

5. Melville, "The Pastoral Economy," ch. 2, ns. 25-30.

6. *Papeles de Nueva España*, Francisco del Paso y Troncoso, ed. (*Madrid and Mexico, 1905-48*) [hereafter cited as PNE] vol. I, 219, 220; vol. VI, 22.

7. PNE I, #533; AGN, Tierras, vol. 1486, exp. 2, fol. 3v; AGI, Justicia, legajo 207 #2, ramo 3, fol. 55r; AGN. Mercedes, vol. 4, fol. 122r.

8. AGN, *Mercedes*, vol. 6, fol. 456; vol. 7, fol. 87; AGN, *Tierras*, vol. 2697, exp. 11; PNE I, 217-18; PNE VI, 33.

9. For a complete discussion of the ecological changes, see Melville "The Pastoral Economy," ch. 5; also, Elinor G.K. Melville, "Environmental and Social Change in the Valle del Mezquital, Mexico, 1521-1600," Comparative Studies in Society and History vol. 32:1(1990), pp. 24-53. The names given in the documents for these plants were: mesquite, lechuguilla, espinos, nopal, tunal, yucca and cardón. Since ocotillo is commonly found in such plant associations, I have translated espinos as occotillo. "Cardón" is currently used to refer to either scotch thistles or a type of nopal cactus.

10. Melville, The Pastoral Economy, ch. 5; and Melville, "Environmental and Social Change".

11. See Leader-Williams, N., *Reindeer of South Georgia* (New York, 1988) for a discussion of the results of his own research into the irruption of reindeer into an island in the Antarctic; and a summary of the New Zealand literature.

12. Ibid., pp. 215-243.

13. Ibid., p. 20.

14. Ibid., p. 20.

15. The earliest records we have of grazing in the Valle del Mezquital are for the 1530s, although it is probable that horses, cattle, sheep and goats were introduced into the region in the 1520s.

16. See Melville, "The Pastoral Economy," ch. 3 for a discussion of the expansion of pastoralism and land takeover; and ch. 5 for a discussion of changing stocking and grazing rates.

17. For a complete description of the ecological changes referred to in this paper, see Melville, "Pastoral Economy," ch. 5; and "Environmental and Social Change".

18. These numbers are not as outrageous as they may appear:  the animals had access to an area of large far in excess of the 780 hectares legally encompassed by the station through the customs of common grazing, grazing public lands, and fallow grazing.

19. PNE VI, p. 181.

20. Lesley B. Simpson, *Exploitation of Land in Central Mexico in the Sixteenth Century* (Berkeley, 1952) p. 13.

21. Climatic changes associated with the Little Ice Age can be proposed to explain this phenomenon, but the little information we have about weather patterns at the end of the sixteenth century do not support this explanation. It is more likely that the catchment value deteriorated and caused a drop in the water table as a result of deforestation and sheet erosion.

22. Leader-Williams, Reindeer, p. 21.

23. When the Great Cocolistle epidemic (1576-81) decimated the Indian agriculturalists.

24. See Richard J. Salvucci, *Textiles and Capitalism in Mexico. And Economic History of the Obrajes, 1539-1840* (Princeton, 1987).

25. These new stations appear to represent new flocks--most of the new land-owners do not appear in earlier documents and all are running sheep.

26. AGI, México, legajo 111, ramo 2, doc. 12.

27. Spaniards did take up agriculture in this region, but sheep grazing dominated in the last quarter of the sixteenth century,

Problems of Forest Conservation:
A Feasible Mechanism for Biodiversity Conservation

Silvia del Amo R.
Gestión de Ecosistemas A.C.

Most of the world's biological diversity is looked after by peasants, who recur to ancient land use practices (Oldfield and Alcorn, 1987). These complex agroecological systems, generally associated to centers of genetic diversity of cultivars. These not only include crop races, but also wild plants and animals which constitute a potential genetic resource for all humanity. Conservation efforts must be integrated by the peasants' efforts toward development, at least in developing countries, devoting special attention to land use practices.

Traditional systems, according to Altieri (1989), present 2 fundamental advantages. On the one hand, they help preserve cultivation practices, which are part of the cultural heritage; on the other hand, peasants who follow traditional systems play an important role in preserving and increasing genetic diversity of cultivars.

In traditional systems, genetic biodiversity is maintained both through cultural intervention and through natural selection.

On the other hand, traditional cultivars have a twofold value that make them worth preserving: as germplasm repositories that can eventually be inserted into modern crops, and as a complex genetic bank adapted to specific environment conditions. To people living in developing countries, these cultivars have an additional value-- to allow production for self-subsistence where modern crops are either not available or not adequate.

## A Possible Solution

Most Third World countries have an economy that cannot subsidize large biotic reserves, unless by means of tourism, by trading off the external debt, or through special support. Obviously, conservation strategies in Third World countries must be different from those implemented in the "first world": restricted conservation must be substituted by a use of natural resources compatible with survival of local groups and a sustained use of the majority, if not of all species (Westman, 1990).

154

## The Problem

It is a widely documented fact that Mexico is one of the countries benefited by a greater diversity in plants and animals. Unfortunately, every year around 500,000-600,000 ha. of temperate, tropical, and semiarid woods are lost in Mexico. Also, it is recognized that 80 percent of the national territory suffers from accelerated erosion (García, 1983; Mass y García, 1990 and Toledo, 1988). According to these data Mexico's most serious problems are loss of biodiversity and erosion.

## A Way to Tackle the Problem

Biodiversity maintenance can be reached by 2 methods: conservation "in-" and "ex-situ". In the first, natural processes are incremented, whereas in the second one, human intervention is favored. However, this classical bipartition is old-fashioned, since among the "in-situ" conservation areas, traditional systems are not included--this designation being exclusively applied to protected areas. These systems, where man's intervention is great and fundamentally of cultural nature, are the result of millenary natural and human selection and constitute the third option in resources preservation (see Table 1).

One way to accomplish this is to establish a network to preserve germplasm. In this case a pragmatic orientation is imposed upon conservation. This kind of "in-situ" conservation has not been favored and has even been underestimated and neglected by official conservation programs, although the World Conservation Strategy mentions the sustained use of resources and production systems.

## Objective

The objective of the project, Biodiversity Maintenance, is to establish an "in situ" germplasm network essentially based on the maintenance of traditional farming systems. Special attention will be devoted to varieties of cultivated species and indigenous races, their wild relatives, species of particular management.

The network will expand within an area which has been previously studied, but whose knowledge has not been systematized; this area is the inventory of traditional cultivars. The network can at length be transformed into an efficient mechanism to transmit knowledge, practices, and techniques. It is necessary that the fieldwork be backed up by a data bank, which will hold all network case-studies.

**Table 1.** The Third Option of In Situ Conservation -- Examples of Management Systems to Maintain Biological Diversity

| Traditional System Maintenance | Species Management Germoplasm Repositories | | Interaction with Social System* |
|---|---|---|---|
| Policultives | Species with agronomical value | Crop species | - Participation of social system, especially, local communities |
| Orchard | | Trap species | - Pragmatic approach to conservation |
| Secondary vegetation Management | | Natural fertilizer | - Conservation of biodiversity and also plurality of appropiation forms |
| Kitchen garden | | Natural insecticide | - Maintenance alive the oral history |
| Hill side management | Species with ecological value | Tolerate species | - Maintenance of pride feeling |
| Raised field | | Promoted species | |
| Chinampas | Species with cultural value | Races and varieties | |
| Camellones | | Indigenous species | |
| etc... | | Medicinal | |
| | | Non-conventional foods | |
| | | Condiment species | |
| | | Other uses | |

Increasing cultural human intervention ▲

Increasing emphasis on natural processes ▼

* This social element is not included in the other types of conservation.

The arrows at the bottom of the table indicated the two principal processes involved in the two types of conservation recognized until today. The natural processes are present in the "in situ" conservation and the human intervention is increasing in the "ex situ" conservation. In the new proposal presented in this paper, we can see that both processes have a priority role, but the human conservation is of a cultural nature, unrelated to the advances of new technology. A new approach of this proposal is the integration of social systems as the key element for the success of conservation (see third column).

This project means for Mexico not only the recovery of biodiversity, but also of cultural aspects in natural resource use and management. Particular attention will have to be paid to actual crop species, those that have been abandoned, and to wild species with an of ecological and cultural significance. The proposed network includes 4 levels, in which diversity loss takes place: 1) ecosystem or agroecosystem level (traditional system); 2) species diversity level; 3) genetic level; and 4) molecular level (Chart 1).

## Methodology

Goal: Preserving cultivated species and their varieties or races, semidomesti-
    cated and wild species, whether tolerated or fomented.

Strategy: Organizing an information network, stressing the worth of and
    promoting these cultivation systems, and forming a data bank.

Tactics: Identifying species and varieties of plants and agricultural or
    horticultural practices in each traditional system.

The incorporation of new members to the network is achieved through visits to the community to establish contact with the future members of it. Once they have accepted, a survey to identify which species are preserved in their cultivation systems is sketched out with their help. Herbarium specimens of the species, races, and varieties are collected and the whole information is stored in the data bank.

The network is to be conceived on 2 levels: researchers and institutions on the one hand and peasants on the other. The first is important because it provides the means to get to know the researchers who have worked in the different communities in natural resource management, and they can introduce us to the community and to the informants. Besides, publications and reports by these researchers constitute our initial back-up literature. The second level, that of the peasants, is the most important one, since they are the active members of the network and on them falls the burden of the actual preservation of the resources.

## Traditional Systems as Fundamental Units of the Network

The WCS points out 4 valid arguments in defense of the conservation of biological diversity: the economic, the ecological, the social, and the aesthetic argument. In the present work, 2 more are added: the cultural and the productive. All of them are intimately connected with traditional systems.

**Chart 1.** Example of Levels in Which Diversity Loss Takes Place in a Traditional System.

"Chinampas," Mixquic, Distrito Federal

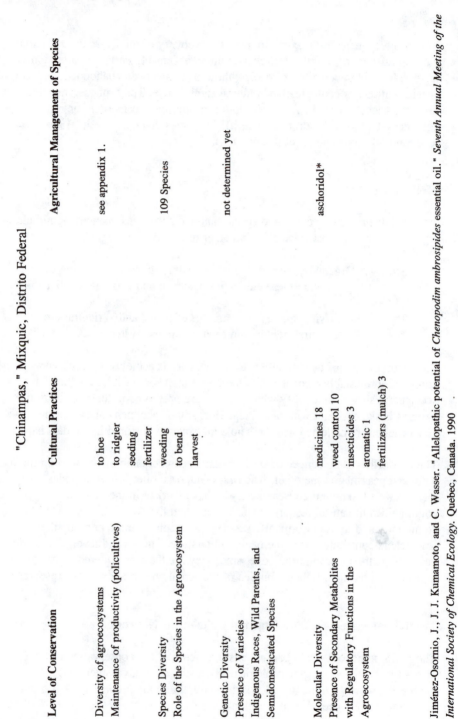

| Level of Conservation | Cultural Practices | Agricultural Management of Species |
|---|---|---|
| Diversity of agroecosystems<br>Maintenance of productivity (policultives) | to hoe<br>to ridgier<br>seeding<br>fertilizer<br>weeding<br>to bend<br>harvest | see appendix 1. |
| Species Diversity<br>Role of the Species in the Agroecosystem | | 109 Species |
| Genetic Diversity<br>Presence of Varieties<br>Indigenous Races, Wild Parents, and<br>Semidomesticated Species | | not determined yet |
| Molecular Diversity<br>Presence of Secondary Metabolites<br>with Regulatory Functions in the<br>Agroecosystem | medicines 18<br>weed control 10<br>insecticides 3<br>aromatic 1<br>fertilizers (mulch) 3 | aschoridol* |

Jiménez-Osornio, J., J. J. Kumamoto, and C. Wasser. "Allelopathic potential of *Chenopodim ambrosipides* essential oil." *Seventh Annual Meeting of the International Society of Chemical Ecology.* Quebec, Canada. 1990 .

## Cultural Argument

Maintenance and conservation of a plurality of agricultural forms also implies that cultural conservation and culture are the sum of material and spiritual creations and of the historical heritage of a social group. In the field of natural resources, this would lead to thinking of the types of management, as a product of human activity, as subjective and environment dependent. This automatically leads to diversity in production and to the value of plants and animals, according to geographical, ecological, and cultural conditions of each human group (del Amo, 1990). The conservation of traditional systems cannot be reached if these are isolated from the preservation of the culture of the local people (Altieri and Merrick, 1987).

## Production Argument

One of the positive characteristics of traditional systems is their cultivation diversity. This peculiarity is being affected by today's tendency toward monoculture. The strategy of market diversification, a well-known one in the economic field, is not accepted in the case of agricultural production; the dominant tendency on the market of different products is reverted precisely in the case of agriculture, where diversification obeys specific geographical, ecological, and cultural situations, and where, in spite of all, the dominant tendency still remains monoculture.

Some of the prioritary requisites of sustained exploitation, as pointed out by the WCS, are exactly accomplished in traditional systems; i.e. 1) resources are not exploited to the point of making their recovery impossible, rather they are managed taking into account the relationship between the species, 2) the exploitation of the resource is regulated, 3) the fortuitous extraction of a species is reduced to a minimum, since they all have a certain value, if not a direct one, at least an ecological or cultural one.

In traditional systems, the strategic principles suggested by the WCS are all attained by: 1) integration; 2) preservation of options; and 3) and 4) accomplishment of a double task of preserving and preventing. A fifth principle neglected by the WCS is public participation, especially of the local communities. Fulfilling these principles guarantees the success of a conservation strategy.

## How Can this Type of Conservation be Boosted?

Much has been said about the positive nature of this type of conservation "in situ"; however the fundamental problem of its maintenance has not been solved yet, due to the influence of modern technology. How can it be supported and allowed to continue existing? In our opinion it is necessary to:

1. Assess the value of traditional cultivation systems, not only as efficient production units, but also as germplasm conservation units, since in these resources preservation has an evidently practical sense.

2. Bring back to the community the scientific and technical information obtained by specialists from these systems, since some of their ideas may be adopted by the peasant to improve the system itself.

3. Support the existence of this kind of production unit, because of the various diversity levels that are concentrated in them: biological, ecological, cultural, diversity of management, and of production.

4. Develop stimulation programs, prizes, and other events among the peasants who maintain the most diversified systems.

5. Give an economic boost to those species, which due to their potential or esthetics are locally important.

6. Increase the experience exchange among peasants of different communities.

7. Maintain the social-cultural organization of the local people, since the traditional system itself acts as a cohesive element.

## Some Final Considerations

The World Conservation Strategy points out the importance of the task of getting to know the variability in natural systems. However, in traditional systems, this task is made easier by the richness of empirical knowledge embodied by the community itself. It only needs to be recorded and made more systematic to be of scientific character.

In WCS preservation of genetic diversity is considered as an investment and an insurance, being necessary to boost agriculture, forestry, and fishing. It allows keeping future options open, is a buffer against negative environmental changes, and also is prime material for scientific and technical innovations.

Wild races constitute an important element in any biological conservation program. Individuals of one species are not identical; there exists individual variation in them, and they are subject to apparently innocuous changes in management practice. As Berry (1983) points out, biological conservation depends on the degree of comprehension of biological variability, both in time and space. This fact assigns a special value to traditional cultivation systems, where one of the main pivots rests precisely on variability.

Biodiversity maintenance does not depend only on the introduction of new and modern technologies. No technology can ever recreate the vast panorama of lost ecosystems, species, and genes. However, maintenance and preservation of different traditional systems are a good and tested approximation.

A third option for "in situ" conservation is proposed: (cf, Table 2 and Figure 1). Under this option--traditional systems conservation--a greater emphasis is placed on natural processes, as in the case of classical "in situ" conservation based on protected areas. At the same time however, human intervention is increased, as in the case of "ex situ" conservation, based on botanical and zoological gardens and where the efforts to store germplasm require heavy expenditures and great technological progress. In this case, human intervention is of a cultural nature; cultivation practices play a very important role in species maintenance and evolution in these traditional systems.

Although at the outset the establishment of this network may appear irrelevant, we can assume with certainty that this is the level where problems can be identified, contacts made, information exchange workshops among peasants set up, etc. This will undoubtably contribute to a correct evaluation of the efficiency of these systems as conservation units of our genetic heritage. On the other hand, through this network it will be possible to compile the inventory of traditional cultivations.

Lastly, it is important to point out that the proposal of this option represents a mechanism in conservation comparable to an underground economy. As a matter of fact, just as underground economy means a safety valve to counteract the economic crisis, in the same way the type of conservation proposed here will allow to attenuate the tendencies of the ecological crisis by counteracting the loss of species. This kind of silent, subterranean but effective conservation constitutes, both on a short and on a long term basis, a feasible biodiversity conservation mechanism for developing countries. Given the span of the country and the enormous diversity of the systems in use, only a representative group is included in the network, but the sample is representative enough to propose the development of similar network as on efficient alternative for germplasm conservation.

**Table 2.** The Third Option of In Situ Conservation -- Management Systems and Conservation Objectives

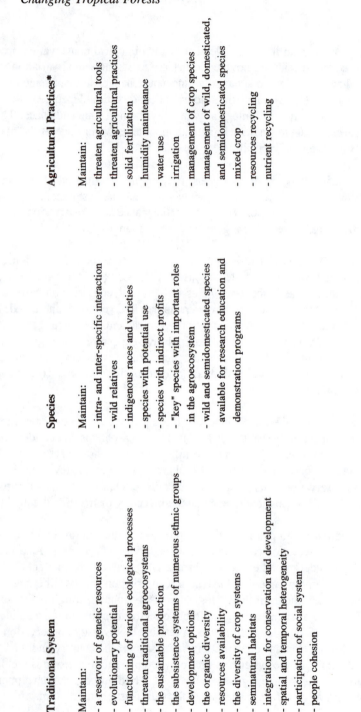

| Traditional System | Species | Agricultural Practices* |
|---|---|---|
| Maintain: | Maintain: | Maintain: |
| - a reservoir of genetic resources | - intra- and inter-specific interaction | - threaten agricultural tools |
| - evolutionary potential | - wild relatives | - threaten agricultural practices |
| - functioning of various ecological processes | - indigenous races and varieties | - solid fertilization |
| - threaten traditional agroecosystems | - species with potential use | - humidity maintenance |
| - the sustainable production | - species with indirect profits | - water use |
| - the subsistence systems of numerous ethnic groups | - "key" species with important roles in the agroecosystem | - irrigation |
| - development options | - wild and semidomesticated species available for research education and demonstration programs | - management of crop species |
| - the organic diversity | | - management of wild, domesticated, and semidomesticated species |
| - resources availability | | - mixed crop |
| - the diversity of crop systems | | - resources recycling |
| - seminatural habitats | | - nutrient recycling |
| - integration for conservation and development | | |
| - spatial and temporal heterogeneity | | |
| - participation of social system | | |
| - people cohesion | | |

* This is a new element not included in the other conservation systems giving a practical sense to conservation.

**Figure 1.** Biological and Cultural Processes Maintained in Traditional Systems

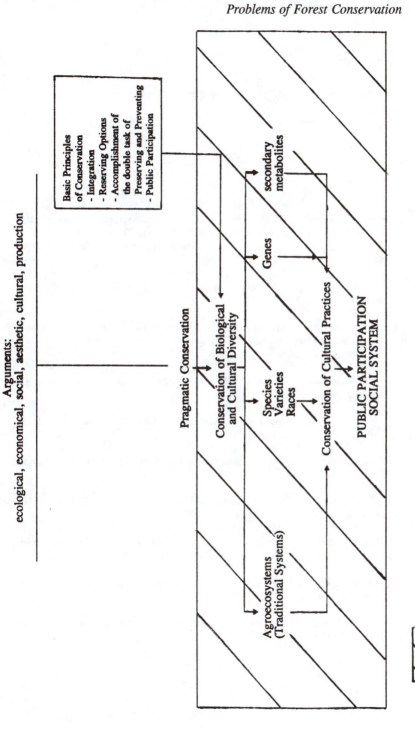

**Arguments:**
ecological, economical, social, aesthetic, cultural, production

**Basic Principles of Conservation**
- Integration
- Reserving Options
- Accomplishment of the double task of Preserving and Preventing
- Public Participation

Pragmatic Conservation

Conservation of Biological and Cultural Diversity

Species
Varieties
Races

Genes

secondary metabolites

Conservation of Cultural Practices

Agroecosystems (Traditional Systems)

PUBLIC PARTICIPATION
SOCIAL SYSTEM

Maintenance of Biodiversity, Continuity, and Change.

## Acknowledgments

Project "Maintenance of Biodiversity" is sponsored by the Rockefeller Foundation.

## References

Altieri, M.A. & L.C. Merrick. 1988. *Agroecology and "in situ" Conservation of Native Crops Diversity in the Third World.* In: Altieri, M.A., M.K. Anderson, & L.C. Merrick. 1987. Peasant agriculture and the conservation of crop and wild plant resources. *Conservation Biology*, 1(1): 49-58.

Altieri, M.A. 1989. Rethinking Crop Genetic Resources Conservation: A view from the South. *Conservation Biology* Vol. 3 March (1): 77-79.

Amo, R.S. del 1990. El uso sostenido: la diversidad biológica, cultural de manejo y de producción. In: R. Rojas (Coor.). En busca del equilibrio pérdido. El uso de los recursos naturlaes en México. Ed. Universidad de Guadalajara.

Berry, R.J. 1983. Genetics and Conservation. In: *Conservation in Perspective*. Warren, A. y F.B. Goldsmith Eds. John Willey & Sons.

Garcia, L.B. 1983. Diagnóstico sobre el estado actual de erosión en México. *Terra* (1): 11-14. México.

IUCN. 1980. *World Conservation Strategy*. International Union for Conservation of Nature and Natural Resources, Gland Switzerland.

Mass, M.J.M. & Garcia, O.F. 1990. La conservación de los suelos tropicales. El caso de México. *Ciencia y Desarrollo* 15(90): 21-36. CONACyT, México.

Oldfield, L.M. & J.B. Alcorn. 1987. Conservation of Traditional Agroecosystems. Can age-old farming practices effectively conserve crop genetic resources? *BioScience* Vol. 37(3): 199-208.

Toledo, V.M. 1988. La diversidad biológica en México.Ciencia y Desarrollo. No.81. Año XIV:17-30.CONACyT, México.

Westman, Walter E., 1990. Managing for Biodiversity: unresolved science and policy questions. *BioScience*, Vol. 40 N(1): 26-32.

# Urban Growth, Economic Expansion, and Deforestation in Late Colonial Rio de Janeiro

Larissa V. Brown
Michigan State University

Deforestation in central and southern coastal Brazil during most of the colonial period was selective. The cost and difficulty of transportation, the small size of the population, and the limited markets tended to constrain extensive damage to coastal forests. Forests near the coast were picked clean of brazilwood, a dyewood whose value had long been recognized and which had been reserved as a royal monopoly. Sugar plantations near coastal ports consumed wood in their boiling houses and packing chests, but the plantations were designed to include forest reserves for wood resources. *Senhores de engenho* long perceived the woods as inexhaustible. Toward the end of the colonial period, however, economic and demographic changes produced serious indications of the inexorable deforestation processes which would accelerate during the nineteenth century.

The existing documentation cannot give us precise information on deforestation and consumption of wood in this period. The Portuguese Crown had long claimed a monopoly of brazilwood and preferential access to the *madeiras de lei*, hardwoods used for naval shipbuilding. But by the late eighteenth century, brazilwood was hard to find in the coastal forests of the captaincy of Rio de Janeiro, and the crown found it very difficult to enforce its preferential claims over ship timbers. Otherwise, wood was not included under the fiscal and regulatory legislation which covered other commodities. Lumbermen did not pay the tithe or other royal duties on wood.[1] All vessels carrying foodstuffs from ports outside Guanabara Bay were required to register with the city clerk of Rio de Janeiro on entry to give the municipal government knowledge of both the amount and price of food in the city.[2] Ships carrying lumber were not required to register, although many did. Firewood collected in suburban and rural areas around the city, and wood rafted down rivers into the bay were not subject to direct regulation or any taxation. We can only learn implicitly about increasing pressure on the forests, from official responses to requests for more ship timbers, from local complaints about shortages, and from legal conflicts over access to woods.

In the 3 decades before Brazilian independence in 1822, the city and captaincy of Rio de Janeiro underwent a process of urban growth and economic expansion, accelerating the incremental destruction of the captaincy's virgin forest, which had expanded when the Haitian Revolution destroyed the sugar and coffee economies of Saint-Domingue. The population of the city's urban core and surrounding hinterland nearly doubled between 1789 and 1822, principally because of the arrival in Rio of the Portuguese royal family fleeing Napoleon's armies. Some 15,000 nobles, bureaucrats, and other Portuguese accompanied them, but the establishment of the imperial capital in Rio de Janeiro also attracted Luso-Brazilians and other Europeans.[3]

Wood was needed for a variety of purposes in colonial Rio de Janeiro: the sugar industry, shipbuilding, construction, and domestic uses. The sugar industry was a prodigious consumer of firewood to fuel the fires in the boiling houses and of boards to make the sugar chests used to package sugar for transport. As early as the end of the 1770s, sugar mills were being abandoned in the new plantation region of the Campos dos Goitacazes northeast of the city of Rio de Janeiro "por nao ter lenhas"--for a lack of firewood.[4] Shipyards in Rio and other coastal ports built all kinds of ships: small coasting and fishing vessels, transatlantic merchant-men, and even some naval warships, though Bahia appears to have been a more important center of naval shipbuilding.[5] Wood was of course important for housing both in town and in the countryside. After 1808 a construction boom took place in the urban and suburban parishes of the city.[6] Wood was not only used as a construction material itself, but as firewood it was necessary for the manufacture of lime from seashells, for making bricks, and in other artisan trades. The lime was used as mortar, cement, plaster, and whitewash. An indication of the construction boom is the trend in prices for lime and bricks. The cost of lime in Rio de Janeiro rose significantly in the 1790s and the first decade of the nineteenth century, suddenly doubling between 1808 and 1810, and rising even higher to peak in 1811. The price fell somewhat in the subsequent decade, but never back to the prices even of the 1790s. The price of bricks also took a sudden leap in the years 1809-1812, peaking in 1811 to double the price of 1808.[7] Finally, every household consumed wood in the form of charcoal or firewood in the daily preparation of meals. Thus, the forests and woods were under attack from engenho owners, owners of lime and brick kilns, domestic users everywhere who needed firewood for cooking, and the construction industry.

The city of Rio de Janeiro was established on a small peninsula jutting out from the western shore of Guanabara Bay in 1565 and in the ensuing centuries settlers fanned out into the alluvial plains, marshes, and hills surrounding the bay. In 1697, a second municipality, the town of Macacu, was established at the north-eastern corner of the bay. No further subdivisions of the city's municipal jurisdiction took place until 2 towns were founded in the late eighteenth century, the most important being Mage in 1789, to the west and north of Macacu, and then between

1814 and 1820 4 new municipalities were carved out of Rio de Janeiro. Macacu and Mage would prove to be important sources of wood for the city.

Royal officials in Rio de Janeiro became particularly aware of the difficulty in acquiring lumber when they were asked to send wood to Lisbon for the naval arsenal. In 1802, Jose Caetano da Lima, the naval intendant in Rio, received orders to send 150 planks of *peroba* or *vinhatico* to Lisbon. In July of that year he wrote that

> The vinhatico in this region is not as unique or solid as that in Bahia: some merchants send for it from there to build their ships because experience has shown that it is much superior: of that which exists in the forests of this captaincy, the nearest is from Rio de Sao Joao, [and] some will come from there. Peroba is the most singular wood that exists in the Campos dos Goitacazes for planks and boards for ship ribs, it is already very far from the rivers and therefore very difficult to transport because so much has been cut over so many years to be sent to Lisbon and used in this port. The same thing has happened to construction lumber; the largest timbers are in the same circumstances of distance from the seaports and great difficulty of transport.[8]

In September he wrote again, noting that in the Rio area most lumbering was concentrated in the forests of Macacu and Posse (the Tapacora area near Macacu, I believe), that the naval squadron had used up a lot of wood in the last 5 years, and that "the forests of this port are not [as] full of wood as those of Bahia, Pernambuco, and Para, because of the distance in which they are located."[9] The same dilemma resurfaced in 1805 and 1806 when the naval intendant found it difficult during the sugar harvest to persuade *senhores de engenho* to assign oxen to bring large timbers out of the woods some 20 km to Porto das Caixas on the Macacu River, whence they would be transported to Rio in large boats and rafts. In 1805 the viceroy declared that there was not enough wood to fill the transport ship which had been sent to collect timbers even if it stayed in port several months. Thus, he planned to send it on to Bahia to complete its cargo.[10]

The captaincy of Rio de Janeiro was already an important sugar and cane brandy producing region when the collapse of the plantation economy in Saint-Domingue led to a decade of rising sugar prices in the 1790s. In the subsequent 2 decades, sugar prices fluctuated, but rising world prices for coffee stimulated the expansion of the new export into hilly land not suitable for sugar production.

The effect on forests of the export boom is in some ways obvious. Although sugar planters in particular were well aware of their need for firewood, any new plantations entailed the clearing of some land. Latifundism had always been at least partially an economic strategy to permit exhaustion and recuperation of lands and woods on a rotating basis within the same estate. As export prices became increasingly remunerative, planters began to give up their commitment to self-sufficiency, using as much of their land as possible to plant sugar cane. They stopped planting food crops for their plantation households, preferring to go to the market to buy manioc for example. In 1798 Viceroy Rezende complained that "in the hope of greater profits, the *senhores de engenho* for the most part have totally abandoned the planting of Manioc, finding it less convenient to plant it, than to buy *farinha* [manioc flour] for the subsistence of their families and their slaves," in spite of his repeated orders to the contrary.[11] The effect of this practice was to encourage greater slash and burn production of manioc flour for the market by specialist food producers, many of whom were located in the municipalities of Mage and Macacu. In the same year, however, royal recommendations to follow the example of Caribbean sugar producers in burning *bagasse* (sugar cane trash) rather than wood met with the declaration that bagasse did not create enough heat in the furnaces. The royal government also wanted planters to take up the plow, but, significantly for our concerns, they declared that each year there was newly cleared land full of tree trunks, vines, and roots which made using plows impossible.[12] Much of the Campos dos Goitacazes region had turned into a veritable sugar-making machine by the end of the eighteenth century, to the detriment of most other economic activities. Some observers discounted the problem of firewood scarcity, asserting that new woods were growing up all the time, but others were less sanguine, and of course, the mature forest was cut down in any case. By 1820 many planters had to purchase firewood, and it was transported relatively far in carts or rafted down rivers, which suggests a growing scarcity. In 1828 the only part of the province of Rio de Janeiro where bagasse was burned in the *engenhos* was in Campos, an improvement adopted by necessity, since in other respects the Campos industry was as technically backward as everywhere else.[13]

Coffee production for export in Rio de Janeiro was still in its infancy at the beginning of the nineteenth century. By independence however, coffee had spread into the Tijuca hills west of the urban core of the city, across the bay to the parish of Sao Goncalo, and into other rural parishes.[14] The French traveler Freycinet recommended that potential investors set up coffee plantations on the hillsides near the city.[15] After 1808 coffee production also began to displace sugar production in small coastal ports west of the city of Rio de Janeiro such as Mangaratiba, Angra dos Reis, Paraty, Ubatuba, and Sao Sebastiao. Coffee production doubled in Angra dos Reis between 1809 and 1818.[16] Planters near many of the small ports of the coast had invested in sugar production simply because it was the principal export crop, but when coffee appeared on the scene many were willing to switch. The French botanist Auguste de Saint-Hilaire observed this phenomenon when in 1818 he passed through Macae, a district between the Campos dos Goitacazes and the city of Cabo Frio in eastern Rio de Janeiro.[17] Approximately 12,500 arrobas (1 arroba = 32 pounds) of sugar were registered entering the port of Rio de Janeiro from Macae in 1799, while in the late eighteen-teens, that amount had been reduced by half or more. In contrast, until 1815, registered entries of coffee from Macae never exceeded 648 arrobas, but by 1822, over 2,000 arrobas of coffee were arriving in Rio de Janeiro.[18]

Coffee, however, was destined to remake the interior of the captaincy of Rio de Janeiro. Much of the captaincy's interior was mountainous and inappropriate for the sugar production which dominated the coastal plain. Until coffee planting took hold, the interior of Rio de Janeiro, including the Paraiba River basin, was very sparsely populated and still covered with virgin forest. Without a lucrative product and a means of transport to get it to market, clearing and establishing estates, not to speak of buying slave laborers, was hardly worth the expense. Coffee production in the interior began at the roadside settlement of Sao Joao Marcos and at the only natural clearing in the Paraiba basin, where the town of Resende emerged. The Sao Joao Marcos-Resende region in 1789 had a population of only 3,734, but by 1821 it had increased to 22,089.[19] After 1808 many *sesmarias* (land grants) were given out in the interior of the captaincy and several new roads were opened from the city into the region which would become the center of coffee production in the decades after independence.[20]

The expansion of the export sector and, in particular, the tendency for producers to abandon the ideal of plantation self-sufficiency in foodstuffs and wood increased the pressure on forests resulting from population growth in the city of Rio de Janeiro. Lumber, firewood, and charcoal were increasingly sought from relatively distant sources as the region around the city became more densely populated and its woods more in demand. The mangrove swamps along the shores of the bay and the rivers and streams which emptied into it were important resources. The bark of the trees was used for tanning hides, the wood for fuel and house construction, and mollusks whose shells were burned in lime kilns congregated in the matted roots of the mangroves on tidal flats. Since 1760 only mangrove trees already stripped of their bark could legally be cut down for firewood, but in 1793 the city council was complaining that indiscriminate cutting of mangroves for the lime kilns had produced a scarcity of the bark.[21] Big landowners and other rural people quarreled over access to the woods. In 1814 some inhabitants of the parish of Iraja and "adjacent islands" who earned their living from gathering firewood and making lime complained about the administrator of a *fazenda* in Iraja who seized slaves and created other obstacles to the exploitation of certain stands of mangrove. While the firewood gatherers tried to claim that all mangrove woods were open to public use, the owner of the fazenda successfully asserted her legal right to exclusive use of mangroves along a stream which was entirely within her property.[22]

Firewood became increasingly expensive in the city after 1808, though observers tended to focus the blame on middlemen or confiscatory demands by the government. As late as 1820 an *almotace* (market inspector) ascribed the high price of firewood completely to "the collusion of middlemen, who lay down a law of price to the public."[23] When the police intendant, Paulo Fernandes Viana, looked into the question of firewood scarcities in 1809, he focused on the effect of government requisitioning policies. In rural districts a bundle of firewood cost 20 reis and transport cost another 10 reis, making the lowest profitable selling price in the city 40 reis (10 reis was the smallest coin). Following a plan established forty years before, the government only paid 30 reis per bundle and late at that, after forced requisition. Firewood sellers either stayed at home at the requisitioning times at the end of the month, or bribed the soldiers to let them pass unmolested. The entire system produced scarcities and higher prices, as well as increasing the incentive for firewood sellers to deal with middlemen. They already sold to middlemen because they wanted to unload their wood, buy the things they needed, and return home by midday. Despite the fact that Viana believed that the shortages would disappear if the government paid fair prices, he also acknowledged that "firewood now comes from farther away, and the population has increased."[24] Viana was also proud of having improved the road from the city to Tijuca, which then became a source of foodstuffs and, in his words, "much charcoal."[25]

A variety of evidence, therefore, indicates growing difficulty in obtaining wood, and scarcity in areas near the city. Rings of cultivation surrounded the urban core and the bay. Although to a newly arrived European the city's immediate hinterland appeared filled with vast and only partly cultivated estates, in Brazilian terms it was densely and intensively cultivated. By the second decade of the nineteenth century the city's suburbs up to about 8 km from the urbanized area were taken up with horticulture and forage crops. The hills surrounding the city were increasingly planted in *capim de Angola* a forage grass brought from Africa which was sold as hay for the city's growing population of horses and donkeys.[26] Some 15 km from the city the plantations began to appear, many of which still had their own jealously, but not very effectively, guarded woods for firewood and charcoal. In 1818 Freycinet believed that a few stands of virgin forest remained available along the rivers flowing into the bay where it could be profitable to set up sawmills to produce wooden boards.[27] Much earlier, however, in 1805, Rio's naval intendant complained that

> there isn't a single inch of land in the surrounding suburbs of this city and port which does not have an owner, because it has all been given out in sesmarias, and those who possess land have built mills to grind sugar cane and plant the same cane and reserve pastures for their own oxen.[28]

As it became necessary to go farther afield in search of wood, the cost of transport was a determining element in access to forests in the complex topography of the Rio de Janeiro region. Only where it was possible to transport lumber and firewood by water was the trade viable and lumbering became a pioneer activity up the river basins of the city's hinterland. The principal sources of charcoal, firewood, and lumber within the drainage basin of Guanabara Bay were the bay islands and the municipalities of Macacu and Mage. On Governador and Paqueta islands, many people made their living by exploiting the mangroves, both through the manufacture of lime and through gathering firewood for sale in the city.[29] As early as 1778 the parishioners of Macacu were said to own 40 charcoal boats and 300 firewood boats to take these commodities to the city.[30] The parish of Santissimo Trindade, up the Macacu River, was a center of logging and the inhabitants rafted logs and boards down the river to the town of Macacu, whence the lumber was taken by boat to Rio de Janeiro.[31] The parishes of Mage were also traditional suppliers of firewood and lumber to the city. The parish of Guapemirim in 1778 contained 30 charcoal boats and 100 firewood boats, and towards 1820 it received more than 2:000$000 in annual income from firewood alone.[32] Small ports on the Mage and Iriri Rivers in the eighteen-teens were said to send more than 40 boats a day laden with foodstuffs and firewood.[33]

Of course, the city did not have to depend on wood and lumber solely from within its immediate hinterland. Although the documentation is uneven, it is clear that Rio de Janeiro increasingly depended on wood shipped from coastal ports to make up the deficit in its immediate vicinity. The city council's register of ships carrying foodstuffs which entered Rio de Janeiro from coastal ports also indicates sources of wood. Although ships carrying wood were not required to register, they often did. Rio de Janeiro received lumber from as far south as the captaincy of Santa Catarina; from the coastal ports of the captaincy of Sao Paulo, especially Paranagua and Iguape; from the captaincy of Espirito Santo to the north; and from a number of small coastal ports within the captaincy of Rio de Janeiro.[34] There were even shipments to Rio of firewood as well as other lumber from the major Sao Paulo port of Santos in the eighteen-teens and early 1820s.[35]

The most important sources of wood outside the drainage basin of the bay were the Macae and Cabo Frio regions of the captaincy of Rio de Janeiro to the northeast of the city. From the earliest days of Portuguese contact, this part of the captaincy of Rio de Janeiro was an important source of dyewoods and other lumber. Even after the establishment of farms and plantations producing sugar, rice, and staple foodstuffs, lumbering continued to be one of the primary occupations of the inhabitants of Macae and the nearby port of Rio de Sao Joao, and, to a lesser degree, Cabo Frio. Rio das Ostras, a small port between Macae and Rio de Sao Joao also began making regular shipments of lumber from at least 1814.[36] In the late 1790s an anonymous memorialist extolled the forest resources of this region, mentioning the "many dozens of boards" exported to Rio de Janeiro and listing 33 different types of trees. He ended however by writing: "All these woods are almost extinct in the forests nearest to the sea."[37]

In the Macae district, lumbering was pushed farther up the Macae River and one of its tributaries, the Sao Pedro River, as plantations were established along the river banks. Rio de Sao Joao and Rio das Ostras were fundamentally exporters of wood to the capital, along with some rice and other foodstuffs. Rio de Sao Joao was probably the most important coastal supplier of lumber to Rio de Janeiro. In spite of the fact that ships with cargoes of lumber were not required to register with the city council, between 1799 and 1822 Rio de Sao Joao was often among the most common ports of origin for coastal vessels entering Guanabara Bay, accounting for 5 to 10 percent of ship entries. This is significant given the fact that 50 percent or more of ship entries were routinely accounted for by Campos and Rio Grande do Sul together.

During the last thirty years before Brazilian independence the forces of export expansion and demographic growth in the Rio de Janeiro region combined to create increasing pressure on the forests of this part of Brazil. The expansion of sugar production not only meant greater consumption of wood in the processing and packaging of sugar itself, but through the tendency to abandon food crop produc-tion on the plantation, had the secondary effect of stimulating a group of specialist food producers for the local market who also made demands on the forests. The coffee economy took hold in Rio de Janeiro in this period as planters moved into the Tijuca hills and the interior, *serra acima*, of the captaincy. New roads and land grants prepared the way for the nineteenth-century coffee assault on the forested Paraiba Valley. Within the city, population growth led to a construction boom in urban and suburban parishes, increasing the demand for charcoal and firewood, and the planting of forage crops on the city's hillsides. The mangrove swamps of the streams and bayshore were more intensively exploited, and as the sources of firewood, construction lumber, and particularly, large ship timbers, became more distant, transport became more difficult and more costly. The city increasingly received wood from outside its own immediate hinterland, contributing to deforestation along the coast from Espirito Santo to Santa Catarina. In practice, most coastal Brazilians still acted as if their forests were inexhaustible, but by the time of independence, the processes which would eventually destroy all but a tiny part of Brazil's Atlantic forest were underway.

## Notes

1. Viceroy Conde dos Arcos to Visconde de Anadia, 21 September 1806, Rio de Janeiro, Arquivo Historico Ultramarino (AHU, Lisbon), Rio de Janeiro caixa 233, doc. 35, describing and commenting upon a *parecer* by the naval intendant of the captaincy of Rio de Janeiro.

2. "Registro do Edital das Posturas de 22 de Agosto de 1795," *Archivo do Districto Federal: Revista de Documentos para aHistoria da Cidade do Rio de Janeiro*, 2, no. 7 (July 1895), 325-331.

3. In 1789 the 4 urban parishes of the city contained 39,147 people, but the combined population of the city and its immediate hinterland, roughly equivalent to what is today called the *baixada fluminense*, had a population of 122,659. By 1821 the now 5 urban parishes contained 67,702 people, while 221,000 lived in the city and *baixada* together. This enumeration was made after King Joao VI had returned to Portugal with some of his retinue. Sources: "Memorias Publicas e Economicas da Cidade de Sao Sebastiao do Rio de Janeiro para uso do vice-rei Luiz de Vasconcellos por observacao curiosa dos annos de 1779 ate o de 1789," *Revista do Instituto Historico e Geografico Brasileiro (RIHGB)*, 68 (1884), 27-28 (incorrect sums were recalculated); "Mappa da populacao da Corte e provincia do Rio de Janeiro em 1821," *RIHGB*, 40 (1870), 137-140.

4. "Relacao do Marquez de Lavradio, Parte II," *RIHGB*, 127 (1913), 340.

5. See F. W. O. Morton, "The Royal Timber in Late Colonial Bahia," *Hispanic American Historical Review* 58, 1 (February 1978), 41-61.

6. Larissa V. Brown, "Internal Commerce in a Colonial Society: Rio de Janeiro and Its Hinterland, 1790-1822," (Ph.D. dissertation, University of Virginia, 1986), 59-61.

7. Harold B. Johnson, Jr., "A Preliminary Inquiry into Money, Prices, and Wages in Rio de Janeiro, 1763-1823," in Dauril Alden, ed., *Colonial Roots of Modern Brazil* (Berkeley: University of California Press, 1973), 248-49, 268-70.

8. Jose Caetano de Lima to Visconde de Anadia, Rio de Janeiro, 21 July 1802, AHU, Rio de Janeiro caixa 201, doc. 43.

9. Lima to Anadia, Rio de Janeiro, 27 September 1802, AHU, Rio de Janeiro caixa 202, doc. 38.

10. Lima to [Anadia?], Rio de Janeiro, 14 July 1805, AHU, Rio de Janeiro caixa 225, doc. 92; Viceroy Fernando Jose de Portugal to Anadia, Rio de Janeiro, 16 November 1805, AHU, Rio de Janeiro caixa 225, doc. 25; Lima to Anadia, Rio de Janeiro, 17 November 1805, AHU, Rio

de Janeiro caixa 224, doc. 92; Conde dos Arcos to Anadia, Rio de Janeiro, 21 September 1806, AHU, Rio de Janeiro caixa 233, doc. 35.

11. Viceroy Rezende to D. Rodrigo de Souza Coutinho, 21 April 1798, Arquivo Nacional, Rio de Janeiro (ANRJ) cod. 69, vol. 8, ff. 54-55. Also in Biblioteca Nacional, Rio de Janeiro, Secao de Manuscritos, (BNRJ) 5, 3, 9.

12. Rezende to D. Rodrigo de Souza Coutinho, Rio de Janeiro, 12 November 1798, ANRJ cod. 69, vol. 8, ff. 104r-104v; "Registo da Carta que escreveo o...Vice Rey...ao Senado em 20 de Marco de 1798," Arquivo Geral da Cidade do Rio de Janeiro (AGCRJ) cod. 16-4-2, f. 85r; Copy of Acordo de Vereanca, 5 May 1798, ANRJ caixa 500, pac 1.

13. Jose Carneiro da Silva, "Memoria Topographica e historica sobre os Campos dos Goitacazes," Rio de Janeiro, 1819, ANRJ cod. 807, vol. 16, f. 96v; Auguste de Sainte-Hilaire, *Viagem pelo Distrito dos Diamantes e Litoral do Brasil* (Belo Horizonte: Itatiaia, 1974), 200, 200 note 47; Jose de Souza de Azevedo Pizarro e Araujo, *Memorias Historicas da Provincia do Rio de Janeiro* (Rio de Janeiro: Ministerio da Educacao e Saude, Instituto Nacional do Livro, 1945), vol. III, 103; Antonio Moniz de Souza, "Viagens e Observacoes de Hum Brasileiro," *Revista do Instituto Geografico e Historico da Bahia*, no. 72 (1945), 115.

14. Affonso de Escragnolle Taunay, *Historia do Cafe no Brasil* (Rio de Janeiro: Departamento Nacional do Cafe, 1939), vol. 5, 15, 241.

15. Louis C. D. de Freycinet, *Voyage autour du monde* (Paris: Chez Pillet Aine, 1827), vol. 1, 256-258.

16. Production increased from 22,000 arrobas in 1809 to 50,000 arrobas in 1818 and exports grew from 15,675 arrobas in the first year to 48,820 arrobas in the second year. "Tabellarische Ubersicht der Production, Consumtion und Exportation des Districts von Ilha Grande em Jahre 1809," Wilhelm C. von Eschwege, *Journal von Brasilien* (Weimar, 1818), vol. 2, facing p. 44; "Mapa comparativo das Producoens do Distrito da Villa de...Angra dos Reis e Seu Termo...1818," Francisco Claudio Alvares d'Andrade to Thomas Antonio da Villa Nova Portugal, 1 February 1819, BNRJ II-34, 23, 12.

17. Saint-Hilaire, *Diamantes*, 184-185.

18. AGCRJ Embarcacoes...Entradas, cod. 53-3-5, 53-5-6, 57-3-9 to 12, 57-3-15 to 16.

19. Alberto Ribeiro Lamego, *O Homem e a Serra* (Rio de Janeiro: IBGE, 1963), 97-99, 102-105, 114; Taunay, *Cafe*, vol. 2, 141- 142, 259-260; Pizarro, vol. IV, 170 and vol. V, 38-45; Johann E. Pohl, *Viagem no Interior do Brasil* (Belo Horizonte: Itatiaia, 1976), 62-63; Viceroy to Camara of Rio de Janeiro, 7 April 1802, AGCRJ cod. 16-4-2, ff. 144v-145; Auguste de Saint- Hilaire, *Segunda Viagem do Rio de Janeiro a Minas Gerais e a Sao Paulo* (Belo Horizonte: Itatiaia, 1974), 102-103. Sources for population statistics, see footnote 3.

20. Brown, 239-244.

21. Alvara, 9 July 1760, *Collecao Chronologica de Leis Extravagantes* (Coimbra: Real Imprensa da Universidade, 1819), vol. 4, 324-325; Edital, 18 September 1793, AGCRJ cod. 16-4-21, f. 41.

22. Paulo Fernandes Viana to Visconde de Aguiar, 6 March 1814, ANRJ cod. 323, vol. 4, ff. 11-12v.

23. 9 December 1820, AGCRJ cod. 58-3-27, f. 6v.

24. Viana to Conde de Linhares (D. Rodrigo de Souza Coutinho), 11 July 1809, ANRJ cod. 323, vol. 1, ff. 100v-102.

25. Paulo Fernandes Viana, "Abreviada demonstracao dos trabalhos da policia," *RIHGB*, 55 (1892), 374-376.

26. Jean-Baptiste Debret, *Viagem Pitoresca e Historica ao Brasil* (Sao Paulo: Martins, 1972), vol. 1, 180.

27. Freycinet, vol. 1, 256-258.

28. Jose Caetano do Lima to [Anadia?], Rio de Janeiro, 14 July 1805, AHU, Rio de Janeiro caixa 225, doc. 92.

29. Pizarro, vol. IV, 81; Manuel Aires de Casal, *Corografia Brasilica* (Belo Horizonte: Itatiaia, 1976), 191.

30. Lavradio, "Relacao...II," 292.

31. Pizarro, vol. II, 215-16; vol. V, 130-131; Petition of Fructuozo de Bastos Monteiro and accompanying documents, despacho in Rio de Janeiro 20 October 1802, AHU, Rio de Janeiro caixa 202, doc. 69.

32. Lavradio, "Relacao...II," 290; Pizarro, vol. III, 205.

33. Pizarro, vol. III, 137-139.

34. Brown, 172, 325-28, 331, 335, 337, 340, 344, 359, 361-363.

35. Mappa da importacao e exportacao do Porto de Santos no anno de 1813, ANRJ caixa 178, pac. 2; "Mappa da Importacao e Exportacao do Porto de Santos no Anno de 1816," *Boletim do Departamento do Arquivo do Estado de Sao Paulo* (BDAESP), 2 (1942), 74; "Mappa da Importacao e Exportacao do Porto de Santos no Anno de 1818," *BDAESP*, 3 (1943), 54; Joao de Souza Pereira Bueno to Provisional Government of Sao Paulo, 4 June 1822, transmitting Santos import-export tables for 182, *BDAESP*, 4 (1943), 71.

36. AGCRJ Embarcacoes...Entradas.

37. "Memoria Historica da Cidade de Cabo-Frio e de todo seu distrito...Anno de 1797," *RIHGB*, 66, tomo XLVI, Parte I (1883), 223-225. The trees listed are: vinhaticos, araribas, cedros, caixetas, ceregeiras, canelas, oleos de cupahiba, sapucaias, pinhoans, parobas brancas e vermelhas, caiunas, jacarandatans, guaraens, guarabu, massarandubas, ipes, arcos de pipa, gorotans, paos-ferros, goraunas, gorapiapunhas, oleos vermelhos, secupiras, oleos pardos, oitis, marendibas, piquia, tatagibas, louros, iricuranas, angelins, cahubins, and pau brasil.

Forest Management in Brazil:  A Historical Perspective

Sebastião Kengen
Brazilian Institute for Environment
and Renewable Natural Resources

Brazil followed a pattern of disregard for its forest resources that characterized the colonization of the other countries of the so-called "New World". The pioneer cleared away forest for agriculture and to supply his wood requirements. In a territory endowed with luxuriant and diverse vegetation, its forest resources seemed to be endless. However, this reason *per se* does not explain the deforestation which followed Brazil's colonization and has marked contemporaneous Brazil.

Considering that the basic aim of forest management is to exploit forest resources on a sustained yield basis, it is fair to assume that Brazilian forest resources have been devastated rather than managed. Therefore, this paper deals with the exploitation of the Brazilian forest resources within the context of its process of development and will provide an insight into the interaction of the social, political, and economic structures.

**A Brief Account of the Management of Tropical Forests**

Forest management is the science and art for planning forest activities to achieve specific goals at operational levels without compromising the biological sustainable limits of the resources to be managed. In other words, management implies a purposive use of natural resources in a sustainable context and presupposes a level of control to achieve acceptable level of regeneration after timber exploitation.

The British, Spanish, and Portuguese had a distinct historical interest in the forest resources of their former colonies. In short, it can be said that to the British the interest in forestry developed much earlier and with strong emphasis on reserve forest land.

In fact, the history of tropical forest management can be traced to more than a hundred years ago in former British colonies, such as India and Myanmar (formerly Burma) still under the British administration. For example, in Myanmar natural forest resources have been utilized according to working plans, designed to manage them on a sustained yield basis, since the mid-1800s. This marked the beginning of a formal Forest Service organization which was the first established in the British Commonwealth and the oldest outside of Europe (Kengen, 1991).

In Portuguese and Spanish colonies, the emphasis was much greater on land for agricultural purposes that was privately owned. In tropical America it is fair to assume that there has been virtually no forest management, with few exceptions, such as in Trinidad and Tobago and small experimental areas in Costa Rica, Surinam, and Colombia (Masson, 1983).

In Brazil in particular, forests were not protected, while the unlimited right of private land ownership allowed the settlers to cut and use fire at any time without any legal restriction. It was only in 1934 with the promulgation of the new Brazilian Constitution that forestry received major attention in Brazilian legislation (Volpato, 1986). However, this did not necessarily mean a better utilization of forest resources.

Even forestry, as an activity organized on a professional basis, is relatively new. Until the 1960s, agronomists, some specializing in silviculture, performed the forester's tasks. Without disparaging the work of these pioneers, it can be said that Brazilian forestry did not become properly established until after the graduation of the first Brazilian trained foresters in 1964. These facts support the above assumption regarding to the difference in the emphasis put on forest resources by the British and Portuguese.

However, this lack of major interest in rational utilization of forest resources cannot be only attributed to difference between Portuguese and English colonization. As pointed out by Masson (1983), these efforts of the establishment of a tropical forest management under a sustained yield basis carried out elsewhere:

> initially made comparatively little impression on other tropical forest administrations and only at the beginning of this century were earnest efforts made to tackle the problems of tropical silviculture in other areas such as Malaysia, Ghana, Nigeria and Trinidad and Tobago. Today only 4.4 percent of tropical closed forest is reported to under management and even this does not normally entail the use of intensive high yielding systems (Masson 1983:1).

In fact, the management and conservation of tropical forest resources have proven a primary challenge throughout the tropics. A series of constraints have been faced by those in charge of these activities. The source of these constraints are of physical nature, e.g. unpredictable weather, complex ecosystems which are not fully understood yet, a high diversity of species of which only a few are marketable, etc. Another source of constraints are of a socio-economic and political nature, such as competing land use, the historical perception of the forest as being a constraint to development instead of being a way to reach it, government policies, etc. Finally, problems of economic forest management are quite difficult in a tropical forest. A tropical forest is a complex ecosystem that has challenged the attempts of almost all forest managers in developing a sound economic method for tropical forest management. There can be some exceptions, but always applicable only under limited circumstances and over a limited area (Masson, 1983; Grainger, 1987).

The complexity and diversity of the tropical forest ecosystem does not allow a single model of forest management applicable all over the tropical countries. It requires specific forest management models. However, to develop these models requires amongst other things political commitment, expertise, time, and money which are, in general, scarce "goods" in the majority of the Third World countries.

**The Brazilian Process of Development:  An Overview**

Brazil was "discovered" in 1500 by the Portuguese. The occupation and settlement of its territory and the other American lands constituted a chapter in the commercial expansion of Europe. Political pressure by other European countries, especially France, England, and Holland, increased Portugal's desire to establish a plantation colony, which could supply highly valuable tropical or mineral commodities as a way of beginning Brazilian colonization.

The early economic formation of Brazil was marked by a series of economic cycles, each associated with a particular commodity. There are no precise beginning and ending dates for these production and trade cycles. However, a reasonable approximation of time periods for these cycles was attempted by Robock (1975), as follow:

> Brazilwood cycle 1500 to 1550
> Sugar cycle 1550 to 1700
> Gold cycle 1700 to 1775
> Rubber and Coffee cycles 1850 to 1930

A mercantile capitalist economy characterized the Brazilian economic formation economy. Its economy was dominated by the production of primary products - agricultural and mining - for export. The period lasted for about 4 centuries, i.e. from the sixteenth century to early 1930s.

From the 1930s Brazil gradually started its industrialization. The emphasis of the Brazilian development model moved from an agricultural economy to a more industrial economy. This process was expanded in the 1950s and from 1964 onwards its present form evolved.

Post-1964 governments adopted policies to promote economic growth at any cost. Thus, policies were launched to create the required climate to attract investments. To do so, extremely generous terms and conditions were offered to both foreign and domestic investors through a fiscal incentives system for programs of regional development. Emphasis was on the Northeast and the Amazon regions and the development of specific economic sectors, such as forestry, fishery, and tourism (Kengen, 1985). Although there were some pitfalls of the fiscal incentives scheme for reforestation, it contributed to an enormous expansion of large-scale industrial forest plantations, especially in the Southeast region. On the other hand, the program of regional development contributed, amongst other things, to the expansion of Brazil's agricultural frontier towards the Center-West region and the Amazon basin. The grant of fiscal incentives encouraged many entrepreneurs to establish cattle ranches in the Amazon basin. This has stimulated a process of deforestation which has been subject to a worldwide discussion.

In response to an increasing rate of deforestation, an internal and external conservation lobby emerged. It has exerted pressure on the government to adopt measures on its programs, plans, and projects which lead to minimize ecological damages and promote a rational use of the region, i.e. avoiding the exhaustion of the country's resources. This led the government to suspend the concession of fiscal incentives to establish farms and cattle ranches in the Amazon basin.

Despite these measures, Brazil's search for development continues. Its great challenge is how to conciliate its development with a rational utilization of its resources. In a territory which is still endowed with virgin land and diverse resources, this is a difficult paradox to resolve.

**Historical Exploitation of Forest Resources**

**Vegetation Types:** Given the diversity of Brazilian edapho-climatic conditions, its vegetation is very diverse, and a detailed analysis of Brazilian vegetation would be beyond the scope of this paper. Thus, for the purpose of this paper the vegetation has been grouped into 3 broad types namely:

    i.   forests - given the diversity of climate and soils they range from tropical rainforest to sub-tropical forest and Araucaria forest.

ii.  cerrado (savanna) - is a plant community composed of tall tufted grasses, and usually tree or shrubs are also present.

iii.  caatinga (thornbush) - is a plant community dominated by thorny trees or shrubs called thorn scrub. It is a xerophytic vegetation typical of dry environments.

**Forest Development:** A major international political and economic integration took place along with the Brazilian colonization. Also the forest continued to be devastated to provide land for agricultural expansion, especially for growing of export crops (Tucker & Richards, 1983). The climax of this occurred during the nineteenth century with the expansion of coffee production under the plantation system.

The Portuguese Crown made some attempts to protect the environment and forest resources. However, despite these attempts to preserve, or at least to avoid over-exploitation of the forests, the mercantilism prevailed (Delson & Dickenson, 1984).

As referred to above, it was only in the 1930s that forest resources received a major attention in Brazilian legislation, even became a constitutional matter.

After the Second World War and, especially in the 1950s, development became a major concern throughout the world. Within this context, the forestry sector was perceived as able to contribute to development. So, FAO and other international agencies concentrated their efforts in order to help Brazil and other underdeveloped countries to explore and to exploit their forests. This task was not difficult given the great willingness of the government in promoting economic development. One example of this was the request for FAO to send a mission to carry out a survey of part of the Amazon forest and to help the establishment of a center to train workers of sawmills in the Amazon basin (INP, 1959).

Underlying these efforts it was the "demonstration effect" of the contribution that the forestry sector had made in the process of development of some developed countries. This approach marked the 1950s and even in the 1960s. It was assumed that the model could be copied by any underdeveloped country regardless of its social, political, and economic structures. Allied to this it was assumed a worldwide shortage of timber and wood-based products which would consequently be supplied by the production of Brazil and other underdeveloped countries.

Although considerable progress in all aspects of forestry could be noted throughout the world, it was in North America, Europe, and in the USSR that it mainly occurred. In Third World countries production had not risen as much as expected, which was a great disappointment of the 1950s (Proceedings of the 5th World Forestry Congress, 1960, vol. 1).

This disappointment can be attributed, among other things, to trying to solve the underdeveloped countries' problems by extrapolating from First World experiences. The argument does not imply anything at all about whether similar effects can or will arise in the underdeveloped country. Therefore, the use of historical data from developed countries seems to be inappropriate.

As far as the Amazon is concerned, in the early stages of the Brazilian colonization the need for labour led the Portuguese settlers to penetrate the Amazon basin to capture indians. These incursions led to knowledge of the forest and exploitation of some forest products, such as cacao, cinnamon, clove, and aromatic resins (Furtado, 1982). As referred to above, in the 1950s some efforts were made to develop a suitable forest management plan for the Amazon basin, but it was difficult to maintain continuity. But a great expansion towards the Amazon basin occurred from the 1970s onwards when the Brazilian government started to direct policies stimulating the expansion of the agricultural frontier, especially the establishment of cattle ranches. However, this was conducted without regard to any forest management plan. In fact, what took place, in the name of progress, was a disorganized expansion of the agricultural frontier characterized by the mismanagement of the forest resources. It is fair to assume that this mismanagement contributed to worsen the situation of underdevelopment than to any actual development (Camino, 1987).

This has caused the Amazon basin to become the focus of attention from the international community, due to the deforestation which "accounts for half of all tropical deforestation", according to Grainger (1987). Although this great rate of deforestation and the region "contains half of all remaining tropical rainforests, these have so far been little exploited for industrial wood" (Grainger 1987:1). In fact, most of the wood has been simply burnt as waste, rather than commercially exploited.

This is quite different to what has happened to the tropical forests of Africa and Asia. In these regions, the main causes for deforestation can be attributed to logging for commercial purposes and to grow food for an ever increasing population.

This mismanagement has greatly contributed to a process of deforestation. However, this mismanagement cannot be viewed in isolation. It follows a rational pattern which cannot be simply ascribed individual malevolent action as has been usually claimed, particularly in the case of shifting cultivation. This can be attributed to as part of the overall "economic and political structures (which themselves obey a systemic logic) that establish the constraints and the incentives of individual action" (Schmink, 1987: 11).

In response to an international pressure, the Brazilian Government, in 1988, has suspended all fiscal incentives to the establishment of cattle ranches in the Amazon basin. This suspension was again confirmed in the economic reforms formulated by the new government which assumed office in 1990.

Another matter of concern regarding deforestation in the Amazon basin has been the establishment of pig iron plants in the area of influence of the iron mine of "Carajás". These plants will use charcoal as a reducing agent. Some research has been conducted in Buritucupu in the State of Maranhao (Thibau *et al.*, 1987) and Marabá in the State of Pará in order to determine which will be the best method of forest management. The goal is to avoid the increasing demand for charcoal that would in turn promote complete devastation. However, it will take sometime before the results can be effectively evaluated.

**Conclusion**

Brazil's search for economic growth and the utilization of its forest resources has been characterized by the paradox between a tendency toward exploitation and attempts to conserve. However, as one can note the exploitation has prevailed, responding to a systematic logic established by economic and political structures. This mismanagement has greatly contributed to aggravate the situation of underdevelopment and, consequently, to decrease the overall standard of living. However, in a territory which is still endowed with virgin land and diverse resources it is not easy to find a solution to the paradox referred to above.

This example of Brazil suggests that the discussion regarding mismanagement of forest resources which has led to an increasing rate of deforestation cannot be simply conducted under an universal view, as well as considered a single factor. On the contrary, it should be analysed case by case under a historical perspective, taking into account the social, political, and economic structures and their inter-relationship.

# References

Camino, R. de. 1987. "Algunas considraciones economicas en el manejo de bosques tropicales". In: *Management of the Forests of Tropical America: Prospects and technologies.* pp. 175-188. Proceedings of a Conference held in San Juan, Puerto Rico. September 22-27, 1986. Rio Piedras: Institute of Tropical Forestry, Southern Forest Experiment Station, U.S.D.A. Forest Service.

Delson, R. M. & Dickenson, J. 1984. "Conservation tendencies in colonial and imperial Brazil: An alternative perspective on human relationships to the land". *Environmental Review* 8(3), Fall, 270-283.

Furtado, C. 1982. *Formaçao Econômica do Brasil* (18th ed.). Sao Paulo: Companhia Editora Nacional.

Grainger, A. 1987. "The future environment for forest management in Latin America". In: *Management of the Forests of Tropical America: Prospects and technologies.* pp. 1-10. Proceedings of a Conference held in San Juan, Puerto Rico. September 22-27, 1986. Rio Piedras: Institute of Tropical Forestry, Southern Forest Experiment Station, U.S.D.A. Forest Service.

Instituto Nacional do Pinho (INP). 1959. *Formaçao de técnicos madeireiros para a Amazônia.* Anuário Brasileiro de Economia Florestal 15, 83-107.

Kengen, S. 1985. Industrial Forestry and Brazilian Development: A social, economic and political analysis, with special emphasis on the fiscal incentives scheme and Jequitinhonha Valley in Minas Gerais. Unpublished Ph.D. thesis. Canberra: Department of Forestry. The Australian National University.

Kengen, S. 1991. Myanma Forestry Sector: An economic review selected issues. National Forest Management and Inventory Project (MYA/85/003). Field Document No. 10. Unpublished.

Masson, J. L. 1983. Management of Tropical Mixed Forest: Preliminary assessment of present status. Fo: MISC/83/17. Rome: FAO. mimeo.

Proceedings of the 5th World Forestry Congress held in the United States of America in 1960.

Robock, S. H. 1975. *Brazil: A study in development progress.* Lexington Lexington Books, D.C. Heath and Company.

Schmink, M. 1987. "The rationality of forest destruction". In: *Management of the Forests of Tropical America: Prospects and technologies.* pp. 11-30. Proceedings of a Conference held in San Juan, Puerto Rico. September 22-27, 1986. Rio Piedras: Institute of Tropical Forestry, Southern Forest Experiment Station, U.S.D.A. Forest Service.

Thibau, C. E.; Jesús, R. M. de & Menandro, M. S. 1987. "Sustained Yield in the Amazon Region". In: *Management of the Forests of Tropical America: Prospects and technologies.* pp. 199-206. Proceedings of a Conference held in San Juan, Puerto Rico. September 22-27, 1986. Rio Piedras: Institute of Tropical Forestry, Southern Forest Experiment Station, U.S.D.A. Forest Service.

Tucker, R. P. & Richards, J. F. (eds.) 1983. *Global Deforestation and the Nineteenth-Century World Economy*. Duke Press Policy Studies. Durham: Duke University Press.

Volpato, E. 1986. Análise da Administraçao Florestal Brasileira. Preliminary version in Portuguese. Unpublished Ph.D. thesis. University of Freiburg.

Indigenous History and Amazonian Biodiversity

William Balée
Tulane University


A recent report to the National Science Board, entitled "Loss of Biological Diversity: A Global Crisis Requiring International Solutions" made the point that the modern extinction spasm we are living through "has been caused by a single species . . ." (NSB 1989:1). The human species itself, regardless of profound societal differences within it, is frequently seen to be the juggernaut of modern extinctions on earth. Several rather pessimistic conservationists apparently assume that humankind, by its very nature, is incompatible with maintaining biological diversity, in terms of current numbers of animal and plant species, on the planet. I understand this pessimism, though I would not implicate the entire human species.

A look at ecological change in Amazonia reveals that humans have disturbed natural landscapes in at least 2 radically different ways. The first kind of disturbance originated in Indian societies that relied mostly on foraging or horticulture. The second kind comes from state societies that depend on fossil fuels. These 2 types of societies differ greatly in their capacity to convert forests and to alter other basic environmental conditions, including climate. For example, Amazonian Indians probably through the centuries have released only relatively minute amounts of $CO_2$ into the atmosphere by the burning of their swidden fields, in comparison with societies associated with modern states, many of which are also involved in tropical deforestation, eutrophication of estuaries, acid rain and forest dieback, toxic waste deposits or other overt threats to the integrity of natural ecosystems. In other words, if global warming is really taking place, because of increasing emissions of $CO_2$ and other greenhouse gases, Amazonian Indians have had but a negligible input (indigenous Polynesian societies, however, may have been associated with higher outputs of $CO_2$--Leslie Sponsel, personal communication, 1991). The quantity of carbon dioxide in the earth's atmosphere rose from 315 to about 340 parts per million only within the last generation (NRC 1983:1; cf. Tirpak 1990:41). Indigenous Amazonian societies, however, reached their climax, in terms of forest burning and use of energy, long before this, that is, prior to the Conquest. Even if modern nation-states profess a conservationist "ethic", as is evident in the increasing profusion of conservationist NGO's, state ministries/secretariats of the "Environment", agencies such as and including the US EPA, and environmental advisory sectors of major world lending institutions, the only solid evidence for recent species extinctions is attributable directly to the same societies.

Although it is plausible that during the terminal Pleistocene (ca. 10,000 to 8,000 years ago) the hunting technologies of pre-agricultural Amazonian Indians, as with Upper Paleolithic peoples elsewhere, contributed to the extinction of large animals, such as giant ground sloths, natural climatic change was probably also involved (Peter Mann Toledo, personal communication, 1990). There is no evidence yet, moreover, for extinctions in Amazonia during and after the Neolithic (the development of settled village life and agriculture), which probably began some 6,000 to 5,000 years ago. A state sponsored conservationist ethic in Western societies came into being only after extinctions of species, such as the Carolina parakeet, the passenger pigeon, and the great auk (Matthiessen 1987), combined with profound alterations of many of the world's habitats, had already become obvious.

Amazonian indigenous societies, in contrast, usually do not possess an explicit policy of conservation nor do they have voluntary associations devoted to preserving biodiversity, perhaps for the simple reason that their economic activities did not produce a need for such. They never had a state. State societies, with their high population densities, elevated levels of per capita energy consumption, and technologies capable of habitat conversion everywhere on the planet are solely responsible for the emerging and justifiably alarming trend of major biotic depletions, not the human species *per se*. There is yet hope for avoiding some depletions, perhaps only as long as non-state societies, such as those of Indians of Amazonia, continue to exist.

This is not to say that agricultural Indians of the Amazon have not altered the environment in significant ways. They have, though the evidence points not to extinctions, but to an increase in biological and ecological diversity. This apparent diversification extends from Neolithic times to the present. Most notable is the suite of domesticated and semidomesticated species. Archaeobotanists (e.g., Pickersgill and Heiser, 1977) have estimated that the number of domesticated neotropical plants exceeds 100. Several appear to be from Amazonia and environs, including pineapple, passion fruit, peanut, manioc, yam, cocoyam, papaya, arrow cane, a bromeliaceous fiber plant (*Neoglaziovia variegata*), annatto, and numerous tree crops. Tree crops, several of which are from the Upper Amazon, according to Clement (1989), include peach palm, guava, Brazil nut, cashew, cacao, inga, cupuassu (cacao family), bacuri (clusia family), abiu (sapote family), sapota (bombax family), Amazon tree grape (mulberry family), and biriba (annona family). Most of the tree crops, in fact, are not completely domesticated: they occur not only cultivated in swiddens and door yard gardens throughout Amazonia, but also in a feral or semi-domesticated state. They tend to be dominant species, however, only where indigenous agriculture has taken place.

Brucher (1989:12) appropriately observed that "It was the Indians who domesticated and improved in quantity and quality the yields of numerous crop plants which today enrich the daily diet of highly developed industrial nations, who know little of their neotropical origin. Those contributions from Middle and South America are superior to other continents . . ." When some conservationists plead for preserving the biological and ecological diversity of Amazonia, wittingly or not they do so at least partly on behalf of the ancestral Indian cultures that contributed to this diversity.


**High Forests vs. Indigenous Fallows in Eastern Amazonia**

Between 1985 and 1990, I carried out inventories of 8 separate hectares of forest in the habitats of the Ka'apor, Guajá, and Tembé Indians of extreme eastern Amazonian Brazil. These hectares span a linear distance of 150 km between the left bank of the Rio Gurupi and the left bank of the Rio Pindaré. The methods used were comparable for each of the 8 hectares: 1) all trees greater than or equal to 10 cm in diameter at breast height (dbh) on each plot were measured, tagged, collected, and identified; 2) all plots were subdivided into 40 sampling units (or subplots) of 10m X 25m, in order to sample relative diversity; 3) all plots were narrowly rectangular in dimension, being either 10m X 1000m or 20m X 500m; 4) all plots were situated on *terra firme* sites near indigenous villages, but none were the objects, at the time of study, of agricultural activity. These inventory methods are identical to those of many other studies in Amazonia (e.g., Campbell *et al.* 1986; Boom 1986; Salomao *et al.* 1988; Gentry 1988). For all 8 hectares, and any combinations thereof, one can calculate total basal area, and, for individual species, basal area, relative frequency, relative density, and relative dominance. In addition, one can estimate the ecological importance value for each species, a derived measure involving the sum of relative frequency, relative density, and relative dominance (see Greig-Smith 1983:152; Campbell *et al.* 1986, Salomao *et al.* 1988).

Of the 8 hectares, 4 (sites 1, 2, 5, and 7) are fallows (sites that were used between 40 and ca. 150 years ago for agriculture but which have not been disturbed by fire since) and 4 (sites 3, 4, 6, and 8) are high forest (a forest of the *terra firme* which has evidently not been disturbed for agriculture within the last 200 or 300 years, if ever). Regardless of the distinctions between these forest types, satellite imagery to date portrays both as being "high forest," i.e., primary vegetation.

That the plots denoted as "fallows" are vestiges of agricultural settlements and not high forest is readily attested in several ways. First, charcoal and potsherds are found to depths of 80 cm at all sites, which are not present on the high forest sites. Although lightning strikes or campfires of nonagricultural peoples could have been responsible for the charcoal, potsherds in Amazonia tend to be strictly associated with agricultural peoples. The charcoal findings are probably artifacts of hearths and/or swidden burning by agricultural peoples. For example, when I asked Guajá Indians near the P.I. Guajá on the Rio Turiaçu as to who were the primordial inhabitants of their fallow (site 2), the ecologically most important species of which was the babassu palm, they replied the "kamarar", which is the Guajá name for the horticultural Ka'apor Indians, their neighbors and former enemies to the northeast. At a Ka'apor fallow (site 7) dominated by the bacuri fruit tree (*Platonia insignis*) near the village of Gurupiuna, some 60 km distant from the P.I. Guajá, on the other hand, the original inhabitants were probably refugee Afro-Brazilian slaves who had lived in a maroon settlement here since before the 1870s. The Ka'apor only arrived in the region around 1874, fleeing state militia who were invading their villages some 200 km to the north. Historical records and Ka'apor lore indicate that the Ka'apor expelled the Afro-Brazilians and took over their lands in about 1878 (Balée 1988). In any event, the Afro-Brazilians who occupied the region *did* have agriculture and would have burned the surrounding forest for swiddens. Site 1, which is a fallow dominated by the babassu palm near the P.I. Awá in the Pindaré basin and today utilized by Guajá Indians, was evidently a former village of the agricultural Guajajara Indians--many potsherds and large charcoal samples were encountered here. Site 5, which is a fallow dominated by wild papaya (*Jacaratia spinosa*) near a Ka'apor village in the Turiaçu basin, had a wellknown and fairly recent history of occupation by the Ka'apor.

If satellite imagery does not yet distinguish between fallow and high forest of the region, what, if any, are the ground truth criteria for maintaining a distinction between the 2? In other words, if fallows looks like high forest, are there any reasons for separating them conceptually? One possible criterion for maintaining the distinction concerns indigenous classification. The Ka'apor call fallow forests taper and high forest ka'a-te. A difficulty arises, however, if one is depending exclusively on indigenous knowledge. This concerns relative shortness of historical memory, which is widespread in these small-scale, egalitarian, nonliterate societies (see Goody, 1977; Murphy, 1979). The Araweté and Asurini, for example, called old fallow forests of the Xingu River basin by the terms ka'a-hete and ka'a-ete, which mean "high forest" (Balée and Campbell, 1990). And the Guajá of the P.I. Awá on the Rio Pindaré described the fallow forest (site 1) that was inventoried as ka'a-ate ("high forest"). The implication is that if a site has been fallowed long enough, the Indians tend to consider it to be high forest, not recognizing that it was once occupied by agricultural peoples. Yet this does not mean that they perceive and encode linguistically a temporal succession from fallow to high forest--indeed, the Ka'apor, at least, evidently do not (Balée and Gély, 1989). The Araweté and Asurini considered stone axeheads and potsherds found in their villages to be the work of divinities, not past agricultural peoples (Balée and Campbell, 1990).

An important physiognomic difference between the 4 fallows and the 4 high forests concerns basal area. The fallows average 24.2 m$^2$ whereas the 4 high forest sites average 29.3 m$^2$. Fallows are typically within a range of about 1824 m$^2$ whereas high forests range from 25-40 m$^2$ (cf. Pires and Prance, 1985; Balée and Campbell 1990; Boom 1986; Saldarriaga and West 1986:364).

In identifying a plot of forest as fallow, a great deal also typically hinges on the presence of "disturbance indicator" species, which may be long-lived. For Van Steenis (1958), disturbance indicators were biological nomads. They occurred as isolated or even rare individuals in a primary forest until a disturbance, such as fire in the service of agriculture, opened space for them. Yet Van Steenis proposed no quantitative measure to determine when a biological nomad becomes an indicator of human activity. In fact, to date, quantitative measures for disturbance indicators are largely lacking (see Brown and Lugo, 1990).

Some species, by their presence on the fallow sites alone, appear to indicate agricultural activity in the past. These include ecologically important species such as the babassu palm, the tucum palm (*Astrocaryum vulgare*), wild papaya (*Jacaratia spinosa*), hog plum (*Spondias mombin*), inajá palm (*Maximiliana maripa*), and the enormous leguminous trees, *Platypodium elegans* and *Apuleia leiocarpa* (see Table 1). There is no evidence that these species ever occur in high forest of the *terra firme* habitats of the Ka'apor, Guajá, and Tembé Indians (although a few, such as *Spondias mombin* and *Maximiliana maripa*, may be encountered in seasonally inundated forests, usually on the banks of rivers).

There are good floristic reasons for separating the high forest hectares from the fallow hectares. The 8 hectares of fallow and high forest in this sample yield 28 pairs of hectares. The similarity of these pairs can be systematically compared using the Jaccard coefficient, which is simply the number of species/total number of species in the sample (i.e., the sum of the total number of species of each plot minus the shared species) expressed as a percentage (Greig-Smith, 1983:151). On average, the coefficient of similarity for pairs of hectares of high forest/fallow is only 10.9 percent. In contrast, the average coefficient of similarity for pairs of hectares of high forest/high forest is 22.8 percent, which is very significantly higher. The average coefficient of similarity for pairs of hectares of fallow/fallow is 17.2 percent, which is also very significantly higher than the fallow/high forest average but not significantly lower than the high forest/high forest average. In

**Table 1.** Comparison of Thirty Ecologically Most Important Species from Fallow (4 ha.) and High Forest (4 ha.) Sites in the Habitats of the Ka'apor, Guaja, and Tembé Indians (Eastern Amazonian Brazil).

| High Forest Species | I.V.* | Fallow Species | I.V.* |
|---|---|---|---|
| Eschweilera coriacea | 37.83 | Jacaratia spinosa | 11.40 |
| Lecythis idatimon | 14.53 | Gustavia augusta | 10.41 |
| Sagotia racemosa | 12.67 | Orbignya phalerata | 9.37 |
| Tetragastris altissima | 11.60 | Astrocaryum vulgare | 7.76 |
| Protium trifoliolatum | 7.76 | Spondias mombin | 6.53 |
| Protium decandrum | 7.07 | Neea sp. 1 | 6.26 |
| Protium pallidum | 6.78 | Pisonia sp. 2 | 6.25 |
| Carapa quianensis | 5.69 | Pouteria macrophylla | 5.71 |
| Couepia quianensis | 5.07 | Maximiliana maripa | 5.40 |
| Pourouma minor | 4.54 | Platypodium elegans | 5.02 |
| Taralea oppositifolia | 4.51 | Platonia insignis | 4.32 |
| Mabea caudata | 4.06 | Simaba cedron | 4.26 |
| Pourouma quianensis | 3.28 | Hymenaea parvifolia | 4.17 |
| Dodecastigma integrifolium | 3.10 | Trichilia quadrijuga | 4.06 |
| Couratari quianensis | 2.77 | Lecythis pisonis | 3.56 |
| Oenocarpus distichus | 2.72 | Dialium quianense | 3.32 |
| Sterculia pruriens | 2.65 | Astrocaryum munbaca | 3.31 |
| Bagassa quianensis | 2.65 | Eschweilera coriacea | 3.19 |
| Cecropia obtusa | 2.60 | Theobroma speciosum | 3.11 |
| Newtonia psilostachya | 2.47 | Lindackeria latifolia | 3.05 |
| Chimarrhis turbinata | 2.40 | Tabebuia impetiginosa | 2.85 |
| Simaruba amara | 2.39 | Myrciaria obscura | 2.75 |
| Euterpe oleracea | 2.37 | Neea sp. | 2.64 |
| Lecythis chartacea | 2.25 | Hymenaea courbaril | 2.61 |
| Parkia pendula | 2.23 | Protium heptaphyllum | 2.59 |
| Protium polybotryum | 2.22 | Tetragastris panamensis | 2.56 |
| Apeiba echinata | 2.19 | Apuleia leiocarpa | 2.53 |
| Fusaea longifolia | 2.18 | Mouriri guianensis | 2.49 |
| Protium giganteum | 2.12 | Cupania scrobiculata | 2.40 |
| Tachigali myrmecophila | 2.11 | Pouteria bicocularis | 2.34 |

| Total I.V. | 166.81 | | 136.22 |

* I.V. = Relative importance value (sum of relative density, relative frequency, and relative dominance); the total of relative importance values for any sample of vegetation is always 300

addition, no tendency exists for nearby plots of high forest and fallow to be more similar than more distantly separated hectares of the same type (i.e., fallow/fallow, high forest/high forest). For example, if one compares high forest site 6 to the fallow site 7 (the 2 plots, which are near the Ka'apor village of Gurupiuna, are separated by about 2.5 km), the coefficient of similarity is only 10.5 percent. Likewise, comparing the high forest site 3 with the fallow site 2 (the 2 plots, which are near P.I. Guajá on the Rio Turiaçu, are separated by only about 1.5 km), the coefficient of similarity is only 11.6 percent. Sites 6 and 7 (near Gurupiuna), on the one hand, and sites 3 and 2 (near P.I. Guajá), on the other, are separated by a linear distance of 60 km. It is interesting, therefore, that the fallow sites 2 and 7 have a similarity coefficient of 19.3 percent, which is very significantly higher than the those of the 2 pairs of nearby hectares. The high forest sites 3 and 6, with a similarity coefficient of 25.1 percent, are also much more similar to each other than either is to its nearby fallow forest. The overriding factor that accounts for divergence in floristic composition between forest stands in this sample is not linear distance between stands, but rather past agricultural perturbation.
When plotted on species-area curves (best-fit curves) as separate forest parcels, the fallow forests and the high forests accumulate diversity at similar rates (see Figure 1). These curves and the evidence of floristic composition support the familiar notion that for secondary forests, "within a span of 80 yr or less, the number of species approaches that of mature forests" (Brown and Lugo 1990:6). The data from my sample show that not only does fallow "approach" high forest in species diversity, the diversity between the 2 forest types is insignificantly different: the 4 fallows had a total of 360 species, whereas the 4 high forests had a total of 341 species. This alone is very significant evidence that these fallow forests represent indigenous reforestation, insofar as species richness of high forests is being replaced by an equivalently rich secondary forest, although the major species, indeed, are different between the 2 forest types.

The most surprising difference, in terms of species, concerns ecologically important species. In comparing the 30 ecologically most important species between the 4 fallows on the one hand and those of the 4 high forest plots on the other, the 2 forest types share but a single species, *Eschweilera coriacea* (see Table 1). This yields a coefficient of similarity of only 1.7 percent for the 30 most important species (number of shared species [1]/total number of species on the 2 plots [(30 + 30) 1] X 100). This difference is extremely significant. The average coefficient of similarity for the 30 most important species of pairs of high forest is 16.4 percent. The equivalent measure for the 30 most important species of pairs of fallow is 11 percent. Yet the average coefficient of similarity for the 30 most important species of pairs of high forest/fallow plots is less than 2 percent. This permits one to conclude that the important species between fallow and high forest are significantly and predictably different.

For reasons that militate against received wisdom, the fallow forests are actually *less* dominated by a few species than are the high, presumably primary, forests. For example, the average total of the 2 ecologically most important species (with a total possible value of 300) of the 4 fallows is only 40.3, whereas the comparable figure for the high forest hectares is 60.3, which is significantly higher. The data I have collected offer only partial support for the statement: "A large number of species in mature forests is due to the presence of rare species. In contrast, secondary forests are usually composed of common species" (Brown and Lugo 1990:7). For analytical purposes, I consider a species to be rare if it occurs only once, regardless of whether it is on one of the fallow hectares or one of the high forest hectares. By this criterion, the high forest has 199 rare species (or 58 percent of the total). Fallow forest has 139 rare species (or 39 percent of the total). Both forest types, in other words, harbor significant quantities of rare species. Many disturbance indicator species, moreover, appear to be *unique* to fallow.

Aside from the differences in age, basal area, overall floristic composition, and ecologically important species between fallow and high forest, another difference, being strictly related to utilitarian concerns, deserves mention. Fallows are indigenous orchards, whether consciously planted or not. Of the 30 ecologically most important species of fallow, 14 are significant food species for 1 of the 3 indigenous peoples of the region, whereas for the 30 ecologically most important species of high forest, only 6 are important food species. Significant food species of the fallow include babassu palm, hog plum, tucum palm, inajá palm, bacuri fruit tree, and copal trees (*Hymenaea* spp.). These orchards would not exist, obviously, had it not been for agricultural activities of the Indians.

**Amazonian Indians and Biodiversity**

It is clear that indigenous agriculture changed the forest profile of Amazonia. At the same time, it is important to recall that high forests still endure in many indigenous areas, coexisting with fallow forests. In many zones of recent penetration by civilization, of course, neither forest type has endured. The activities of horticultural Indian societies probably did not so much perpetuate as fairly ignore many primary forests, hence the continuing survival of these forests today in indigenous areas.

Indigenous old fallows contrast markedly with pastures and monocultural swiddens, not only for remote sensing but also in terms of the ground truth. Old fallows represent the logical opposite of deforestation--they are, in other words, reforested patches of vegetation, even if the most important species are different from the aboriginal forest and even if the species in these fallows were not always actually planted. They are the result of indigenous, non-state agroforestry practices, whether intentional or not. Slash-and-burn agriculture and "sustainable agriculture" are, by this evidence, not necessarily opposed categories. Yet the official US EPA position is that one tropical forestry measure for reducing sources of greenhouse gases would be to substitute "slash-and-burn" with "sustainable agriculture" (see Tirpak 1990:42). State societies may not exhibit sustainable slash-and-burn regimes, but indigenous, non-state societies of Amazonia evidently did and continue to do so.

Pre-Columbian horticultural peoples also altered the land itself, as, for example, the evidence of Roosevelt (1987, 1989) suggests: enormous mounds, earthworks, and differential burials on Maraj; Island at the mouth of the Amazon and near Santarém, at the mouth of the Tapajós River. Anthrosols (*terra preta do indio*-- "Indian Black Earth"), which occur widely throughout the Amazon Basin, also evince pre-Columbian human manipulation of the natural landscape (Smith 1980). Many landscapes, soils, and forests today in the Amazon, then, intimate a very old human factor, specifically, one that was not associated with a state society dependent on fossil fuels.

In some ways, modern Amazonian Indians use and manage the forest differently from their pre-Columbian forebears. Present-day hunter-gatherer societies of the forests of eastern Amazonia, such as the Guajá, who do not fell and burn forest for swidden fields, exert less influence on forest composition than did the pre-Columbian chiefdoms of Amazonia, some of which may have been on the road to becoming state societies at the time of the European Conquest. Modern hunter-gatherers also demonstrate less effect on forest composition than modern horticultural village societies, such as the Ka'apor.

It is a remarkable yet well-documented fact that many of them lost the art of fire-making. The hunting and gathering Guajá Indians of extreme eastern Amazonia, for example, until they came into contact with Brazilian government authorities in 1974-75, did not know how to make fire from a drill and hearth, an art that is otherwise known widely to horticultural peoples of the Amazon basin. They used torches from the combustible sap of the *Manilkara huberi* tree (maçaranduba, sapote family) to travel from camp to camp and maintained camp fires burning with slow-burning woods, such as *Sagotia racemosa* of the spurge family, known to them as the *miriko-'i* ("woman-tree"). Unlike their horticultural neighbors, the Ka'apor, they did not recently fell and burn forests for swiddens. Yet in the remote past, the ancestral culture of the Guajá Indians certainly had a suite of domesticated crops.

The foraging Guajá speak a language within the Tupi-Guarani family of languages. The plant vocabulary of the mother language, proto-Tupi-Guarani, which was spoken about 1800-2000 years ago (Migliazza, 1982; A.D. Rodrigues, personal communication), contained numerous words for neotropical domesticates, including maize, manioc, yam, sweet potato, pineapple, peanut, gourd, calabash, cashew, annatto dye plants, and bromeliaceous fiber plants (cf. Lemle 1971, Rodrigues 1988). Thus, Guajá is descended from a proto-language associated with a horticultural society. The Guajá lost horticulture and their plant domesticates probably because of introduction of Old World diseases and attendant marked depopulation that followed the Conquest. Severe depopulation destabilized the society and led it to become increasingly nomadic (Balée, in press). In the tropics, full-time nomadism appears to be incompatible with horticulture.

Even if modern hunter-gatherers exert a negligible effect on forest composition, their ancestors did manipulate many of the forests often erroneously understood to be "primary" today (Balée 1989a, 1989b). It would be mistaken, however, to consider these manipulations of the past as "forest conversion," since the artifactual fallows the Indians left behind are a kind of forest, too. In any case, the pre-Columbian chiefdoms of the past, together with the hunting and gathering and horticulturalist Indian peoples of today in the Amazon have more in common with each other than any of them do with modern nation states, in terms of mutually distinctive effects on biological and ecological diversity. Among these diverse types of societies, only the state type is associated directly with massive biotic depletions of today. The pre-Columbian chiefdoms, however more dense and growing their populations and technologies before the Conquest, had a greater similarity to modern Amazonian Indians than to state society also in terms of their languages, marriage customs (for example, infant betrothal), and religious beliefs (for example, polytheism). Indian cultures of today, in other words, have in large part descended from remote pre-Columbian ancestors, in spite of recent Western influences that many of them have experienced. Most of the crops and the species that modern Indians exploit in primary forest and fallow are also neotropical, another connection to the pre-Columbian past. Most modern Amazonian Indians continue to be like their forebears--they are still Indians--certainly in terms of their plant resources and the ways in which they use and manage these.

The resource management practices of the indigenous farmers and foragers of Amazonia of today are less destructive of the environment, by any measure, than our rapacious nation-states with economies based on the burning of fossil fuels. Myers (1988) recently identified 10 "hot spots" in certain tropical forests worldwide, including one in western Amazonia, which are high in endemic species and which are immediately threatened with habitat conversion and species extinctions. Myers predicted that unless forceful conservation measures are implemented soon, more than 17,000 or about one-half of the some 34,200 endemic plant species in these hot spots will be extinct by around the year 2000. This figure accounts for more than 13 percent of all tropical forest plant species worldwide. Myers's estimate for the number of animal species immediately facing extinction in these areas is 350,000. This coming extinction "spasm", which will be like nothing seen on earth since mass extinctions of plants and animals in the late Cretaceous, some 65 million years ago, is but an artifact of state societies, not the human species in general. Not a shred of evidence connects modern Amazonian Indians, whose societies are by definition non-states, to the increasingly apparent scenario of major biotic depletions.

Yes, pre-Columbian chiefdoms and modern horticulturalist village societies did and continue to alter the "natural" environment. But to the archaeology of the future, indigenous burial mounds and other earthworks (such as stockades) will represent a qualitatively different kind of environmental manipulation than mega-projects such as the hydroelectric dams of Tucuruí, Balbina, and, if ever constructed, Kararaô (on the Rio Xingu). The trails and highways that linked indigenous Amazonian villages, some of which, such as the Tapajós Indian village at the mouth of the Rio Tapajós, were incipient urban centers (Roosevelt, 1989), will never rival the Belém-Brasília, Belém-Sao Luís, and Trans-Amazon highways in terms of habitat conversion. The carbonized plant remains from prehistoric Indian villages, swiddens, and fallows will suggest far more species of trees than the strata indicative of cattle pastures and monospecific stands of rice. Finally, there is really no indigenous equivalent for rivers and lakes poisoned by mercury in the modern Amazon gold rush.

The time has come for the conservationists of state societies to rethink their premises. Without the state, would their activities still be needed, if even meaningful? *With* the state, which is a given condition of most of the world's population, can conservationists slow down or reverse the trend of species extinctions in our time? If modern states cannot protect the remaining Indian villages and non-state societies of the world, will they ever be able to emulate them in terms of resource use, management, and biological and ecological diversification? We may know the answers to these difficult questions in the very near future.

# References

Balée, W. 1988. The Ka'apor Indian wars of lower Amazonia, ca. 1825-1928. Pages 155-169 *in* R. R. Randolph, D. M. Schneider, and M. N. Diaz (eds.), Dialectics and gender: Anthropological approaches. Westview Press, Boulder, CO.

___. 1989a. Cultura na vegetaçao da Amazônia brasileira. Pages 95-109 *in* W. A. Neves (ed.), Biologia e ecologia humana na Amazônia: Avaliaçao e perspectivas. *Coleçao Eduardo Galvao*, Museu Paraense Emilio Goeldi, CNPq, Belém.

___. 1989b. The culture of Amazonian forests. *Advances in Economic Botany* 7:121.

___. in press. People of the fallow: An historical ecology of foraging in Lowland South America. *In* K. H. Redford and C. Padoch (eds.), Conservation of neotropical forests: Building on traditional resource use. Columbia University Press, New York.

___. and D. G. Campbell. 1990. Evidence for the successional status of liana forest (Xingu River basin, Amazonian Brazil). *Biotropica* 22(1):36-47.

___. and A. Gély. 1989. Managed forest succession in Amazonia: The Ka'apor case. *Advances in Economic Botany* 7:129-158.

Boom, B.M. 1986. A forest inventory in Amazonian Bolivia. *Biotropica* 18(4):287-294.

Brown, S. and A.E. Lugo. 1990. Tropical secondary forests. *Journal of Tropical Ecology* 6:132.

Brucher, H. 1989. Useful plants of neotropical origin and their wild relatives. Springer-Verlag, Berlin.

Campbell, D.G., D.C. Daly, G.T. Prance, and U.N. Maciel. 1986. Quantitative ecological inventory of terra firme and várzea tropical forest on the Rio Xingu, Brazilian Amazon. *Brittonia* 38(4):369-393.

Clement, C.R. 1989. A center of crop genetic diversity in Western Amazonia. *Bioscience* 39(9):624-631.

Gentry, A.H. 1988. Tree species richness of upper Amazonian forests. *Proceedings of the National Academy of Sciences* USA (Ecology) 85:156-159.

Goody, J. 1977. The domestication of the savage mind. Cambridge University Press, New York.

Greig-Smith, P. 1983. Quantitative plant ecology. Berkeley and Los Angeles, University of California Press.

Lemle, M. 1971. Internal classification of the Tupi-Guarani linguistic family. Pages 197-199 *in* D. Bendor Samuel (ed.), Tupi Studies I, Summer Institute of Linguistics Publications in Linguistics and Related Fields, Publication no. 29. Summer Institute of Linguistics, Norman, Oklahoma.

Matthiessen, P. 1987. Wildlife in America. Viking Penguin, New York.

Migliazza, E.C. 1982. Linguistic prehistory and the refuge model in Amazonia. Pages 497-519 *in* G.T. Prance (ed.), Biological diversification in the tropics. Columbia University Press, New York.

Murphy, R. 1979. Lineage and lineality in lowland South America. Pages 217-224 *in* M. Margolis and W. Carter (eds.), Brazil: Anthropological Perspectives, Essays in Honor of Charles Wagley. Columbia University Press, New York.

Myers, N. 1988. Threatened biotas: "Hot spots" in tropical forests. *The Environmentalist* 8(3):187-208.

NRC (National Research Council). 1983. Changing climate: Report of the carbon dioxide assessment committee. National Academy Press, Washington, D.C.

NSB (National Science Board). 1989. Loss of biological diversity: A global crisis requiring international solutions. National Science Foundation, Washington, D.C.

Pickersgill, B. and C.B. Heiser, Jr. 1977. Origins and distribution of plants domesticated in the New World tropics. Pages 803-835 *in* C.A. Reed (ed.), Origins of agriculture. Mouton, The Hague.

Pires, J.M. and G.T. Prance. 1985. The vegetation types of the Brazilian Amazon. Pages 109-145 *in* G.T. Prance and T. Lovejoy (eds.), Key environments: Amazonia. Pergamon Press, New York.

Rodrigues, A.D. 1988. Proto-Tupi evidence for agriculture. Paper presented at the First International Congress of Ethnobiology, Belém.

Roosevelt, A. 1987. Chiefdoms in the Amazon and Orinoco. Pages 153-185 *in* R.D. Drennan and C.A. Uribe (eds.), Chiefdoms in the Americas. University Presses of America, Lanham, MD.

___. 1989. Lost civilizations of the lower Amazon. *Natural History*, February, pp. 74-83.

Saldarriaga, J.G. and D.C. West. 1986. Holocene fires in the northern Amazon basin. *Quarternary Research* 26:358-366.

Salomao, R.P., M.F.F. Silva, N.A. Rosa. 1988. Inventário ecológico em floresta pluvial tropical de terra firme, Serra Norte, Carajás, Pará. *Boletim do Museu Paraense Emilio Goeldi*, sér. Bot., 4(1):1-46.

Smith, N.J. 1980. Anthrosols and human carrying capacity in Amazonia. *Annals of the American Association of Geographers* 70(4):553-566.

Tirpak, D. 1990. The intergovernmental panel on climate change (IPCC): The U.S. position. Pages 41-43 *in* Proceedings of the Conference on Tropical Forestry Response Options to Global Climate Change. University of Sao Paulo, Brazilian Institute for the Environment and Renewable Natural Resources (IBAMA), and US Environmental Protection Agency, Washington, D.C.

Van Steenis, C.G.G.J. 1958. Rejuvenation as a factor for judging the status of vegetation types: the biological nomad theory. Proceedings of a Symposium on Humid Tropics Vegetation, pp. 212-215. UNESCO, Paris.

Como El Hombre Blanco Se Aposó De La Tierra Indígena En El "Pontal Do Paranapanema" En El Estado De "São Paulo" - Brasil

J. Régis Guillaumon
Instituto Florestal

## Introducción

La región enfocada en el presente estudio está localizada en el extremo Oeste del Estado de "São Paulo" y se denomina "Pontal do Paranapanema". Comprende el espacio entre los ríos "Paraná", "Paranapanema" y el parteaguas entre los ríos "Peixe", "Feio" y riacho "Coimbra" y entre las coordenadas geográficas 21°30'a 23°00'de Latitud Sur y 49°30'a 53°30'de Longitud W Gr.

## La Civilización Guarany Y Las "Entradas Paulistas"

Antes del siglo XIX, la región fue escenario del confronto de blancos y indios, con enfoques bastante antagónicos. Por un lado, através de los jesuitas, ya que tres reducciones, Santo Ignacio, o "Ipaumbuzu" Loreto o Nuestra Señora de Loreto y San Pedro, localizadas en el "Paranapanema" a cerca de 130 km de su embocadura, constituyeron puntos avanzados en la organización de la república de los "Guaranies", al penetrarem selva a dentro, esquivandose de la esclavización por los españoles del Plata. Por otro lado, a través de los "bandeirantes paulistas" que veían en los aldeamientos indígenas, inclusive en las reducciones, puestos fáciles para aprisionamiento de los indios como mano de obra esclava en la fase anterior a la importación de negros en la colonización portuguesa del Brasil.

Lugon (1977) estudió muy bien la cuestión de la organización de los "guaranies" en lo que poderia considerar como la primera economia planeada y orientada que se tiene noticia en la historia.

Las primeras reducciones se establecieron en la cuenca del Plata, en el comienzo del siglo XVII. Del Paraguayc, los jesuitas partieron por tierra y el viaje terminó en barco por el "Paranapané", en el punto de confluencia del "Pirapé". Nuestra Señora de Loreto fue fundada en los primeros dias de Julio de 1610, en la ribera del río "Piraga", bajo la dirección de los curas Cataldino y Maceta. El núcleo de la reducción fue constituido por doscientas familias de indios bautizados algunos años antes por los curas Ortega y Filds. A 10 de Septiembre la vida social funcionaba ya en la normalidad y en poco tiempo, Loreto se quedó superpoblada. Fue entonces, una legua y media de distancia, surgindo Santo Ignacio-Mini, que acogió, desde los primeros dias, várias centenas de familias. Poco después, dos otras reducciones fueran aun creadas en la misma región y otras tantas aparecieron en el "Paraná". Diferentes tribus fundieronse en la nueva comunidad más amplia formada por la reducción. Con el surgimiento de los servicios administrativos para aplicación de la regulación, surgieron numerosas funciones nuevas, acabando por debilitar la autoridad de los caciques que, protegidos por ciertos curas, pasaron a constituir una pequeña nobleza, más o menos decorativa. En general, las reducciones contaban con un número de caciques entre treinta y cincuenta. Santo Ignacio llegó a contar con cincuenta y siete, aunque por ocasión de su liquidación no contase con más de una quincena de ellos. Ocupaban cargos más o menos importantes de "tenientes", "mayordomos", etc, en el conjunto de la república. Por toda parte los ocupantes españoles encontraron en esta república consejos municipales electos por el pueblo; el cura Mastrielli ya habla de un "cabildo" o consejo electo, compuesto de "alcaides, fiscales y otros ministros".

Desarrollaron pacientes ensayos de cultura y reproducción de la "yerba-mate", uno de suyos principales productos de comercialización con los colonos del Plata; sus plantaciones fueron coronadas de éxito y extensas áreas fueron consagradas al su cultivo en las cercanías de cada reducción, uno de los mayores sucesos agronómicos de los jesuitas. En general, todos los productos de las Misiones tenían superioridad sobre los otros porque la suya preparación era racionalizada y salia de la rutina. Adoptaban una orientación más general de las profesiones y una especialización por reducción, de acuerdo con las condiciones del suelo, clima, gusto de los habitantes y necesidades del conjunto de la República. Cada reducción cultivaba, particu- larmente, un ramo de la indústria: tonelería, artesanías en cuero, tejeduría, etc. San Francisco Xavier producia tapizes de alto precio, San Juan tenia los mejores fabricantes de instrumentos musicales, Apóstolos fundia las campanas. La grande especialidad de Loreto eran las estatuas y esculturas. En el campo de las artes, fueron grandemente incentivados por el cura Berger, que partió para el Paraguay en 1616, donde pasó toda su vida y, com su guitarra, convertió multitudes de infidelis. Enseñó a los "Guaranies" la música instrumental y vocal. Siendo también escultor y pintor, ejerció un ministerio extremamednte fecundo. Fue sucedido por el cura Jean Baes, que dió un caracter aun más artístico a las execuciones musicales de los "Guaranies". Baes fuera antes un maestro renombrado en Europa, como músico de la Corte del archiduque Alberto, después junto de la infanta Isabel y, posteriormente, como maestro de la capilla en la Corte de Carlos V. Los "Guaranies" aprendieron com Baes la notación más perfecta de la época. Conserva- ron y transmitieron preciosamente, sus repertorios y composiciones personales. Baes se murió en 1623, en Loreto, al servicio de los apestados. Cura Sepp, llegado casi un siglo después, és también presentado como músico, "virtuosi" en todos los géneros de instrumentos y compositor. Muchos de los curas jesuitas que para acá vinieron eran dotados de grandes méritos en el dominio científico; mientras unos se dedicaban a la colonización, otros ocupabanse en estudiar las lenguas indígenas, la historia natural y las explotaciones; tenian una predilección muy especial por la Botánica, quiere por la materia en si quiere por sus relaciones con la medicina.

Elaboraron cierto número de catálogos muy atrayentes sobre las plantas medicinales de que ellos habian experimentado el uso, con los nombres guaranies y españoles, sobresaliendose el cura Montenegro y el cura Asperger. La primera oficina de impresión instalada en el Plata fue, por iniciativa de los jesuitas de la República Guarany, a fin de imprimir los libros en lengua guarany (Lugon, 1977). Según el cura Gay (Giovannetti, s.d.), el cura Montoya escribió su obra magistral sobre la "Arte de la Lengua Guarany" en N. Sra. de Loreto, libro que fue impreso en la reducción de Santa Maria, situada un poco al norte de Iguazu. Fue el primer libro impreso en América Latina. Montoya hiciera imprimir un catecismo, una gramática y un diccionário guarany en Madrid, en 1639. Caracteres tipográficos especiales habian sido creados para transmitir las particularidades de la pronuncia. Un volumen intitulado "Temporal y Eterno" salió de la imprenta de Loreto en 1705, después un diccionário. El guarany llegó a ser elevado a la dignidad de lengua escrita y impresa, habiendo sido enseñada en la Universidad de Córdoba. El trabajo empezado por Maceta y Cataldino tuvo tal desarrollo que Montesquieu escribe en "L'Esprit des Lois" calificando la obra de los jesuitas de grandiosa, mismo que se hubiese reduzido apenas al establecimiento de la industria. La República Guarany, perdida en médio de la pampa y del bosque virgen fue en la época el único estado industrial de Suramérica (Lugon, 1977).

Las reducciones de Santo Ignacio y Loreto llegaron a rivalizarse con las más esplendorosas poblaciones del Paraguay. Loreto, después de Real Ciudad del Guaira y Vila Rica, fue la tercera capital de la región de las Misiones. Por tanto, Loreto, Santo Ignacio y San Pedro, esta situada en la ribera "paulista" representaron en el siglo XVII las centinelas avazadas del progreso en el inmenso "sertão do Paranaspanema" (Giovannetti, s.d.).

Por otro lado, la codicia y la incomprensión de colonizadores españoles y portugueses siempre ha estado inquieta y rondó las reducciones, interesados en el aprisionamiento y esclavización del indio. En 1643, dieciséis misionarios ya habian sido asesinados, sin citar los incontables sacrificios de indios recém convertidos (Lugon, 1977). Leme (1980) relata que Antonio Preto, hijo del afamado Manoel Preto, fundador y primer patrón de la capilla N. Sra. de la Expectación, llamada de "Ó", hoy barrio de la ciudad de "São Paulo", participó de varias "entradas" de los "paulistas" a los "sert^aes" del río "Grande", llamado "Paraná" por los españoles, y a los del río Uruguay y conquistó tantos indios que llegó a contar en su hacienda de la capilla del "Ó" con 999 indios de arco y saeta. De el hace odiosa mención, en el libro sobre Simon Maceta y Francisco Dias Tanho, misionarios en la Provincia de Paraguay, impreso en Pamplona en el año de 1687, cuando describe la "entrada" que Manoel Preto hizo, asaltando la ruducción de Santo Ignacio en 1623 y 1624, cuando era superior el cura Simon Maceta, y de la de Loreto, los curas Antonio Ruiz y José Cataldino. Dias Tanho califica los "paulistas" de "mamelucos, gente atrevida, belicosa y sin ley, que solo tienen de cristianos el bautismo y son más carniceros que los infieles". Afirma que la tropa de los "paulistas" se componia de 800 "mamelucos" (estos son los blancos) y 3.000 "tupis" (estos son los indios administrados por los "paulistas", con armas de fuego y otros instrumentos de guerra); que las reducciones fueron saqueadas, las iglesias profanadas, los indios y los curas presos, heridos o muertos.

Por ocasión del ataque de la "Bandeira" de Manoel Preto y Antonio Raposo a las reducciones, el número de indios aldenados subia a más de 100 mil. Cuando esta "Bandeira" embestió sobre Real Ciudad del Guaira, que era la capital del inmenso imperio teocrático de los jesuitas, hubo el éxodo y todos se refugiaron en las reducciones del río "Paranapanema" (Giovannetti).

**La Conquista Del "Sertão Do Paranapanema"**

Bruno Giovannetti (s.d.), que trabajó como ingeniero en la construcción del ferrocarril "Sorocabana", en el inicio deste siglo, colectó importantes informaciones sobre la región en contacto con la populación local (Le Selva, 1952).

La ocupación antigua del valle de "Paranapanema", que habia bajado de las cabeceras a través de la penetración para el interior, promovida durante el ciclo de la mineración en el valle de "Ribeira" - 1600 a 1697 (Petrone, 1966), estacionó por muchos años en las cercanias de la grande estrada del Sur y se estendió ya en el primer cuartel del siglo XIX, hasta el río "Pardo" y su afluente "Turvo"; la ocupación de las tierras más allá del río "Pardo", en dirección al río "Paraná", solo tuvo inicio en la mitad del siglo pasado (Sampaio, 1890).

En mediados del siglo XIX, mientras cada vez más se poblaba el interior del Estado de "São Paulo", la región conocida como "Sertão do Paranapanema" continuaba a constar en los mapas oficiales como zona desconocida; la línea avanzada de la colonización estaba a la altura de "Botucatu" (fundada en 1844 y elevada a villa en 1855), "Avaré" (fundada en 1862 y elevada a Distrito Policial en 1867), "Lenóis" (fundada en 1867), "São Domingos" (fundada en 1858 y transferida por decreto de 1867 para la ribera del río "Pardo" dando origen a la villa de "Santa Bárbara do Rìo Pardo") (Giovannetti, s.d.).

Fue en la segunda mitad del siglo XIX que los primeros rozadores pisaron el suelo de la región, actualmente conocida por "Alta Sorocabana", vagando por la selva y afrentando los indígenas. En 31 de mayo de 1856, José Theodoro de Souza registró su posesión en la parroquia de "Botucatu", en la cual declaraba poseer todas las tierras desde el río "Turvo" hasta la barra del "Tibagi" con el río "Paranapanema", a partir del año de 1847 (Giovannetti s.d.). En 1850, la Ley n° 601, de 18 de Septiembre, tornaba nulas las ocupaciones de tierras por otro título que no fuera el de compra; solamente en 30 de Enero de 1854, todavia, fue promulgada su reglamentación, en la tentativa de salvar el pratrimonio "devoluto" de la Provincia (Giovannetti, s.d.; Botelho, 1905).

La ocupación del "Pontal do Paranapanema", en el extremo oeste del Estado de "São Paulo", está relacionada con la decadencia de la mineración en "Minas Gerais", que tuvo su apogeo entre 1741 y 1770 y con la guerra del Paraguay (1865 a 1870). Toda la inmensa región situada en ángulo de convergéncia de los ríos "Paraná" y "Paranapanema" pertenció a los primeros ocupantes : José Theodoro de Souza y Francisco de Paula Moraes (Giovannetti, s.d.).

José Theodoro de Souza, nacido en "Pouso Alegre" Estado de "Minas Gerais", entre 1805 y 1815, penetrando por la cuenca del río "Pardo", pasó por "Trés Ranchos", "São Domingos", "Lenóis", "Santa Cruz do Río Pardo" y atravesó el río "Turvo" a la altura de la villa de "São Pedro". A la ribera del río "Novo", en la barra del riacho "Barraca", puzo un crucero como señal para más tarde fundar un poblado y seguió, atravesando el río "Pari" y alcanzando las vertientes del río "Capivara". Volvió, entonces , al suyo Estado natal para traer material humano y poblar la región descobierta. No trajo esclavos, pero labradores libres, hombres acostumbrados a la dura y áspera vida de la selva (Giovannetti, s.d.).

En 1868, José Theodoro fundó la villa de "São José do Rio Novo", hoy "Campos Novos Paulista", con escritura hecha en "São Domingos", Comarca de "Agudos". Este desbravador, motivado por la ocupación de la tierra, buscaba una nueva riqueza. Atrás, en las vertientes del río "Pardo", habia dejado el coronel Francisco Dias Baptista, José Marques do Vale, Jorgino Marques, João Gonalves Ferreira, José Ferreira Maciel, Carlos Bernardino de Souza, Joaquim Luiz Dutra, José Pinto Cardozo y otros. Con la fundación de "São Pedro" y "Campos Novos", el centro de acción fue transferido de las vertientes del río "Pardo" para "Campos Novos" (Giovannetti, s.d.). El número de habitantes de la villa de "Campos Novos" aumentó rapidamente en función de la guerra del Paraguay; moradores de "Pouso Alegre" para allá migraron a fin de encontrar refugio y calma (Enciclopédia dos..., 1957).

Con la llegada de nuevos labradores y comerciantes, que iban se estableciendo a la ribera del río "Novo", la villa de "São José do Río Novo" iba se desarrollando. José Theodoro, con residencia fija en esta villa aun no se contubo y seguió rumbo oeste, donde travó lucha con los indios que intentaban obstruir el camino. Después de atravesar el río "Capivara", penetró las vertientes del riacho "Alegre" y seguió rumbo a "São Matheus" abrindo los primeros caminos para el território incógnito, creando nuevas células de poblamiento, nuevos centros, de los cuales, niervos y arterias se extendian para regiones aun no trilladas. En 1874, la villa fue asaltada de sorpresa por aproximadamente 1.000 indios "Coroados" que pretendían la muerte del aventurero José Theodoro. En ella se establecieron, aun antes de su muerte (1875), Nicolau De Mayo, con origen en la nobleza italiana y que primero se habia establecido en "Sapecado MG", Domingos Ursaia, su cuñado, José Justino Ferreira y Geraldo Gerdulli, provenientes del Estado de "Minas Gerais", que constituyeron la élite en la época (Giovannetti, s.d.).

Las gigantescas posesiones de José Theodoro y Francisco de Paula Moraes, luego se fraccionaron en latifundios menores, pero siempre con areas mucho grandes. El elemento humano fijóse en los campos, en la zona de vegetación baja, rarefacta, de "cerrados" que margeaban los ríos "Paraná", "Paranapanema" y "Peixe". "Campos Novos" se tornó el centro que abrigó, por muchos años, los exploradores de la propiedad territorial inculta. Fue una época de esplendor para su vida comercial, cuya fama llegó a todos los rincones del Estado (Giovannetti, s.d.).

**El Fraccionamiento De La Propriedad**

En la conquista de la nueva tierra, los aventureros con capacidad de mando supieron hacerse señores poderosos llevando vida nababesca con las fabulosas rentas de un pedazo de papel falsificado en la casa del compadre jefe político. Los primeros ocupantes hacian ventas por documento privado, de cierta "agua" o de una hacienda compuesta de "campo y bosque" y, no habiendo texto impreso prohibindo la firma "a rogo" todas las escrituras otorgadas por los ocupantes analfabetos traían firmas de "terceras personas". Várias de estas "terceras personas" hacian ventas sin la concordancia del proprietario, el cual generalmente ignoraba por completo las reducciones ejercidas sobre sus posesiones (Giovannetti, s.d.). Posteriormente, toda esa inmensa mesopotamia fue reconocida y fraccionada por los ingenieros, conquistandolas al gentío en duros embates. Solamente tres localidades representaban puntos de progreso: "Campos Novos", "Platina (ex Saltinho)" y "Conceião de Monte Alegre". De "Campos Novos" partían los "paulistas" y "mineros" selva adentro, en busca de la tierra prometida. La marcha acelerada para el oeste representaba el início del proceso de las grandes divisiones de tierras, que se acclcró con el ferrocarril "Sorocabana". "São José do Rio Novo" fue elevada a Districto Policial en 24 de Junio de 1878 y, en 13 de Abril de 1880 pasó a Districto de Paz. La ciudad pasó a ser el centro donde residían los abogados y ingenieros envueltos en el proceso de ocupación. Fue transformada en Municipio, en 10 de Marzo de 1885 y llegó a abarcar toda la extensión territorial comprendida entre el riacho "Coimbra", las barrancas del "Paranapanema" y "Paraná" y el parteaguas "Peixe-Paranapanema". En las vertientes del río "Novo" con el "Taquaral" y "Pari" el desarrollo demográfico era animador, teniendo entonces inicio el deslocamiento de agricultores para el oeste. También, en 1885, Nilo Diodati hizo vários servicios topográficos en el río "Turvo" y intentó penetrar en el río "Peixe", habendo sido rechazado por los indios "Coroados" que, en numerosas tribus, habitaban toda aquella importante cuenca hidrográfica (Giovannetti, s.d.).

Sobre los asaltos y muertes provocados por los indios "Caingangs" entre 1880 y 1886, referese Ihering (1907) como hecho que creó dificuldades en el poblamiento del "Pontal do Paranapanema".

Em 1886, fue trazada la linea divisoria del grande inmoble "Monte Alvão" que, partiendo de la cabecera del riacho "São Matheus", en la hacienda que pertenció a Maia Guimarães ("Quatá"), seguia hasta encontrar el río "Peixe". Enfrontando los indios de la región, fue en este año fundada, por Pedro Pocai, la villa de "Salto Grande" (Giovannetti, s.d.).

Data del mismo año la expedición emprendida al valle del "Paranapanema" por la Comisión Geográfica y Geológica del Estado de "São Paulo". Narra esta Comisión, que el poblamiento regional era originario, basicamente, del sur de "Minas" y compuesto de creadores y de agricultores que se deslocaban con toda su familia y con todos sus haberes, abundancia de ganado y de capitales. La población nueva que llegaba a los "sert[a]es do Paranapanema" (o de "Campos Novos") iba se estableciendo diseminadamente, desde el "Turvo" hasta allende del "Laranja Doce", formando una série de establecimientos de agricultura y de ganaderia al longo de la única carretera abierta a través de los campos con trazado sinuoso, de forma a ligar muchas fincas familiares y haciendas, cuyos fondos llegaban a la orilla del "Paranapanema". Con respecto a la demografía regional, así se expresa aquella Comisión: "Según el apadronamiento de 1872, la populación establecida en el valle del "Paranapanema" era de 57.460 habitantes; por el apadronamiento más nuevo, hace poco publicado por la "Comissão Central de Estatística", este guarismo elevase ahora a 89.840 habitantes, acusando así un aumento de 32.443 almas en el período de 14 años, aumento que, por lo menos en un tercio representaba la populación inmigrada en los últimos años (Sampaio, 1890).

El período de mayor intensidad de ventas de tierras fue de 1890 a 1891, por parte de los herederos de los primeros ocupantes. En 1891, los ingenieros Nilo Diodati y

Simão Levy, en compañia de Jacob Molitor, organizaron una expedición para proceder al relevamiento de un trecho del río "Peixe", arriba de la barra del riacho "Panela", pero tuvieron su entrada impedida por la acción de los indios (Giovan-netti, s.d.). Una escritura privada, de 12 de Julio de 1892, por ocasión de la venta del inmoble denominado "Ribeirão do Morro Vermelho", en la "Fazenda do Rio do Peixe", Município de "Campos Novos", és uno de los testimonios de la expedición de Diodati (Campos Novos Do Paranapanema, 1892).

En 1890, la Comisión Geográfica y Geológica recomendaba que el poder público deberia intervenir para poner fin a las luchas entre colonos y indios, dando protec-ción a los contendores con medidas indirectas. Según la misma Comisión, la catequesis era la mejor medida para la pacificación destos "sertões". Ponderaba, que esta medida no deberia tener en vista la transformación del aborigen en agente de una civilización que el no comprendía, que no era lícito esperar del indio domesticado un obrero como lo requeria nuestra civilización; reconocia que el indígena amansado o domesticado por la palabra del misionario perdia toda la nobleza salvaje sin ganar la capacidad o la nobleza moral; en el contato con la raza más fuerte, que lo subyugaba, el solamente tenia a perder. Afirmaba que, "reducir el indio por la mansidón, lo protegendo contra el exterminio que la conquista de la tierra fatalmente lo condena, tornar mismo posible la asimilación de una y de otra raza, he todo cuanto puede dar la catequesis y lo cuanto bastará para la civilizaci'on seguir su camino". Recomendaba también el proseguimiento de la carretera que venía de "Botucatu", a través del "cerrado", haciendola llegar hasta el "Paraná" (Sampaio, 1890).

El ferrocarril, por su vez, solo vá avanzar por el valle del "Paranapanema" en el siglo XX. En el período de 1886 a 1900 el ferrocarril "Sorocabana" continuará el trecho que empezaba en "Laranjal", alcanzando "Botucatu", "Lenóis", "Avaré" y "Cerqueira Cesar", allende del ramal para "Itapetininga" (Saes, 1981) y, en su paásage rumbo a "Presidente Epitácio", irá desviar su rota de la ciudad de "Campos Novos", por presión de los propios políticos locales, que temian perder las posiciones de mando (Le Selva, 1952).

**La Retomada De La Catequesis Y La Defensa De Los Indios**

La región era habitada por tres tribus indígenas: los "Coroados", (Caingangs), que ocupaban las vertientes del río "Peixe" en aldeas esparcidas; los "Cayuás" (Guaranies), que ocupaban las vertientes del río "Paranapanema", originarios de "Jataí", en el "Paraná", donde ya habian sido catequizados por el Fray Timóteo de Castelnuevo; los "Xavantes" ocupaban la meseta central, en una faja de campo y "cerrado" que iba de "Avaré" al riacho "Rebojo", afluente de la orilla derecha del

"Paranapanema" y la región conocida como llanura de los "Agudos", que los desbravadores consideraban como el nido de la "bugrada" no sujetada (Giovannetti, s.d.; Sampaio, 1890).

A pesar de las consideraciones de la Comisión Geográfica y Geológica, la catequesis era vista por los que detenían el poder como una forma de "amansar" e subyugar el indio para que las tierras pudiesen ser ocupadas en este proceso de conquista y avanzo de las fronteras de la "civilización".

La catequesis de los indios nómades del "Sertão do Paranapanema" cupo a los capuchinos franciscanos. En 22 de Mayo de 1888, llega a "Campos Novos", Fray Mariano de Bagnaia, convidado por el comisario de la ordem para organizar la catequesis, al qual vino juntarse Fray Francisco de Alatri, a 7 de Mayo del mismo año. En esta retomada, este puede ser considerado el primer ciclo de la catequesis de indios del "Pontal do Paranapanema", que se interrompe con la muerte de Fray Mariano, siendo retomada solamente en el inicio del siglo XX, cuando llega el cura Bernardino de Lavalle, a 23 de Julio de 1901. Los "Caingangs", de la tribu de los "Coroados", que vivian en la ribera del río "Peixe", solo van ser contactados en el siglo seguiente, en 1904 (Giovannetti, s.d.).

**Los Movimientos En Defensa Del Indio**

Ihering (1907) estimaba en 10 mil el número de índios en el Estado de São Paulo en 1904.

En la realidad, los primeros resultados concretos en la defensa del aborigen, con la conciencia y el reconocimiento de una otra cultura, solo surgieron en el inicio del siglo XX, como resultado de intensa campaña emprendida por los intelectuales de la época, de manera especial por el "Centro de Sciencias, Letras e Artes de Campinas", uno de los más activos grupos de presión, de inspiración positivista. Las actas de sus sesiones ordinarias, de 1908 a 1912, son punteadas de informaciones sobre los indígenas, con denuncias de malos tratos, agresiones y matanzas, y de acciones concretas en defensa del gentio (Sessões do Centro; ..., 1913; Idem, 1913a; Idem, 1913b; Idem, 1913c; Idem, 1914; Idem 1914a; Idem, 1914b; Idem, 1914c; Idem, 1914d; Idem, 1914e; Idem, 1914f; Idem 1914g; Idem, 1914h; Idem, 1914i; Idem, 1914j; Idem 1914l; Idem, 1914m; Idem, 1914n; Idem, 1914o; Idem, 1914p; Idem, 1914q; Idem, 1914r; Idem, 1914s; Idem 1914t; Idem, Idem, 1915u). El estopín desta acción fue la tesis presentada en 1904, por Hermann Von Ihering (1907), director del Museo Paulista, en la Exposición Universal de Saint Louis - U.S.A., según el qual "los actuales indios del Estado de São Paulo no representan un elemento de trabajo y progreso. Como también en otros Estados de

Brasil, no se puede esperar trabajo sério y continuo de los indios civilizados y como los Caingangs salvajes son estorbos para la colonización de las regiones del "sertão" que habitan, parece que no hay otro medio, de que se pueda lanzar mano, sino su exterminio. La converción de los indios no tiene dado resultado satisfactorio; aquellos indios que se unieron a los portugueses inmigrados solo dejaron una influencia maléfica en los hábitos de la populación rural. és mia conviccíon de que és debido esencialmente a esas circunstancias, que el Estado de São Paulo és obligado a introducir millares de inmigrantes, pues que no se puede contar, de manera eficaz y segura, con los servicios de esa populación indígena, para los trabajos que la labranza exige".

La campaña daquel Centro fue iniciada con la publicacíon de un hojuelo, que penetró en todos los rincones del País, formando opinión y haciendo repercutir sus vozes en el 1º Congreso Brasileño de Geografia de "Rio de Janeiro" y en la Sociedad Científica de "São Paulo" (Stevenson, 1915).

En 17 de octubre de 1908, Tito de Lemos concitaba el Centro de Ciencias, Letras y Artes para que se dirigise a las sociedades de Geografia de "São Paulo" y de "Rio de Janeiro", convidandolas a sumaren fuerzas contra la teoria de exterminación de los indios sustentada por Ihering, (Sessões do Centro; ..., 1913; Ihering, 1907). Relata Stevenson (1915), en su discurso publicado en la revista del "Centro" que, por iniciativa de Tito de Lemos, aquella Institución, tomara la "resolución de trabajar por todos los medios a su alcanze, para hacer cesar las horripilantes cazadas humanas de las orillas del "Paranapanema" atrayendo para ese asunto de vital importancia, las vistas de los poderes públicos, la acción de las asociaciones científicas, de la prensa y del pueblo". En octubre daquel año, el "Centro" convoca las sociedades geográficas de "São Paulo" y de "Rio de Janeiro" para campaña conjunta contra los asesinatos de indios "Coroados" en conflicto con trabajadores del ferrocarril "Noroeste do Brasil", con base en denúncias hechas por el "Correio da Manhã" donde se noticiaba que "hicieron del asesinato del indio una nueva espécie de sport con que mucho se divierten", exigiendo que la acusación fuese apurada y si constatada aplicado el artículo del Código Penal (Sessões do Centro; ..., 1913a). Contra la tesis de Ihering, consiguen también la adhesión unánime de la congregación del Museo Nacional, que apoya la campaña del "Centro" (Sessões do Centro; ..., 1913a). Aun en 1908, referese Horta Barbosa al facto de que, según Ihering, solamente en los estados de "Paraná" y "Rio Grande do Sul" se estaba dando alguna atención al problema del indio, encontrandose estes desprotegidos en "São Paulo" y en "Santa Catarina" (Sessões do Centro;..., 1913b).

En 26 de Diciembre de 1908 és lida en la sesión daquel "Centro" minuta de representación para ser encaminada contra el martirio de los indios del Estado, solicitando providencias enérgicas y eficazes y presentando sugestiones, entre las quales, el policiamento por la "Forca Pública" en los locales que los blancos estuviesen en contacto con el indio, "no se admitindo, sin proceso y julgamento la alegación de legítima defensa, de la vida, de la honra o de defensa de la propriedad y de que el Gobierno dejase de fornecer armas de fuego a los civiles empleados en ferrocarriles y otros, y que el mismo gobierno no más considerase como "devolutas" las tierras ocupadas por indios a fin de distribución (Sessões do Centro; ..., 1913c). La tesis de defensa del indio, apoyada por el "Centro", fue entonces, por Gustavo Enge, llevada al 1º Congreso Brasileño de Geografia, realizado en "Rio de Janeiro" en 7 de Septiembre de 1909 (Stevenson, 1915; Sessões do Centro; ..., 1914a).

En 1º de Enero de 1910, el "Centro" decide por la manifestación de protesto al Ferrocarril "Noroeste", solicitando providencias para harmonizar los salvajes con los trabajadores (Sessões do Centro; ..., 1914d).

En Sesión del "Centro" de 15 de Enero de 1910, Domingos Jaguaribe defende la tesis del Gobierno de "São Paulo" destinar 300.000 ha esclusivamente a los indios (Sessões do Centro; ..., 1914e).

Em medio de todas estas acciones, se consigue llevar la preocupación al Gobierno Federal que institui el Servicio de Protección al Indio y establece critério jurídico por el cual el indígena, aunque segregado por la "civilización" era considerado igualmente ciudadano brasileño y adquiria los mismos derechos que los demás brasileños, a la vida y a la comunión social (Stevenson, 1915).

En sesión de 31 de enero de 1910 (Sessões do Centro;..., 1914f) és noticiada la correspondencia de la Asociación de Protección y Auxílio a los Indios de Brasil, recién creada en el "Rio de Janeiro", agradeciendo las congratulaciones del "Centro". En esta misma sesión, Horta Barbosa referese al proyecto del Ministro de la Agricultura, Rodolpho Miranda, para "discriminatória" y defensa de las tierras de los indígenas contra las incursiones de los civilizados y aun manifesta la posición de la Comisión de Defensa de los Indios en el sentido de que se oficiase al gobierno de "São Paulo" manifestación en el sentido de que las tierras ocupadas por los indígenas fuesen respetadas, loando aun la idea que tuvo Domingos Jaguaríbe de reservales 300.000 hectareas.

Todavia, a pesar de combatiren el alculturamiento del indio a través de la catequesis, no se daban cuenta los positivistas que, dentro de un otro enfoque, sus buenas intenciones no los exentaba también del proceso de despersonalización y alculturamiento, ahora motivado en transformar los indios en agricultores y trabajadores, coparticipantes del desarrollo dentro de la ideologia del progreso por ellos pregonada.

Los objetivos de la campaña no son atingidos al nivel del Estado, pero al nivel de la Federación en el 10 de Febrero de 1910 (Sessões do Centro; ..., 1914l) és comunicada en la reunión del "Centro", el nombramiento de Horta Barbosa para el recién creado Servicio de Protección a los Indios y Localización de Trabajadores Nacionales, del Ministerio de la Agricultura, aun en la gestión del ministro Rodolpho Miranda.

Finalmente en 19 de Marzo de 1910 (Sessões do Centro; ..., 1914h), el "Centro" es comunicado por la Secretaria de Agricultura del Estado de "São Paulo" sobre las resoluciones tomadas por el Gobierno Federal con relación a los indígenas, habiendo este Estado puesto a la disposición del Gobierno Federal todas las tierras "devolutas" necesárias para asientamiento de los indios de "São Paulo".

En sesión de 28 de Marzo, proponese que el "Centro" oficie sus cumplimientos al coronel Candido Mariano da Silva Rondon, por su nombreamiento como director de la Comisión Indígena promovida por la Secretaria de Agricultura (Sessões do Centro; ..., 1914i).

De otras acciones de protección al indio también participó el Centro de Ciencias Letras y Artes de Campinas, saliendo de los límites del Estado de "São Paulo". Vale citar, palabras de estímulo y reconocimiento dirigidas aquél "Centro" por el coronel Candido Mariano da Silva Rondon, notório pacificador de los indios brasileños. En respuesta al documento encaminado por aquella institución, este ilustre indigenista, en 3 de Junio de 1911 (Sessões do Centro; ..., 1914r), asi se exprime: "Vosotros que fuistes de los primeros a vibrar golpe mortífero en la orgullosa teoria del exterminio íncola americano, creada a pretexto de su supuesta irreductibilidad a la evolución social, bién mereceis en este momento de expansión fraternal, los aplausos afectuosos de las almas bién nacidas. La vuestra manifestación cívica estimulame a mayores esfuerzos, si tal fuera posible en el cumplimiento de mi deber, de forma a corresponder mejor y más dignamente, a las esperanzas que, en el servicio de protección de los indios y localización de trabajadores nacionales, todos los fervorosos patriotas depositan".

Finalmente, en sesión de 1º de Abril de 1912, es comunicada en el "Centro" la pacificación de los indios "Caingangs", hasta entonces considerados irreductibles, gracias a los esfuerzos de Luiz Bueno Horta Barbosa, inspector del servicio de protección a los indios en el Estado de "São Paulo" (Sessões do Centro; ..., 1914u).

Todo el empeño de los indigenistas de la época, entretanto, no fue suficiente para hacer frente al interés económico en la tenencia de la tierra haciendo con que, como siempre, el lado más débil perdiera la cuestión y las populaciones indígenas desaparecieran del "Pontal do Paranapanema"

## Conclusión

La lección que se quita de esta análise, transpone la simples narrativa histórica de los factos y sirve de alerta para la forma como estas cuestiones seran tratadas en las regiones aun no incorporadas a la economia de mercado. El espacio de la selva, al lado de su aspecto de conservación de ecosistema natural, guarda aun la cuestión de la preservación del espacio indio. La protección destes ecosistemas se mostra intimamente ligada a la preservación de la cultura de los pueblos de la selva, amenazada de extinción por la ampliación del espacio blanco. La lección del pasado puede ser el punto de apoyo para políticos y estadistas en la conducción de los conflictos que se le tocan en la actualidad.

Anexo 1. Situación de la area del estudio

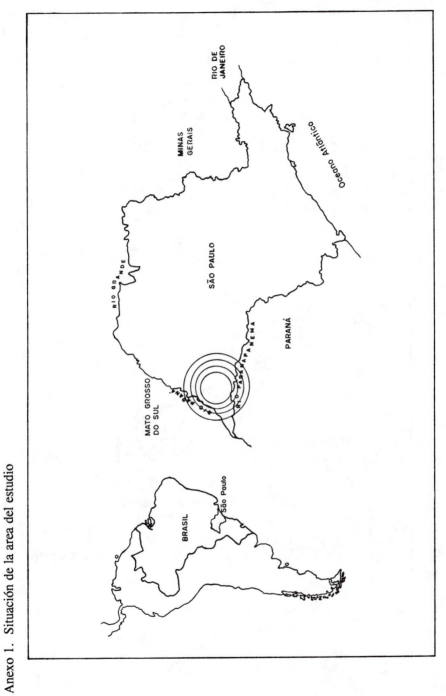

Anexo 2.  Las reducciones rumbo al interior

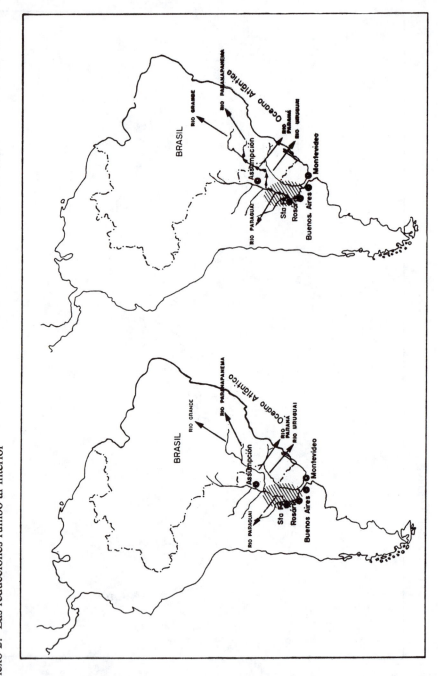

Anexo 3. La ocupación del "Pontal do Paranapanema"

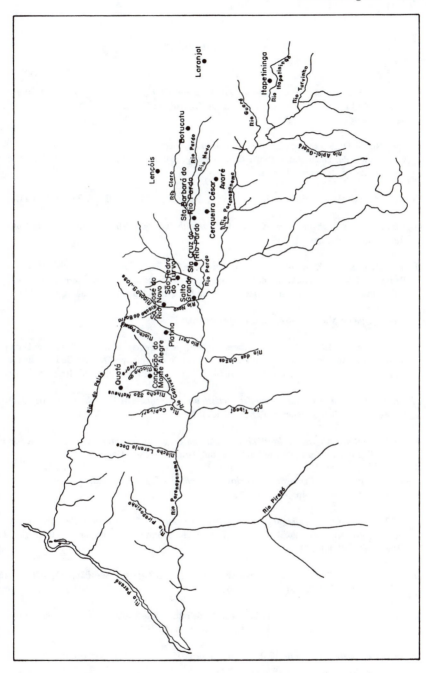

# Referencias

Botelho, C. 1905. *Relat rio apresentado ao Dr. Jorge Tibiriç , presidente do Estado; anno de 1904.* São Paulo, Secretaria dos Neg cios da Agricultura Commercio e Obras P blicas do Estado de São Paulo.

Campos Novos Do Paranapanema. 1892. *Escritura de compra e venda de 12 de julho de 1892.* Outorgante: Pe. Decio Augusto Chefado. Outorgados: Dr. Jeronymo de Cunto; Engenheiro Nilo Diodati e Domingos Ursaia.

Enciclop dia Dos Munic pios Brasileiros. 1957. Rio de Janeiro, IBGE. v. 28. p. 210-212

Giovannetti, B. s.d. *Esboço hist rico da Alta Sorocabana.* São Paulo. Empresa Gr fica da Revista dos Tribunais. 166p.

Ihering, H. von. 1907. A antropologia do Estado de São Paulo. *Revista do Museu Paulista,* São Paulo, 7:202-257. (Traduzido da 2. ed. ingleza deste estudo, que fôra elaborado para ser distribu do a pedido da comissão respectiva, na Exposição Universal de São Luiz, 1904).

Le Selva, L cia de Mayo. 1952. Informaç es verbais.

Leme, T. de A. P. 1980. *Nobili rquia paulista; hist rica e geneal gica.* 5. ed. acrescida da parte in dita com uma biografia do autor e estudo cr tico de sua obra por Afonso de E. Taunay. Belo Horizonte, Ed. Itatiaia. São Paulo, EDUSP. 280p. t. 1 (Reconquista do Brasil, nova s rie, 5).

Lugon, C. 1977. *A rep blica "comunista" cristã dos guaranis; 1610-1768.* Trad. por Alvaro Cabral. 3.ed. Rio de Janeiro, Paz e Terra. 353p.

Petrone, P. 1966. *A baixada do Ribeira; estudos de geografia humana.* São Paulo, Fac. de Fil. Ciênc. e Letras da Univ. de São Paulo. 366p. (Boletim, 284 - Geografia 14)

Sampaio, T. F. 1889. *Exploração dos rios Itapetininga e Paranapanema.* Rio de Janeiro, Imprensa Nacional. 14p.

___. 1890. *Consideraç es geogr phicas e econômicas sobre o "Valle do Rio Paranapanema".* São Paulo, Leroy King Brookwalter. 73p. (Boletim da Commissão Geographica e Geol gica do Estado de São Paulo, 4)

Saes, F. A. M. de. 1981. *As ferrovias de S ~ ao Paulo; 1870-1940.* São Paulo, HUCITEC/INL - MEC. 199p. (Coleção estudos hist ricos)

Sessões do Centro; acta da sessão ordin ria de 17 de outubro de 1908, 306. 1913. *Revista do Centro de Sciencias, Letras e Artes,* Campinas, 12(34):17-72.

___; acta da sessão ordin ria de 12 de dezembro de 1908, 313. 1913a. *Revista do Centro de Sciencias, Letras e Artes,* Campinas, 12(34):76-77.

___; acta da sessão ordin ria de 19 de dezembro de 1908, 314. 1913b. *Revista do Centro de Sciencias, Letras e Artes,* Campinas, 12(34):77-78.

___; acta da sessão ordin ria de 26 de dezembro de 1908, 315. 1913c. *Revista do Centro de Sciencias, Letras e Artes*, Campinas, 12(34):78-80.

___; acta da sessão ordin ria de 3 de julho de 1909, 335. 1914. *Revista do Centro de Sciencias, Letras e Artes*, Campinas, 12(34):47-48.

___; acta da sessão ordin ria de 18 de setembro de 1909, 342. 1914a. *Revista do Centro de Sciencias, Letras e Artes*, Campinas, 12(34):51.

___; acta da sessão ordin ria de 15 de outubro de 1909, 343. 1914b. *Revista do Centro de Sciencias, Letras e Artes*, Campinas, 12(34)51-53.

___; acta da sessão ordin ria de 1 de janeiro de 1910, 352. 1914d. *Revista do Centro de Sciencias, Letras e Artes*, Campinas 13(35-36):96-97, jun./set.

___; acta da sessão ordin ria de 15 de janeiro de 1910, 354. 1914e. *Revista do Centro de Sciencias, Letras e Artes*, Campinas, 13(35-36):98-99, jun./set.

___; acta da sessão ordin ria de 31 de janeiro de 1910, 356. 1914f. *Revista do Centro de Sciencias, Letras e Artes*, Campinas, 13(35-36):99-101, jun./set.

___; acta da sessão ordin ria de 19 de fevereiro de 1910, 359. 1914g. *Revista do Centro de Sciencias, Letras e Artes*, Campinas, 13(35-36):101-102, jun./set.

___; acta da sessão ordin ria de 19 de março de 1910, 363. 1914h. *Revista do Centro de Sciencias, Letras e Artes*, Campinas, 13(35-36):103-104, jun./set.

___; acta da sessão ordin ria de 28 de março de 1910, 364. 1914i. *Revista do Centro de Sciencias, Letras e Artes*, Campinas, 13(35-36):104, jun./set.

___; acta da sessão ordin ria de 28 de maio de 1910, 371. 1914j. *Revista do Centro de Sciencias, Letras e Artes*, Campinas, 13(35-36):106-107, jun./set.

___; acta da sessão ordin ria de 10 de setembro de 1910, 378. 1914l. *Revista do Centro de Sciencias, Letras e Artes*, Campinas, 13(35-36):110, jun./set.

___; acta da sessão ordin ria de 10 de dezembro de 1910, 382. 1914m. *Revista do Centro de Sciencias, Letras e Artes*, Campinas, 13(35-36):111-112, jun./set.

___; acta da sessão ordin ria de 07 de janeiro de 1911, 384. 1914n. *Revista do Centro de Sciencias, Letras e Artes*, Campinas, 13(37):55, dez.

___; acta da sessão ordin ria de 28 de janeiro de 1911, 385. 1914o. *Revista do Centro de Sciencias, Letras e Artes*, Campinas, 13(37):55-56, dez.

___; acta da sessão ordin ria de 20 de maio de 1911, 388. 1914p. *Revista do Centro de Sciencias, Letras e Artes*, Campinas, 13(37):57, dez.

___; acta da sessão ordin ria de 27 de maio de 1911, 389. 1914q. *Revista do Centro de Sciencias, Letras e Artes*, Campinas, 13(37):57-58, dez.

216    *Changing Tropical Forests*

___; acta da sessão ordin ria de 3 de junho de 1911, 390. 1914r. *Revista do Centro de Sciencias, Letras e Artes*, Campinas, 13(37):58-60, dez.

___; acta da sessão ordin ria de 30 de junho de 1911, 393. 1914s. *Revista do Centro de Sciencias, Letras e Artes*, Campinas, 13(37):61-62, dez

___; acta da sessão ordin ria de 10 de fevereiro de 1912. .403. 1914t. *Revista do Centro de Sciencias, Letras e Artes*, Campinas, 13(37):68-69, dez.

___; acta da sessão ordin ria de 1 de abril de 1912, 408. 1914u. *Revista do Centro de Sciencias, Leras e Artes*, Campinas, 13(37):71, dez.

Stevenson, C. 1915. Centro de Sciencias. *Revista do Centro de Sciencias, Letras e Artes*, Campinas, 14(41):37-42, dez. (discurso)

Promise and Performance of Industrial Plantations
in Two Regions of Australia and Brazil

John Dargavel
Australian National University

Sebastião Kengen
Brazilian Institute for Environment
and Renewable Natural Resources

The depletion of the wood resources in the world's natural forests through deforestation and the degradation of their productive capacity has been countered to some extent by sustained yield management and the creation of new resources in extensive forest plantations. (In this paper "plantations" refers to forest trees unless otherwise stated.) However, the long history of managing natural forests to sustain their yields (Steen 1984) has been confined largely to the temperate forests of Europe and North America, and even there future wood yields are now uncertain in the face of air pollution and the transfer of some forests to national parks. The history of tropical forests provides little evidence that attempts to manage them for sustained yield have been successful (Dargavel et al. 1988), and even if they were to be, the yields that they could sustain might be insufficient to meet the demands of increasing world populations and per capita consumption. By contrast, the shorter history of forest plantations demonstrates how resources can be increased. For example, the plantation resources of Australia, Brazil, Chile, New Zealand, and Portugal are already supporting forest products industries and exports. The extensive plantations of fast-growing, often exotic species that are being developed in these and other countries of the tropics and sub-tropics will be major suppliers of the world's industrial wood in the future.

The distinction between natural forests and forest plantations is not always clear, because planting can be used to restore, regenerate, or enrich the resources of existing natural forests as well as to create new ones. John Evelyn (1620-1706) was an early and most enthusiastic advocate of planting to counter the "waste and destruction" of British forests (Evelyn 1664). Although he considered the use of exotic species and fertilisation, he was primarily concerned with restoring the indigenous oak forests that supplied the navy with ship timbers. The idea that large-scale planting could increase forest productivity and foster regional development through industrialisation was most clearly demonstrated in the Landes region of southwest France from the eighteenth century (Barrere et al. 1962; Reed 1954).

State programmes of planting to arrest the spread of coastal sand dunes, and to drain and road the heaths behind them made the otherwise almost useless land suitable for growing the native *Pinus pinaster*. The extensive plantations that were then developed on private and communal lands provided resin, turpentine, pit props, and railway sleepers or ties in the nineteenth century and wood for pulp and paper mills in the twentieth. Given the dominant place of French forestry in the nineteenth century, the Landes development set the idea that forest resources could be created on land too poor for agriculture, and hence without political controversies over its use. Large plantation projects in Australia, New Zealand, South Africa, and elsewhere were expanded during the 1930s, partly with funds for unemployment relief. Their wood provided the resources to support new industries and were successful examples of forest development. Plantations are created also, often on a much smaller scale, for firewood, fodder, amenity, catchment protection, and many other reasons. Such non-industrial plantations have been estimated to comprise 41 percent of the plantation areas in developing tropical countries (Levingston 1984).

In the 1950s and 1960s, industrialisation was widely believed to be the source from which benefits would diffuse or "trickle down" through society. The case that industrial forestry could stimulate such development in underdeveloped countries was put in a seminal paper by Westoby (1962). He believed that the forest sector had strong forward and backward linkages with the rest of the economy, that the demand for forest products would rise sharply and therefore that investment in industrial forestry would realise development objectives. Westoby's arguments were elaborated and repeated for both developed and underdeveloped countries (eg. Gregory 1965; Sartorious and Henle 1968; McGregor 1976) and advocated strongly by the Food and Agriculture Organisation (FAO) of the United Nations in sponsoring the "FAO World Symposium on Man-made Forests and their Industrial Importance" (FAO 1967). However the prevailing trickle-down theory of development on which this rested was increasingly criticised, especially from Latin America where Frank (1969), after studying Chile and Brazil, was to claim that it was "empirically invalid", "theoretically inadequate" and "policy-wise ineffective" for developing countries. He held that the net benefits of industrial investment in underdeveloped countries diffused, not down, but up to the core countries of the world system. By 1978, Westoby too considered that "the famous multiplier effects are missing" and that:

> nearly all the forest and industry development which has taken place in the underdeveloped world has been externally oriented, aimed at satisfying the rocketing demands of or the rich industrialised nations, the basic forest products needs of the peoples of the underdeveloped world are further from being satisfied than ever ... (Westoby 1978).

Such polarized and opposing views have influenced but oversimplify the history of plantations in particular regions. Moreover, the critiques fail to distinguish between different objectives that may have applied and the different manners in which projects have been executed. More recently, the environmental objective of creating plantation resources in order to reduce or eliminate logging from natural forests has to be considered in some countries.

In this paper, we examine the history of plantation developments to create new industrial resources in 2 regions of Australia and Brazil. Although very different in their social and political aspects, they have similarities in their geography and in their plantations. They are countries of similar area and share common latitudes. Climatically, Brazil has found that several Australian eucalypts (*Eucalyptus viminalis, E. saligna, E. grandis, E. europhylla, E. citriodora* and *E. teretecornis*) grow most successfully, while both countries have found that pines from Central America and the Caribbean (*Pinus caribaea varieties hondurensis* and *bahamensis*) and the south of the United States of America (*P. radiata, P. elliottii* and *P. taeda*) grow successfully. In both countries, plantations have been developed mostly as large-scale enterprises, rather than as woodlots by farmers or small-scale landowners. Technologically, both countries have adopted capital and energy-intensive methods of establishment and logging, and are scientifically advanced in aspects such as genetic improvement. Economically, the plantations of both countries, as in many others, have been developed with heavy involvement by the state, though that is now being called into question. Industrially, the forest sector of both countries is tied into the world market to a considerable extent through trade and multi-national enterprises. Indeed, the plantations may be more deeply incorporated in, and are certainly more vulnerable to, the operation of the world economy than the natural forests. Moreover, the products of plantations in Brazil and Chile, for example, will exert a competitive pressure, directly of indirectly, through the world market on the products of plantations in Australia and New Zealand.

## Australia

Australia has been establishing plantations for a little over a century and has 905,000 hectares of which 858,000 hectares are coniferous species. *Pinus radiata* is the main species planted, with *P. elliottii, P. caribaea var hondurensis* and hybrids (including one with *P. tecunumanii*) being used in the tropical areas of Queensland and *P. pinaster* on poor coastal sands in Western Australia. Of the few indigenous conifers only hoop pine, *Araucaria cunninghamii* has been planted to any extent. Only 42,000 hectares of eucalypts (mainly *Eucalyptus regnans,*

*E. globulus, E. grandis, E. saligna* and *E. nitens*) and minor areas of other hardwoods (mostly *Populus* spp.) have been planted.

Australia has very limited natural resources of the easily worked softwood timbers and those which were readily accessible were quickly exploited by the early white settlements. As the population expanded rapidly from the discovery of gold in 1851, domestic production could not meet the demand for softwood timber which was supplied increasingly by imports from the Baltic, North America and later New Zealand (Dargavel 1990). In spite of the plantation programme, Australia still imports 40 percent of its softwood timber requirements.

The commercially usable resources of hardwood timbers from the eucalypt forests were restricted to the higher rainfall areas. At least half these forests were cleared, or severely modified for agricultural settlement and much of the wood destroyed. In the gold fields and in areas close to the growing cities, the readily accessible forests were soon cut out. By 1865 a report to the Victorian parliament had recommended, among other things, that plantations be commenced (Victoria 1865). Various exotic and indigenous species were tried by the colonial, and later State governments in small-scale plantings in Victoria and the other States (Carron 1990). The original arguments for establishing them were based on the physical ability to grow trees and use the wood, especially the softwoods which it was expected could replace imports. This production argument was later married to an argument for land reclamation and, following the Landes example, several of the earliest plantations in Victoria, Tasmania, and Western Australia were established on poor coastal dunes where they subsequently failed.

From the 1920s arguments for plantations and other forestry schemes stressed the advantage of the supposedly healthy employment that they provided. New South Wales and Victoria operated "reformation" and "reforestation" camps for good conduct prisoners, and Tasmania proposed a camp for the "reclamation" of homeless boys from Britain. This association of forestry with moral virtue and social reform was found also in South Africa (Roach 1989). Various relief schemes were introduced during the depression of the 1930s which provided unemployed people with food and shelter in exchange for working on public enterprises, some of which involved planting trees. However the general economic difficulties, followed by those of World War II, meant that the rate of planting was slow. By 1950 Australia had about 115,000 hectares of coniferous plantations and 11,000 hectares of hardwood plantations.

The post-war economic boom in Australia was associated with further industrialisation and proved a successful period of development during which Gross National Product increased at 4-5 percent a year, average real incomes increased and there was virtually full employment. The arrival of 2.5 million refugees and immigrants in the 1950s and 1960s, and a period of high birthrates created a great demand for housing; between 1947 and 1971, 1.8 million homes were built which doubled their number. To meet the housing and other demands, sawlogs were cut from the eucalypt forests beyond their sustainable yield level. Moreover, the small domestic pulp and paper industry expanded rapidly, though not enough to satisfy the total demand. Australia still imports a quarter of its pulp and paper consumption. The ability to make paper pulp from eucalypts by several processes was very largely developed by Australian research that had first started in the 1920s. Although Australian pulp mills drew their wood resources from the natural forests, the technology made fast-growing eucalypt plantations an attractive means of creating new resources for the industry in Brazil and elsewhere.

From the 1950s, the States expanded their plantations financing them with loan funds, and from the 1960s and 1970s with soft loans from the Commonwealth (i.e. federal) Government. The arguments that supported these public investments were based on import replacement to make Australia self-sufficient, the decentralisation of development from metropolitan to rural areas and the employment provided. Some of the industrial companies and a few investment companies established their own plantations, as did a number of farmers and small holders with various incentive schemes. Over the last 20 years these industrial, investment, and small plantations have increased their proportion of the total planted area from 25 to 30 percent.

The plantations were established mainly on state forest land carrying eucalypt forests of low productivity, most of which had been degraded by overcutting. From the mid-1970s, this was criticised for destroying the environmental values, and particularly the fauna habitat, provided by the eucalypt forests. In consequence, governments transferred their new planting first to poor quality farmland and later, as the quantity of this progressively reduced, to land of better quality. In the 1980s this became forcefully opposed by rural communities and some local governments in Victoria and New South Wales on the grounds of the local depopulation and social dislocation it caused.

Two bold but very different proposals to expand the area of Australia's plantations were advanced at the end of 1987 to address the national problems. One, for industrial development put forward by the Forestry and Forest Products Industries Council (FAFPIC 1987), representing the state and commonwealth forest agencies, envisaged planting 530,000 hectares of pines and 65,000 hectares of eucalypts plantations. This, with continued logging in the eucalypt forests, would provide the resources for new mills (9 pulp mills, 34 sawmills and 12 panel mills) which would not only supply the domestic market but produce a surplus for export thus eventually reversing the sector's adverse balance of payments. The other, for conservation put forward by the Australian Conservation Foundation (ACF 1987; Cameron and Penna 1988), envisaged planting 230,000 hectares of pines and 200,000 to 350,000 hectares of eucalypts to supply a scaled-down industry which would have to stop logging in the eucalypt forests over the next 15 to 30 years thus, eventually reversing the environmental impacts of logging in the eucalypt forests.

These proposals prompted the preparation of a comprehensive view of Australia's international and trade setting, historical experience, scientific research, policies, and other matters related to plantations (Dargavel and Semple 1990). Although the objectives and experience of the various States and companies clearly differed and scientific research was well advanced, there was a pronounced lack of economic analysis; few substantial economic analyses of public plantations (three-quarters of the national total) have been conducted and fewer still published. Doubtless this is because the appraisal of efficiency and effectiveness is a relatively new requirement in public sector management in Australia, however it has meant that Australia with a century of practical and scientific experience does not know whether many of its plantations can produce wood at an internationally competitive price. As we have seen, the development of plantations had been supported by a series of arguments and objectives (cheap land, cheap labour, decentralisation, development, employment, balance of payments, conservation, and so forth) that went beyond the strict tenets of economic efficiency. Now, government forest services carry large accumulated debts and it is not clear whether industry will develop in ways that will enable them to be recouped profitably.

## South Australia

The most fully developed and successful of Australia's state plantations are those in the south-east of South Australia; their history has been described by Lewis (1975, 1988), Carron (1985), Boardman (1988) and Geddes (1990). South Australia had very small areas of commercially exploitable eucalypt forests and relied heavily on timber imports from other colonies and overseas from its first white settlement in 1836. In 1870 the government's surveyor-general, G.W. Goyder, called for the reservation of land for forestry purposes, the creation of government plantations, and the encouragement of private tree planting. By the end of the 1870s, 79,000 hectares of reserves had been created, planting commenced and a Scottish forester, J.E. Brown, appointed as Conservator. In 1883 forestry was organised under a distinct government department - the Woods and Forests Department. Brown commenced a series of trials of both exotic conifers and native eucalypts in several regions. These were continued by his successor, W. Gill, and the main species, *P.*

*radiata*, selected as capable of very high growth rates in the higher rainfall regions of the south-east and the Adelaide hills, though the amount of land there was limited. To produce sawlogs, a rotation of 40 years with intermediate thinnings was envisaged.

By the start of the twentieth century, the department produced thinnings from the young plantations, but could neither sell the logs to the few small hardwood mills, nor readily persuade a market used to high quality imports to accept the wood. In response, the department started its own tiny sawmill in 1903 and sold the timber for fruit cases. Milling and marketing in the face of import competition remained a problem and the department expanded its sawmilling activities, building several small mills, kiln seasoning plants and a large sawmill in 1930.

The planting rate was increased with British loan funds, under their Overseas Settlement Scheme, which enabled the 2,000 hectares a year to be planted for 10 years from 1926. However there were a number of problems which caused the future of the plantation project to be scrutinised closely in the 1930s. First was the difficulty of selling the small wood from thinning operations. An initial proposal to build a pulpmill fell through, but a small1to make at least 2,500 tonnes a year was completed in 1941. Second was the fact that many of the plantations had been established on very poor sites, or were growing poorly. A special investigation in 1934 found that 42 percent of the 19,000 hectares was "ineffective and uneconomic". Research in the 1930s and 1940s was to show that some of this was due to trace element and nutrient deficiencies which could be corrected in young plantations by applying zinc, copper and phosphate. Careful site and soil surveying was also introduced to enable the worst sites to be avoided, and studies of growth, yield and thinning were commenced to provide data for scientific management.

The government continued the plantation project by planting at 1,200 hectares a year and building a second large sawmill in 1941. A private company built a veneer plant to peel the best quality logs and the sawn timber became better accepted on the market partly because the quality improved as the logs were older. The department gained experience and better machinery, partly because wartime shortages of timber imports made the local product welcome. By 1975 the Department had 76,000 hectares of plantations, but little land left to plant. A long programme of research into tree breeding, site preparation, fertilisation, and weed control has enabled it to raise the productivity of its stands by about 30 percent. This was offset in 1983 by a fire which destroyed 21,000 hectares. However, private investment was attracted to the sector from the 1920s particularly by investment companies selling area bonds to individual small investors. By 1987 there were 90,000 hectares of public and private plantations in South Australia. This, combined with planting by the Victorian government in adjacent areas just over the border, gradually built up the total resource in what became known as the "Green Triangle" region.

Industrial capacity expanded and diversified from the 1950s. Large modern sawmills were built by the department and by 2 private companies and various treatment and furniture plants opened. The problem of finding a market for thinnings was solved when 2 particle board mills and a tissue mill were built. Several other mills for making laminated beams and other products have been built in the region. Overall, the plantation project has stimulated the development of a thriving, technologically advanced region based on an integrated timber industry. Although the government plantations were debt funded, they have paid interest on the loans, and from 1955 to 1983 paid a surplus to the state. Notably in this era of privatization, the government has retained ownership of its plantations and sawmills.

**Brazil**

Although Brazil was incorporated into the world system 3 centuries before Australia, its major forest plantation development is more recent and far larger.

In contrast to Australia, Brazil has great tropical forests whose current deforestation has attracted world wide attention (eg. Browder 1988). However deforestation and subsequent land degradation is a much older and more widespread phenomenon. For example, the poverty of the North East region has been associated with soil exhaustion by the sugar plantations operated there in the sixteenth and seventeenth centuries. In the South and South East regions which contain the 2 largest cities, Rio de Janeiro and Sao Paulo, much of Brazil's industry and about 60 percent of the population, the natural forests which covered about 75 percent of the land in 1500 had been reduced to about 13 percent by the 1980s mostly to create grazing and farming lands, but also by uncontrolled logging for lumber and charcoal production (Magnanini 1959; IBDF/DE 1983, Dean 1983). Part of the forest land carried the natural softwood, *Araucaria angustifolia* (syn. *A. brasiliana*). The natural cerrado (savanna) of tall tufted grasses with trees and shrubs dominant in some areas and the caatinga (thornbush) which occurred in some of the drier parts was similarly reduced.

Brazil's economy, much like Australia's, rested on its primary agricultural and mining base with cycles in which sugar, gold, rubber, and coffee were major export staples. Industrial development in the nineteenth century was limited and the railway infrastructure, largely built with British capital, served to link the hinterland with port cities, such as Rio de Janeiro. Coffee was grown on sites cleared of the primary forest in southeastern Brazil and wood shortages occurred, particularly in the State of Sao Paulo.

The Paulista Railway Company commenced plantations in 1910 to overcome the shortages in their area. Their forestry section was headed by Edmundo Navarre de Andrade who made a special study of eucalypts on several visits to Australia (Penfold and Willis 1961). The company tested 144 species on different soil types and found that *Eucalyptus alba, E. camaldulensis, E. citriodora, E. saligna,* and *E. teretecornis* grew best. Companies and land owners planted eucalypts and pines throughout the South and Southeast regions so that by 1961 Brazil had 560,000 hectares of plantations and was planting at the rate of about 30,000-35,000 hectares a year.

Brazil adopted the prevailing industrialization model of development of the 1950s, much as enunciated by the Economic Commission for Latin America (ECLA). It aimed for "fifty years of development in five" through an aggressive policy of "Import Substitution Industrialisation" and exports of manufactured goods (Lessa 1983). The state played a crucial role through massive investments in infrastructure, basic industries, backward regions of the North East, and the Amazon, and by providing tax incentives to attract private investment. In what was called the "Brazilian economic miracle", industrial production rose to reach 11 percent a year and Real Domestic Product to reach 6 percent (1956-1961). However this achievement was accompanied by a fall in the real minimum wage between 1940 and 1990 to a quarter of its value, a fall in the share of the national income received by the poorest half of the population from 17 to 13 percent (Pereira 1982), a period of repressive military rule, and a massive burden of foreign debt.

The policy of accelerated industrial development was extended to the forest sector in 1966 by offering tax incentives to individuals and corporations who established forest plantations. Paradoxically similar incentives were offered for clearing tropical rainforests in the Amazon region for cattle ranches. The complex and frequently modified details of the plantation incentives scheme have been described by Kengen (1985), Galvao and do Couto (1984), and others. It enabled corporations and individuals to offset the costs of planting the trees and maintaining them for their first 3 years against their tax payments. Initially at its most generous, the tax savings could equal the cost of the total plantation investment, excluding land costs. Subsequent modifications linked investment in plantations to investment in the North East and Amazon regions and required them to be in industrial or special regions. Size requirements and other provisions favoured corporate investors. The response was spectacular. In a decade, the rate of planting leapt to over 400,000 hectares a year as investors, many of them without previous experience or links with industry, hurried to obtain effectively free plantations on whatever land they owned or could buy cheaply.

The costs to the state, in the taxes foregone, were justified as part of the general development strategy and by a number of specific arguments originating from the trickle-down theory (eg. Beattie 1975; Beattie and Ferreira 1978; Berger 1979; Neves 1979; Nogueira 1980). They posited that: plantations could be developed on otherwise unproductive or marginal land, that rural employment would be created and rural-urban migration would be reduced, that plantation resources would attract industries, that they would start a process of modernisation which would lead to regional development and raise the general standard of living. The financial crisis of the Brazilian state led to the plantation incentives scheme being finally dropped in 1988.

The rapid expansion of plantations provided the wood resources for 2 industrial development programs. The National Pulp and Paper Program, announced in 1974, envisaged not only supplying the domestic market but exporting pulp to the world market. Over 3 million tonnes were exported in 1983 and a target of 20 million tonnes a year was set for the year 2000.

The National Steel Development Program similarly aimed to expand production and exports. However Brazil lacks adequate coal resources and relies on charcoal for almost half of its pig iron production; most of the charcoal steel industry is located in the State of Minas Gerais. The international oil crisis in 1973 made charcoal an attractive source of energy and the government pressed some of the major industries, such as pulp and paper, cement, food, beverage, and textiles to substitute charcoal for at least a quarter of their oil consumption.


**Minas Gerais**

Minas Gerais in the South East region is a major plantation State, second only to Sao Paulo. Its population of 15 million is about the same as Australia's. Its development followed the general cyclical pattern of Brazilian economic development with its forest resources being exploited for the gold, coffee, cotton, cattle, and then charcoal for the pig iron industry.

Charcoal was produced from both the forest and the cerrado to supply the pig iron and steel firms. By 1950 10 of the 12 firms were operating their own forest and cerrado reserves to produce charcoal as well as buying it from third parties. Their reserves ranged from 145,000 to 150,000 hectares and were managed to regenerate naturally for future exploitation in 20 to 25 years time. The larger firms made small plantings of eucalypts, which had been planted successfully in the adjacent State of Sao Paulo. Their fast growth enabled 3 or 4 crops to be produced on 7- to 8-year rotations in the same time that 1 crop could be produced naturally. Industrial expansion in the 1960s led to further planting, so that there were about 60,000 hectares of eucalypt plantations in the State by 1960. One of the largest companies, Companhia Siderurgica Belgo Mineira, then planned an ambitious program to plant 5,000 hectares a year for 18 to 20 years in order to match charcoal demand to its plans to expand steel production (Osse 1961).

Planting expanded rapidly with the incentives scheme and from 1973, Minas Gerais became the leading State in terms of area planted. From the introduction of the scheme in 1967 until 1981, the approved plantation projects totalled an area of 1.4 million hectares, or about one-third of the area approved for the whole of Brazil. Only about 1.2 million hectares was actually planted, and of this, 85 percent was planted with eucalypts. The incentives scheme attracted investors who apparently had no clear idea of the end use of the wood that they were growing. A survey in 1980 showed that while 48 percent were destined for charcoal production, and 9 percent were destined for pulp, firewood and timber production, the end use of the remaining 43 percent was unknown (IBDF/UFRRJ 1983).

**Jequitinhonha Valley**

About one-third of the plantations in Minas Gerais were established in the Jequitinhonha Valley. Their social and economic performances were compared in a case study to the outcomes expected from the arguments put forward for plantation establishment, mentioned earlier (Kengen 1985). The Jequitinhonha Valley is a large area of some 78,000 square kilometres with a predominantly rural population of 824,000 (1980). It is a rather stagnant area that "presents a context of depression in many of its basic aspects, despite the potential of its human and natural resources" (F. CETEC 1980). Being more distant from the industrial areas of the State, its plantation development occurred later when the land prices in the closer areas rose, and with a higher proportion of plantations for which the end use was uncertain. The plantation expansion only reached the Jequitinhonha Valley in the 1970s and it was not until the mid 1970s that the great rush occurred. In the decade 1971-1980, 354,000 hectares were planted (by comparison only 282,000 hectares were planted in Australia during the same period). The rapid and obvious dynamism caused by the arrival of people, machines, and increasing consumption was greeted euphorically as the "medicine" to cure the valley's chronic underdevelopment.

Although the majority of the peasants had been living in the valley for generations, most did not have documents to prove legal ownership of their land. This enabled the State government agency, RURALMINAS, to classify their land as devolutas, or ownerless, and hence belonging to the government and able to be sold to the plantation investors. Plantation investors also purchased land from farmers. While much of the land on the plateaux selected for plantations was of low agricultural productivity in grazing and fallow, it nevertheless provided partial sustenance and medicinal plants for the rural population. An expansion of cattle ranching occurred at the same time and the combined effect was to decrease the proportion of land held by small farmers, increase the concentration of land ownership, and increase the degree of urbanisation within the Valley.

Although the plantation developments did provide employment and some infra-structure services such as health and water supplies for their workforce, the general level of services in the Valley was extremely low. Although some health indicators improved during the period of plantation development, indicators of health problems typical of more urban and industrial centres increased. As the boom period of planting occurred only in the 1975-1980 period, it is perhaps too early to tell whether the resource created will attract industries to the Jequitinhonha Valley, or whether it will remain a supplier of raw materials for processing elsewhere.

**Promise and Performance**

On a world scale, the creation of new wood resources in large plantations is the new face of industrial forestry. With all the advances of species selection and breeding, fertilisation, weed and pathogen control, and a high degree of mechanisation, it promises great increases in physical productivity. But to what ends and to whose benefit will this promise be employed?

Australia and Brazil saw the development of plantations strategically, as part of national development aims. In both cases, the state took a highly interventionist approach directly by establishing plantations in Australia, and indirectly by subsidising landowners to do so in Brazil. A number of social and economic objectives and arguments were marshalled in support of these policies in each country. In South Australia and in the Jequitinhonha Valley, the policies led to plantation development ahead of industrial development, which South Australia was able to overcome only by the state forest service becoming an industrial producer. Repetto and Gillis (1988) have shown that ill-advised government policies for natural forests have in many cases led to their misuse, and much the same may apply to plantations.

Preparing well-advised policies for plantations is not an easy task. Planners have to find a path through the complex web of scientific advances, changing economic circumstances, national and social objectives, environmental requirements and policy measures. To do so, they need to draw on all the experience that has been gained with plantations; documenting and analysing that experience is the task for forest historians.

## Abbreviations

| | |
|---|---|
| ACF | Australian Conservation Foundation |
| FAFPIC | Forestry and Forest Products Industry Council |
| F. CETEC | Fundacao Centro Tecnologico de Minas Gerais |
| FAO | Food and Agriculture Organisation of the United Nations |
| IBDF/DE | Instituto Brasiliero de Desenvolvimento florestal/Departmento de Economia Florestal |
| IBDF/UFRRJ | Instituto Brasiliero de Desenvolvimento Florestal/Universidade Federale Rural do Rio do Janeiro |
| IUFRO | International Union of Forest Research Organizations |

## References

ACF. 1987. *Australia's Timber Industry: Promises and Performance.* Hawthorn, Victoria: ACF.

Barrere, P., Heisch, R. and Lerat, S. 1962. Ch. 6. La region du Sud Ouest. in *France et Demain.* Paris: Press Universite de France.

Beattie, W.D. 1975. An economic analysis of the Brazilian fiscal incentives for reforestation. Purdue University (Unpublished PhD thesis).

Beattie, W.D. and Ferreira, J.M. 1978. *Analise financeira e socio-economica do reflorestemanato no Brasil.* Colecao: *Desenvolvimento e Planejamento Florestal.* (Serie: Estudos perspectivos para o period 1975-1985. Brasilia: IBDF/COPLAN)

Berger, R. 1979. The Brazilian fiscal incentive Act's influence on reforestation activity in Sao Paulo State. Michigan State University (Unpublished PhD thesis).

Boardman, R. 1988. Living on the edge - the development of silviculture in South Australian pine plantations. *Australian Forestry*. 51(3): 135-56.

Browder, J.O. 1988. Public policy and deforestation in the Brazilian Amazon. in Repetto, R. and Gillis, M.(eds). *Public Policies and the Misuse of Forest Resources*. Cambridge: Cambridge University Press.

Cameron, J.I. and Penna, I.W. 1988. *The Wood and the Trees: A Preliminary Economic Analysis of a Conservation Oriented Forest Industry Strategy*. Hawthorn, Victoria: ACF.

Carron, L.T. 1990. A history of plantation policy in Australia. in Dargavel, J. and Semple, N. (eds). *Prospects for Australian Forest Plantations*. Canberra: Centre for Resource and Environmental Studies, Australian National University.

Carron, L.T. 1985. *A History of Forestry in Australia*. Australia: Australian National University Press.

F. CETEC. 1980. Projecto estudos integrados do Vale do Jequitinhonha. Belo Horizonte: F. CETEC.

Dargavel, J. 1990. Place of timber merchant firms in the structure of the industry. Paper to IUFRO XIX World Congress, Montreal, 5-11 August 1990 (S6.07, Forest History Group).

Dargavel, J., Dixon, K. and Semple, N. (eds) 1988. *Changing Tropical Forests: Hstorical Perspectives on Today's Challenges in Asia, Australasia and Oceania*. Canberra: Centre for Resource and Environmental Studies, Australian National University.

Dargavel, J. and Semple, N. (eds). 1990. *Prospects for Australian Forest Plantations*. Canberra: Centre for Resource and Environmental Studies, Australian National University.

Dean, W. 1983. Deforestation in Southeastern Brazil. in Tucker, R.P. and Richards, J.F. (eds) *Global Deforestation and the Nineteenth-Century World Economy*. Durham, N.C.: Duke University Press.

Evelyn, J. 1664. *Sylva, or a Discourse on Forest Trees, and the Propagation of Timber in His Majesties Dominions*. London.

FAFPIC. 1987. *Forest Industries Growth Plan*. Melbourne: FAFPIC.

FAO. 1967. *Unasylva*. 21(3-4) issue devoted to key papers and recommendations.

Frank, A.G. 1969. *Sociology of Development and Underdevelopment of Sociology*.

Galvao, A.P.M. and do Couto, H.T.Z. 1984. Man-made industrial forests in Brazil: prospects and constraints. in Wiersum, K.F. (ed.) *Strategies and Designs for Afforestation, Reforestation and Tree Planting*. Wageningen: Pudoc.

Geddes, D.J. Plantation and industrial development in the south-east of South Australia. in Dargavel, J. and Semple, N. (eds). 1990. *Prospects for Australian Forest Plantations*. Canberra: Centre for Resource and Environmental Studies, Australian National University.

Gregory, G.R. 1965. Forests and economic development in Latin America. *Journal of Forestry* 63(2): 83-8.

IBDF/DE. 1983. *Inventario Florestal Nacional (Sintese dos Resultados)*. Brasilia: IBDF/DE

IBDF/UFRRJ. 1983. Avaliacao economica das florestal da Regio Sudeste: disponsibilidade florestal e consumo industrial. Rio de Janeiro: *UFRRJ* (3 volumes).

Kengen, S. 1985. Industrial forestry and Brazilian development: a social, economic and political analysis with special emphasis on the fiscal incentives scheme and the Jequitinhonha Valley in Minas Gerais. Canberra: Australian National University (Unpublished PhD thesis).

Lessa, C. 1983. *15 Annos de Politica Economica*. Sao Paulo: Editora Brasiliense (4th edn).

Levingston, R. 1984. International policy and action for forestation. in Wiersum, K.F. (ed.) *Strategies and Designs for Afforestation, Reforestation and Tree Planting*. Wageningen: Pudoc.

Lewis, N.B. 1988. The Woods and Forests Department of South Australia. in Dargavel, J.(ed.) *Sawing, Selling & Sons: Histories of Australian Timber Firms*. Centre for Resource and Environmental Studies, Australian National University.

Lewis, N.B. 1975. A hundred years of forestry in South Australia: 1875-1975. Adelaide: Woods and Forests Department (Bulletin no. 22).

Magnanini, A. 1959. Area des grandes formacoes vegetais no Brasil. *Anuario Brasileira de Economica Florestal*. 11: 295-303.

McGregor, J.J. 1976. The existing and potential roles of forestry in the economics of developing countries. in Evaluation of the contribution of forestry to economic development. UK: Forestry Commission: 1-7 (Bulletin no. 56).

Neves, A.R. 1979. Avaliacao socio-economica de um programa de reflorestamento na Regiao de Carbonita, Vale de Jequitinhonha, MG. Universidade Federal de Vicosa (Unpublished MSc thesis).

Nogueira, U.B. 1980. Charcoal production from eucalyptus in Southern Bahai for iron and steel manufacture in Minas Gerais, Brazil. Michigan State University (Unpublished PhD thesis).

Osse, L. 1961. As culturas de eucalyptus da Companhia Siderurgica Belgo Mineira. *Anuario Brasiliero de Economia Florestal*. 13: 102-12.

Penfold, A.R. and Willis, J.L. 1961. *The Eucalypts: Botany, Cultivation, Chemistry and Utilization*. London: Leonard Hill.

Periera, L.C.B. 1982. *Economia Brasileira: Uma Introducao Critica*. Sao Paulo: Editora Brasiliense.

Reed, J.L. 1954. *Forests of France*. London: Fabre & Fabre.

Roach, T.R. 1989. The white labour forest settlement programme in South Africa 1917-1938. Johannesberg: University of Witwatersrand, Unpubl. MA thesis.

Sartorious, P. and Henle, H. 1968. *Forestry and Economic Development*. New York: Praeger.

Sedjo, R.A. 1983. *The Comparative Economics of Plantation Forestry: A Global Assessment*. Washington, D.C.: Resources for the Future.

Steen, H.K. (ed.) 1984. *History of Sustained Yield Forestry: A Symposium*. (USA): Forest History Society.

Victoria. 1865. *Report on the Advisableness of Establishing State Forests*. Votes and Proceedings of Parliament 1864-5.

Westoby, J.C. 1962. Forest industries in the attack on underdevelopment. *Unasylva*. 16(4): 168-201. Reprinted 1987 in *The Purpose of Forests: Follies of Development*. Oxford: Blackwell.

Westoby, J.C. 1978. Forest industries for socio-economic development. Proc. Eight World forestry Congress, Jakarta, 16-28 October 1978 (Paper FID/GS). Reprinted 1987 in *The Purpose of Forests: Follies of Development*. Oxford: Blackwell.

The Environmental History of Amazonia:
Natural and Human Disturbances,
and the Ecological Transition

Leslie E. Sponsel
University of Hawaii

Virtually all the world's tropical forests are populated, usually by indigenous peoples. In order for local, state, or international interests to exploit forest resources, the rights of indigenous groups must be denied and the groups themselves displaced. It is no accident, therefore, that indigenous peoples are disappearing at an even faster rate than the tropical forests upon which they depend. Their own survival is intricately linked with that of their forests. They also represent our best first line of defense against the destruction of the forests (Clay 1989:1).

## The Issue

About one-fourth of the national territory of Colombia is Amazon forest, an area of some 47 million hectares. During the last 3 decades the rate of deforestation in Colombia has accelerated to the point that most of its forest will be destroyed within 50 years. Yet some 25 million hectares of remote forest remain in use by the more than 450,000 indigenes in the Colombian Amazon. In an attempt to conserve the forest for the future of the nation in 1988, President Barco's government decided that the best way to stop deforestation would be to recognize the prior rights of indigenous people to their ancestral lands in the nation's Amazon territory. The government acknowledges that only traditional indigenous societies have developed strategies and techniques for sustainable use of the natural resources of the forests without long term environmental degradation. The hope is that if the government respects the rights, territory, and culture of indigenous societies, then they will prosper and also conserve the forest. This assumes that the indigenes will retain their traditional system of environmentally sound beliefs, values, and practices, in spite of some acculturation including Western education, medicine, and trade (Bunyard 1989).

Recently Redford (1990) criticized the Colombian model by calling into question the assumption that indigenes are necessarily practical ecologists and conservationists, and offering a critique and revisionist view: (1) Although indigenes are

233

knowledgeable about the ecology of their habitat, they have long had a serious impact on their environment. (2) Their environmental impact increases as they are attracted to Western ways to improve some aspects of their life. (3) These societies can not be held as models for economic development and environmental conservation.

In developing his argument Redford (1990) also makes these assertions: (1) Most tropical forests have been severely altered by precontact human activities. (2) Indigenes have the same capacities, needs, and desires as Westerners, and there are no cultural barriers to their exploitation of the natural resources in their habitat to the point of depletion and environmental degradation. (3) The documented cases of sustainable systems of resource use by indigenous societies depend on low population density, abundance of land, and little if any involvement in a market economy, conditions which no longer exist in most of Amazonia. Under the new conditions, indigenous societies will not maintain the integrity of their traditions including sustainable resource use. (For other critics of the environmental impact of indigenes in Amazonia see Colchester (1981), Hames (1979:247-250), and Smole (1976:199-211)).

On the other hand, Reichel-Dolmatoff (1971, 1976) is among those who have argued that many indigenous societies are in harmony with their forest ecosystem in Amazonia. In particular, he has described an elaborate cultural system which is ecologically viable for the Desana of the Colombian Amazon (see pp. 7-8). (For other examples of this position see Clay 1988, McDonald 1977, Posey (1990), Ross (1978), Smith (1983), and Sponsel (1986a)).

But Hames (1988:92-93, 106) counters the position that traditional indigenes are primal ecologists and practical conservationists with the observation that field research has not adequately tested and quantitatively documented the behavioral correlates and corresponding environmental impact of ideas which appear to have conservation functions. This also points to the problem raised by environmental philosophers of the degree of congruency between ideals and actual behavior in relation to the environment (Callicott and Ames 1989).

The emerging debate, over *whether indigenes are noble conservationists or savage destroyers of nature in Amazonia*, parallels an enduring controversy about the environmental history of native peoples in North America and elsewhere (Callicott 1982, Cronon and White 1986, Diamond 1988a, Hughes 1983, Martin 1978, Vecsey and Venables 1980, etc.). In turn the controversy is part of a much broader and deeper series of related subjects--human nature, the human niche, the contrast between "primitive" and "civilized" societies, and whether any of these are basically positive or negative (Barnes 1923, Berkhofer 1978, Vecsey and Venables

1980:3-4). Unfortunately, such controversies are not merely academic squabbles, but they have very serious practical ramifications. For example, the argument that whites could use the land better than indigenes was repeatedly employed as a rationale for justifying the dispossession of indigenes from their ancestral lands in North America (Hagan 1980, Hughes 1983:135, Vecsey and Venables 1980:xix). The same thing has happened and can still happen in Amazonia, which is probably the last frontier on our planet. There is a remarkable persistence of the colonial and neocolonial rationalization for resource exploitation, dispossession of indigenes, and imposed change which variously asserts that they are primitive, anachronistic, backward, inefficient, unproductive, wasteful, destructive, and irrational in their technology, subsistence, and economy (Bodley 1988, 1990).

While this brief paper will not resolve such controversies, some progress may be made by reviewing the environmental history of Amazonia within the framework of (1) a comparison of natural and human disturbances, and (2) the concept of the ecological transition. This new approach reveals that societies range on a continuum from equilibrium to disequilibrium, and occasionally shift positions, albeit usually in the direction of disequilibrium. It describes human-environment interactions in relative rather than absolute terms. By placing these issues in the perspective of the environmental history of Amazonia, it also reveals that *indigenous impacts on the ecosystems were usually comparable to small scale natural disturbances*, whereas Western impacts are without precedent and clearly unnatural. This new approach may prove to be more productive than the usual either-or and all-or-none positions. For instance, there is the common procedure of identifying a few dramatic cases of resource depletion and environmental degradation by indigenous societies in prehistory and history to cast suspicion on all native-environment interactions (Diamond 1988a).

**Natural Disturbances**

In the last 2 decades in Western science, there has been a shift from viewing the Amazon in terms of order and stability to disorder and instability, and also from 'homogeneity to heterogeneity. (These shifts mirror the rise of chaos theory in recent years (Gleick 1987)). Amazonia is far from homogeneous and stable, it is composed of a rich diversity of dynamic ecosystems, and within each there is a complex mosaic of gaps, successions, and mature forest with associated differences in the composition, structure, and function of plant and animal communities. The phenomena contributing to the dynamic temporal and spatial variation and associated biodiversity in Amazonia include but are not limited to climatic fluctuations, fluvial dynamics, tree falls, and fire (Sponsel 1986a).

Natural deforestation may have occurred in prehistoric times on a massive scale albeit gradually over millennia. During the last million years of the Pleistocene epoch, even in Amazonia there were climatic changes which may have in turn triggered major changes in the geographical distribution of plant and associated animal communities in both the highlands and lowlands. During drier periods when rainfall dropped below 1500 mm/year, savannas expanded and forests contracted into restricted "refuge" areas, while the opposite transpired during wetter periods. Also changes in sea level would have increased or decreased flooding along the Amazon River and some of its major tributaries given the remarkable flatness of the basin. These dynamic processes are suggested by studies of changes in terrain, sediments and their pollen at a limited number of widely separated sites (Haffer 1987). However, there is considerable controversy over Pleistocene refugia and related ideas (Colinvaux 1989, Prance 1982).

On a smaller scale, there have also been more moderate regional and local perturbations in climate and weather during the Holocene. For instance, pollen analysis of lacustrine sediments from Kumpak Lake in Ecuador reveal that unusually heavy storms occurred during the last 5,000 years washing sediments in pulses from the shore into the lake. These and other natural disturbances have contributed to the species diversity of Amazonia (Colinvaux, *et al.* 1985, Colinvaux 1989).

The dynamic processes of fluvial geomorphology, especially lateral erosion and channel changes along floodplains, are another major source of natural disturbance in the forest ecosystems of Amazonia as revealed by analysis of Landsat satellite imagery for the upper Amazon in Peru. For instance, the migration of meanders along a floodplain through time creates oxbow lakes and meander scrolls - a sequence of ridges running parallel to the river course. In turn this creates long, continuous, and dynamic strips of successional forests in a sequence of different ages along recent and abandoned river channels. This spatial distribution differs markedly from the scattering of natural tree falls. Analyses indicate that 26.6 percent of the modern lowland forest has characteristics of recent erosional and depositional activity, and 12.0 percent of the lowland forest along rivers is in successional stages. Primary succession on newly deposited riverine soils is a major mode of forest regeneration. By causing high turnover, disturbances and variation in forest structure, these river dynamics may be a major factor creating and maintaining the high between-habitat species diversity characterizing the upper Amazon; that is, closely packed, small-scale habitat mosaics which are highly stable in species composition but relatively short-lived (Salo 1987, Salo, *et al.* 1986).

Light gaps and successional processes triggered by tree falls from wind storms, tree senility, ephiphyte overload, and other factors also contribute to the dynamic variation and biodiversity of Amazon forests. Estimates of tree-fall frequencies from several lowland forests reveal that 4-6 percent of the plant communities are in the first 5 years of forest regeneration in light gaps (Denslow 1987, Hartshorn 1980, Pickett and White 1985, Uhl *et al.* 1989).

Amazon forest ecosystems are adaptable; they have evolved numerous mechanisms to regenerate after disturbance. These include seedlings and saplings present in the forest understory which quickly respond to light gaps, the capacity to produce sprouts from stem bases or roots following destruction of the main stem, germination of otherwise dormant seeds buried in soils (200-1,000 seeds/$m^2$), and the dispersal of seeds from outside the disturbed area through wind and animal agents like birds, bats, and rodents. Tree species have a remarkable adaptability to persist in a semi-arrested state of development and then to rapidly accelerate their growth rates when treefalls create light gaps in the forest (Uhl, *et al.*, 1989).

Fire has probably been a moderate level disturbance in forests occurring infrequently but repeatedly with low intensity for more than 6,000 years. In much of Amazonia it is difficult to find soils without some charcoal fragments (Uhl, *et al.*, 1989:236). A study in the vicinity of San Carlos de Rio Negro in Venezuela discovered charcoal in soil samples from tierra firme and igapo sites, some associated with anthrosols and ceramic potsherds, others not. The oldest archaeological evidence in the area dates through thermoluminescence from 3750 +/- 20 years B.P., whereas charcoal remains have been radiocarbon dated as far back as 6260 +/- 100 years B.P. The circumstance that older dates are associated with charcoal but not with ceramic potsherds suggests that natural as well as anthropogenic fires occurred. (No preceramic occupation by foraging societies is known archaeologically from Amazonia (Meggers 1988)). One catalyst for the natural fires may be increased aridity in forests at the northern margin of the Amazon (Sanford, *et al.*, 1985). However, fires are not necessarily destructive in forest ecosystems. When fires are of low intensity they may stimulate vigorous regrowth as both nutrients and plant colonists are readily available (Uhl, *et al.*, 1989:237).

*In the Amazon certainly forest disturbance is nothing new, and it is not necessarily unnatural or destructive.* Indeed, disturbance can promote a healthy forest and enhance its adaptability and biodiversity, as for example by creating patches of forest at different stages of the successional process which are more resilient than a forest stand of uniform age. Disturbances may help maintain and even enhance biodiversity at various levels, including within and between ecosystems. Since meandering rivers, treefalls, and fire are among the numerous factors which have triggered succession for millions of years in Amazonia, the forest ecosystems have evolved numerous mechanisms for regeneration (Uhl, *et al.*, 1989).

The critical factor in disturbance is its scale in terms of both temporal duration and spatial extent. The temporal duration ranges from decades to millennia. The spatial scale ranges from less than a hectare to 100,000 km². The chances of regeneration are better for disturbances which are shorter in time and smaller in space. As will be discussed later, traditional indigenous disturbances of forest were probably usually comparable to small natural disturbances like tree falls, whereas disturbances in recent decades by Western activities, like conversion to pasture for cattle ranching, are on a scale comparable to some of the larger natural disturbances (Uhl, *et al.*, 1989).

However, there is at least 1 major and crucial difference between large-scale natural and Western disturbances--the temporal dimension. Natural disturbances occur over centuries or millennia in terms of the stress and regeneration, whereas Western disturbances occur over a few years or decades yet require centuries or even millennia for regeneration depending on the particular situation. The abruptness, intensity, and magnitude of disturbances by Western activities is often unnatural compared to climatic, geological, and biological processes which trigger disturbances. Moreover, because they are so devastating, Western disturbances may obstruct natural regeneration. In the highly weathered soils of interior Amazonia, nutrient availability is usually the most important limiting factor on plant growth. Western deforestation often involves extensive areas which expose soils to erosion and nutrient loss, and which destroy or retard many of the mechanisms of forest regeneration (Uhl, *et al.*, 1989).

When anthropogenic disturbances do not mimic or parallel natural disturbances, then they are abnormal and endanger the ecosystem. *Human disturbance of forest ecosystems becomes unnatural when the natural regenerative capacities of the forest ecosystem are retarded or destroyed.* This happens increasingly when Western disequilibrium societies effectively and persistently penetrate the forests. It may also happen in some cases when traditional indigenous societies are radically acculturated and shift from equilibrium to disequilibrium.

## The Ecological Transition

The related concepts of equilibrium, disequilibrium, and the ecological transition are heuristic as they provide a framework for thinking about processes of human ecology, adaptation, and evolution (Sponsel 1987). Societies fall along a *continuum* from equilibrium to disequilibrium which are opposite tendencies. However, these are *relative* rather than absolute conditions, and this is not some rigid typology.

An equilibrium society is one where the population is in a dynamic balance with is resources below the carrying capacity of its environment. An equilibrium society has a population that is small and controlled; needs and wants are minimized by being limited mostly to satisfy basic biological requirements; the society is self-sufficient through exploitation of its local environment; technological capacity is limited; the principal energy sources are animals, human muscle, and fuelwood; and most individuals have daily contact with their natural environment. The society's environmental impact is negligible or minimal (Bennett 1976:13).

In contrast, a disequilibrium society is characterized by a large population which is increasing with little if any intrinsic controls. Resource consumption extends well beyond satisfying basic biological requirements. Moreover, the population becomes increasingly dependent on resources derived from beyond the local region from an extensive network of trade and commerce which exploits distant ecosystems. Energy sources extend beyond solar, muscle, and fire to include water and/or wind, and eventually fossil fuels and nuclear energy. Increasingly numbers of individuals lose daily contact with their natural environment by becoming craft, bureaucratic, and other specialists whose work is independent from subsistence. Environmental impact increases markedly, including resource depletion (e.g., game depletion and local extinction of species for commercial exploitation) and environmental degradation (e.g., deforestation).

A particular society may shift its position through time in either direction, although it appears that the more common direction is from equilibrium to disequilibrium. This more common direction of movement is the ecological transition (Bennett 1976:113-120). In the course of cultural evolution there have been several revolutionary developments--principally the use of fire, agriculture, the state, industrialization, and modernization (Campbell 1983). Each of these revolutionary developments represents new thresholds in the ecological transition with new levels of disequilibrium. Also when a society in equilibrium comes into contact with one in disequilibrium, then the former usually moves in the direction of greater disequilibrium (Colchester 1981, 1989, Bennett 1976:113-120, 1980, Bodley 1990).

## Case Studies

Although numerous case studies from Amazonia could be used to illustrate equilibrium societies as well as anthropogenic disturbances of forest ecosystems, available space restricts coverage to a few selected cases, and those chosen are among the best documented and most striking ones: Desana environmental ecological philosophy and ethics, Yanomami trek hunting, Bora swidden-fallow agroforestry, and Mebengokre ethnoecology of forest fields.

The Colombian Desana live in an area of white sands and black waters, an oligotrophic ecosystem which is poor in nutrients, productivity, and species diversity. Here as well as elsewhere in much of the interior Amazon animal protein may be one important limiting factor on the human population, if not the most important one (Gross 1975, Reichel-Dolmatoff 1971:13, Sponsel 1986a:74-77).

If the Desana population were to increase substantially, or if their game resources were to be depleted, then they could not readily move to new territory since they are circumscribed by neighbors and attached to their homeland through cultural and historic roots. The Desana have an elaborate system of myth, ritual, and symbol which regulates the dynamics of their human and animal populations as they interact in the dynamic predator-prey relationship. A rich and complex cultural system of faunal (food) and sex prohibitions are central in regulating this predation process. Their adaptive strategy is reinforced by a theory of disease which considers illness to be a result of some violation of a prohibition or disturbance of the spiritual world. The shaman monitors the taboos, game abundance, hunting, and other aspects of the cultural ecology of the Desana village as the manager of this equilibrium society. Also the shaman performs rituals to maintain equilibrium and to restore it if it declines. In these and other ways the Desana are supposed to manage their resource exploitation to ensure sufficient quality protein without game depletion (Reichel-Dolmatoff 1971, 1976).

About 15,000 Yanomami live in the mountainous headwater region which divides the Orinoco and Amazon river basins and also forms the border zone between Venezuela and Brazil. The Yanomami represent an ancient and stable adaptation to the interior forest ecosystem. Divergent lines of evidence from genetics and linguistics indicate that the Yanomami have been an independent group for over 2,000 years (Spielman, *et al.*, 1974:643). Although they cultivate bananas, plantains, and other crops in swidden gardens, traditional Yanomami are primarily foragers and only secondarily farmers.

The Sanema, a northern group of Yanomami in Venezuela studied by the author, adapt successfully to the limiting factor of animal protein in at least 5 ways: (1) extensive knowledge of the behavioral ecology of prey species and habitats; (2) highly skilled and efficient use of hunting technology (bow and arrow, curare, etc.); (3) generalized and opportunistic predation usually based on a solitary hunter walking long distances along trails, except for group hunts focused on peccary; (4)

low density and high mobility of their population; and (5) regular movement of hunting, garden, camp, and village sites (Sponsel 1981:198-278). Overnight hunting camps function as a compromise between sedentary and nomadic existence; that is, as a temporary alternative to village relocation as game is depleted in its vicinity (Sponsel 1981:226-228). Faunal prohibitions practiced by the Sanema and other Yanomami (Taylor 1974) may function to conserve some wildlife species and enhance biodiversity (MacDonald 1977, Ross 1978, cf. Colchester 1981).

The recently reported long-term field study by Good (1987, 1989, 1990) emphasizes trekking as the most important and distinctive characteristic of Yanomami subsistence. The Yanomami schedule their planting in gardens to allow periods away from the village when their staple of plantains are not harvestable. During these periods they trek and camp in the forest living off of gathering wild plant foods, hunting game in areas where it is more abundant, and occasional fishing. The community may spend 40 percent or more of the year on such treks where hunting yields are 60 percent higher than in the hunting range adjacent to the village. They rarely spend more than 2 to 4 consecutive months in a village, and may camp in the forest as long as 3 months in a row.

In eastern Peru the Bora carefully manage swidden fallows to the point that their abandonment and forest regeneration are gradual and interconnected processes. Their agroforestry involves a combination of annual and perennial crops together with spontaneous natural forest regrowth. Swiddens range in size from 0.25 to 1.0 ha. Management of swiddens and fallows includes planting, polycropping, crop scheduling and zonation, selective weeding, ash fertilizers, organic insecticides, tree coppicing, tree planting (including species which fix soil nitrogen), and directed succession. While the most productive fallow is between 4-12 years of age, even after 35 years some fallows are still used to some degree for subsistence and market products (Denevan, *et al.*, 1984, Denevan and Padoch 1988).

A fallow plot 19 years old included trees up to 25 m high, clear stratification, and some 233 trees belonging to 82 species (over half the trees in single occurrences). In 1 transect informants identified 22 useful trees. These included construction materials (11 spp., 25 individuals), medicinals (4 spp., 4 individuals), food plants (2 spp., 11 individuals), artisan material (1 sp., 1 individual), utilitarian plants (4 spp., 4 individuals), edible grub trees (2 spp., 2 individuals) (Denevan, *et al.*, 1984:353-354).

Bora swidden-fallows represent an ecologically stable and viable form of resource and land use management. Moreover, such indigenous strategies for forest resource use imply that, *with much larger precontact populations, forests were much more extensively used and disturbed, although apparently usually without degradation and destruction* (Denevan, *et al.*, 1984:346-347, Denevan and Padock 1988:1-2). The Kayapo who call themselves the Mebengokre are an indigenous society of more than 3,500 people living in the Mato Grosso region of central Brazil. The related phenomena of trekking, nomadic agriculture, and island forests provide an example of how this society alters and manages ecosystems in ways which are sustainable, enhance biodiversity, and promote forest regeneration (Posey 1989:241).

The Mebengokre make several treks for a duration of a month or more each year without returning regularly to their village and its surrounding gardens for food provisions. Each village maintains at least 500 km of trails which average about 2.5 m in width. Associated with these trails and adjacent temporary camps are forest fields in the savanna. A sample of 10 of these fields revealed 120 plant species of which 75 percent are probably planted (Posey 1990:19). These resource islands may include yams, sweet potatoes, fruit trees, and other food plants as well as palms with multiple uses, vines for drinking water, and medicinal plants (Posey 1985). Moreover, some of the plants are purposefully introduced as favorite foods of game species such as deer and tapir; thus these resource islands also function as game farms and reserves for periodic subsistence hunting.

These forest islands in the savanna can be considered as a form of restoration ecology. They promote the regeneration of forests by establishing concentrations of trees that in turn attract fauna which disperse seeds from the "natural forests" thereby introducing additional plant species (Posey 1989:243). In this and other aspects of Mebengokre cultural ecology, Western distinctions between such dichotomies as forest and savanna, forest and field, natural and anthropogenic, and wild and domesticated are hard to apply on etic grounds and invalid on emic grounds (Posey 1983:241, 1984:125).

**Discussion**

For millions of years in Amazonia climatic, fluvial, and other disturbances of the forest have long been an important part of the natural processes in its dynamic ecosystems. Accordingly the ecosystems are resilient having evolved mechanisms for regeneration. Indeed *such disturbances may be necessary to maintain biodiversity* (Diamond 1988b, Lugo 1988).

Like natural small scale disturbances, indigenous practices may help to maintain and even enhance biodiversity. The Desana, Yanomami, Bora, and Mebengokre cases are typical of traditional indigenous societies in Amazonia which practice sustainable resource use and conservation without irreversibly depleting resources and degrading the ecosystem. (Also see Clay 1988, Hames and Vickers 1983, Hecht 1982, Jordan 1987, Sponsel 1992). While human antiquity in Amazonia extends back more than 6,000 years, by the year 1500 A.D. *when Western penetration started the Amazon as a biome was not yet endangered*; that only became a reality mostly in the last 2 decades under the influence of Western societies.

Most indigenes have a long-term interest in their habitat and use its natural resources mainly for food and other necessities, and in some cases also to a degree for household income. In contrast most Westerners are transients in Amazonia with only a short-term interest to exploit the forest for quick profit and then escape to civilization to live a rich life. Western exploitation of Amazonia is clearly characterized by arrogance, ignorance, myopia, greed, and destruction. If the social and ecological consequences of development experiments such as Bragantina, Fordlandia, Transamazonia, Jari, and Carajas are considered, then Western disturbances in Amazonia are enormously more destructive than anything indigenes have ever done (Bourne 1978, Fearnside 1989, Hecht and Cockburn 1989, Moran 1981, 1983, Prance 1990, Russell 1987, Stone 1985, Tucker 1990). The environmental impact of indigenous societies, *even cumulatively after thousands of years*, is negligible in contrast to that of Western societies in recent decades.

For example, the typical Western approach to agriculture in Amazonia reflects mid-latitude farming practices and in general is just the opposite of the indigenous one. Westerners eliminate biological diversity and complexity by imposing a narrowly limited and controlled range of specific cash crop monocultures which are especially vulnerable to pests and disease as well as market fluctuations (Posey 1984:101, 104). Cutting and burning forest to prepare short-lived pastures for cattle ranching is also an extremely inefficient, unproductive, and wasteful use of Amazonian forests by irresponsible and greedy land speculators subsidized by government loans and/or tax incentives (Hecht 1989, Hecht and Cockburn 1989). In these and other respects, Western practices are clearly unnatural disturbances of the forest ecosystem leading to massive deforestation, erosion of biodiversity, and other deleterious environmental consequences (Fearnside 1989, Lutzenberger 1982, Repetto 1990, Wilson 1990).

In Amazonia ecologically and culturally sustainable indigenous societies have flourished with low population density, abundant land, and subsistence economies with little or no involvement in markets. Many indigenes societies remain this way, although the situation is changing rapidly with increasing influences and pressures from the West. Some of these changes result in greater levels of indigenous environmental impact such as game depletion under greater population and market pressures (Colchester 1981, 1984, 1989, Gross, *et al.*, 1979, Hames 1979, Johnson 1977, Redford 1990, Redford and Robinson 1987, Saffirio and Scaglion 1982, Smith 1981, Stearman 1990). However, *victims should not be confused as villains*. Increased settlement size and sedentarization, land area decline, and market involvement usually result from Western penetration including missionization, colonization, and government administration (Colchester 1989, Gross, *et al.*, 1979, Stearman 1990).

Any dynamic system changes through time and indigenous cultures and societies are no exception. Traditional societies are adopting some Western ways in an attempt to enjoy the best of both worlds, and they are also changing through new innovations of their own. Certainly indigenes in Amazonia have the same individual physical and mental capacities as any subgroup of the human species. They also share some of the same needs and desires, but not all. Moreover, in terms of resource exploitation and environmental impact, indigenous societies do not have the same demographic, technological, economic, and political capacities and concerns as Western societies. They never created problems like Jari and Carajas. Often they have quite different cultural beliefs, values, and goals, and some of these may function as constraints on Westernization. Some societies such as the Yekuana of the Venezuelan Amazon have experienced contact for centuries, but have been very selective in what they have accepted from the West and simultaneously maintained their ethnic identity and integrity in the process (Arvelo-Jimenez 1971, Sponsel 1986b).

Given the *precedents* of an indigenous society's occupation of an area for centuries or even millennia and its general ecological and sociocultural viability, they deserve the right to occupy and exploit their own land and resources in their own terms without external impositions which ignore their knowledge, needs, and concerns. Most indigenes do not totally reject their own culture in favor of the West. They wish and should have basic rights with regard to such things as their land title, culture, religion, and language. However, indigenous people also wish and should have a basic right to share the benefits of Western health, education, and the like, as they choose.

To blame indigenous peoples for the deleterious results of the ecological transition in Amazonia is to confuse the victim for the villain, because without Western influence most indigenous societies would remain more or less in dynamic equilibrium with negligible or beneficial environmental impact. Indigenous societies which have inhabited a territory for centuries or even millennia without degrading it contrast markedly with the rapidity, intensity, and magnitude of environmental degradation under Western activities.

From the few case studies briefly reviewed here it is clear that much of what has been considered to be primary, natural, or undisturbed forest in Amazonia may be the product of millennia of indigenous resource management and conservation as well as coevolution (Posey 1984:125). Indeed, although apparently invisible to most Western eyes, *Amazonia has already long ago been economically developed by indigenous societies*, but mostly in sustainable ways which do not irreversibly deplete resources and degrade or destroy their ecosystems. Most disturbances of the ecosystems by traditional indigenous societies are no greater than natural disturbances and likewise help to maintain and even enhance biodiversity.

So far Western attempts to develop Amazonia have been such disastrous failures that new directions are clearly needed. At least in part these new directions may come from the heuristic models provided by traditional indigenous societies like the Desana, Yanomami, Bora, and Mebengokre, if only government leaders and bureaucrats, technocrats, bankers, developers, and scientists can transcend their arrogance, ignorance, and prejudice (Appell 1988, Clay 1988, Posey 1990:23, Sponsel 1986a, 1987). Such indigenous models for economic development and environmental conservation can never be exactly imitated, but they can provide a basis for seriously rethinking and redesigning Western societies to be more viable ecologically and socioculturally. It is becoming increasingly clear that the survival and welfare of civilization and the biosphere is at stake as well as that of Amazonia (Myers 1984).

## Conclusion

The survival of Amazonia and indigenes are intimately interdependent. One of the keys to the future of Amazonia is to recognize the ecological knowledge, understanding, wisdom, practices, and rights of the original people of Amazonia as well as the fact that they have their own modes of economic development which in contrast to the Western modes are usually sustainable rather than destructive. The framework of environmental history in Amazonia together with the theoretical concepts associated with the ecological transition and the case studies from cultural ecology help to place these matters in perspective. They also demonstrate that the past is indispensable for understanding the present and coping with the future in Amazonia as elsewhere.

# References

Appell, G.N., 1988. "Costing Social Change," In *The Real and Imagined Role of Culture in Development*, Michael R. Dove, eds., Honolulu: University of Hawaii Press, pp. 271-284.

Arvelo-Jimenez, Nelly, 1971. *Political Relations in a Tribal Society: A Study of the Ye'cuana of Venezuela*. Ithaca: Cornell University Doctoral Dissertation.

Barnes, Harry Elmer, 1923. "The Natural State of Man (An Historical Resume)," *The Monist* XXXIII(1):33-80.

Bennett, John, 1976. "The Ecological Transition: From Equilibrium to Disequilibrium," In his *The Ecological Transition: Cultural Anthropology and Human Adaptation*. New York: Pergamon Press, pp. 123-155.

___, 1980. "Human Ecology as Human Behavior: A Normative Anthropology of Resource Use and Abuse," In *Human Behavior and Environment: Advances in Theory and Culture*, Irwin Altman, ed., New York: Plenum Press, pp. 243-277.

Berkhofer, Robert F., Jr., 1978. *The White Man's Indian: Images of the American Indian from Columbus to the Present*. New York: Alfred A. Knopf.

Bodley, John, ed., 1988. *Tribal Peoples and Development Issues: A Global Overview*. Mountain View, CA: Mayfield Publishing Co.

___, 1990. *Victims of Progress*. Mountain View, CA: Mayfield Publishing Co.

Bourne, Richard, 1978. *Assault on the Amazon*. London: Victor Gollancz, Ltd.

Bunyard, Peter, 1989. "Guardians of the Forest: Indigenist Policies in the Colombian Amazon," *The Ecologist* 19(6):255-258.

Callicott, J. Baird, 1982. "Traditional American Indian and Western European Attitudes Toward Nature: An Overview," *Environmental Ethics* 4(4):293-318.

___, and Roger T. Ames, 1989. "Epilogue: On the Relation of Idea and Action," In *Nature in Asian Traditions of Thought: Essays in Environmental Philosophy*, J. Baird Callicott and Riger T. Ames, eds., Albany, New York: State University of New York Press, pp. 279-289.

Campbell, Bernard, 1983. *Human Ecology*. New York: Aldine Publishing Co.

Clay, Jason, 1988. *Indigenous Peoples and Tropical Forests: Models of Land Use and Management from Latin America*. Cambridge, MA: Cultural Survival, Inc.

___, 1989. "Defending the Forests," *Cultural Survival Quarterly* 13(1):1.

Colchester, Marcus, 1981. "Ecological Modelling and Indigenous Systems of Resource Use: Some Examples from the Amazon of South Venezuela," *Antropologica* 55:51-72.

___, 1984. "Rethinking Stone Age Economics: Some Speculations Concerning the Pre-Columbian Yanoama Economy," *Human Ecology* 12(3):291-314.

___, 1989. "Indian Development in Amazonia: Risks and Strategies," *The Ecologist* 19(6):249-254.

Colinvaux, Paul A., 1989. "The Past and Future Amazon," *Scientific American* 260(5):102-108.

___, *et al.*, 1985. "Discovery of Permanent Amazon Lakes and Hydraulic Disturbance in the Upper Amazon Basin," *Nature* 313:42-45.

Cronon, William, and Richard White, 1986. "Indians in the Land," *American Heritage* 37(5):19-25.

Denslow, Julie S., 1987. "Tropical Rainforest Gaps and Tree Species Diversity," *Annual Review of Ecology and Systematics* 18:431-451.

Diamond, Jared, 1988a. "The Golden Age That Never Was," *Discover* 9(12):70-79.

___, 1988b. "Factors Controlling Species Diversity: Overview and Synthesis," *Annals of the Missouri Botanical Garden* 75:117-129.

Denevan, William M., *et al.*, 1984. "Indigenous Agroforestry in the Peruvian Amazon: Bora Indian Management of Swidden Fallows," *Interciencia* 9(6):346-357.

Denevan, William M., and Christine Padoch, eds., 1987. *Swidden-Fallow Agroforestry in the Peruvian Amazon*, Bronx, NY: The New York Botanical Garden, Advances in Economic Botany 5:1-107.

Fearnside, P.M., 1989. "Charcoal in the Amazon," *Ambio* XVIII(2):141-143.

Gleick, James, 1987. *Chaos: Making a New Science*. NY: Viking.

Good, Kenneth R., 1987. "Limiting Factors in Amazonian Ecology," In *Food and Evolution: Toward a Theory of Human Food Habits*, Marvin Harris and Eric B. Ross, eds., Philadelphia: Temple University Press, pp. 407-421.

___, 1989. *Yanomami Hunting Patterns: Trekking and Garden Relocation as an Adaptation to Game Availability in Amazonia, Venezuela*. Gainesville: University of Florida Doctoral Dissertation.

___, 1990. *Into The Heart*. New York: Simon and Schuster.

Gross, Daniel R., 1975. "Protein Capture and Cultural Development in the Amazon Basin," *American Anthropologist* 77(3):526-549.

___, *et al.*, 1979. "Ecology and Acculturation Among Native Peoples of Central Brazil," *Science* 206:1043-150.

Haffer, J., 1987. "Quaternary History of Tropical America," In *Biogeography and Quaternary History in Tropical America*, T.C. Whitmore, and G.T. Prance, eds., NY: Oxford University Press, pp. 1-18.

Hagan, William T., 1980. "Justifying Dispossession of the Indian: The Land Utilization Argument," In *American Indian Environments: Ecological Issues in Native American History*, Christopher Vecsey and Robert W. Venables, eds., Syracuse, NY: Syracuse University Press, pp. 65-80.

Hames, Raymond B., 1979. "A Comparison of the Efficiencies of the Shotgun and the Bow in Neotropical Forest Hunting," *Human Ecology* 7(3):219-252.

___, 1988. "Game Conservation or Efficient Hunting?," In *The Question of the Commons: The Culture and Ecology of Communal Resources*, Bonnie J. McCay and James M. Acheson, eds., Tucson: Arizona University Press, pp. 92-107.

___, and William T. Vickers, eds., 1983. *Adaptive Responses of Native Amazonians*. NY: Academic Press.

Hartshorn, Gary S., 1980. "Neotropical Forest Dynamics," *Biotropica* (Supplement) 12(2):23-30.

Hecht, Susanna B., 1982. "Agroforestry in the Amazonian Basin: Practice, Theory and Limits of a Promising Land Use," In *Amazonia: Agriculture and Land Use Research*, Susanna B. Hecht, ed., Cali, Colombia: Centro Internacional de Agricultura Tropical, pp. 331-371.

___, 1989. "The Sacred Cow in the Green Hell," *The Ecologist* 19(6):229-234.

___, and Alexander Cockburn, 1989. *The Fate of the Forest: Developers, Destroyers and Defenders of the Amazon*. NY: Verso.

Hughes, Donald J., 1983. *American Indian Ecology*. El Paso: Texas Western University.

Johnson, Allen, 1977. "The Energy Costs of Technology in a Changing Environment: A Machiguenga Case," In *Material Culture: Styles, Organization, and Dynamics of Technology*, Heather Lechtman and Robert Merrill, eds., St. Paul, Minn.: West Publishing Co., pp. 155-167.

Jordan, Carl F., ed., 1987. *Amazonian Rain Forests: Ecosystem Disturbance and Recovery*. NY: Springer-Verlag.

Lugo, Ariel E., 1988. "The Future of the Forest: Ecosystem Rehabilitation in the Tropics," *Environment* 30(7):16-20,41-45.

Lutzenberger, Jose A., 1982. "The Systematic Demolition of the Tropical Rain Forest in the Amazon," *The Ecologist* 12(6):248-252.

Martin, Calvin, 1978. *Keepers of the Game: Indian-Animal Relationships and the Fur Trade*. Berkeley: University of California Press.

McDonald, David R., 1977. "Food Taboos: A Primitive Environmental Protection Agency (South America)," *Anthropos* 72(5/6):734-748.

Meggers, Betty J., 1988. "The Prehistory of Amazonia," In *People of the Tropical Rain Forest*, Julie Sloan Denslow and Christine Padoch, eds., Berkeley: University of California Press, pp. 53-62.

Moran, Emilio F., 1981. *Developing the Amazon*. Bloomington: Indiana University Press.

___, ed., 1983. *The Dilemma of Amazonian Development*. Boulder: Westview Press.

Myers, Norman, ed., 1984. *Gaia: An Atlas of Planet Management*. NY: Doubleday.

Pickett, S.T.A., and P.S. White, eds., 1985. *The Ecology of Natural Disturbance and Patch Dynamics*, NY: Academic Press.

Posey, Darrell A., 1983. "Indigenous Ecological Knowledge and Development of the Amazon," In *The Dilemma of Amazonian Development*, Emilio F. Moran, ed., Boulder: Westview Press, pp. 225-257.

___, 1984. "Ethnoecology as Applied Anthropology in Amazonian Development," *Human Organization* 43(2):95-107.

___, 1985. "Native and Indigenous Guidelines for New Amazonian Development Strategies: Understanding Biological Diversity Through Ethnoecology," In *Change in the Amazon Basin*, John Hemming, ed., Machester; University of Manchester Press, II:156-181.

___, 1989. "Alternatives to Forest Destruction,: Lessons from the Mebengokre Indians," *The Ecologist* 19(6):241-244.

___, 1990. "The Science of the Mebengokre," *Orion Nature Quarterly* 9(3):16-23.

Prance, Ghillean T., ed., 1982. *Biological Diversification in the Tropics*. NY: Columbia University Press.

___, 1989. "Rainforested Regions of Latin America," In *Lessons of the Rainforest*, Suzanne Head and Robert Heinzman, eds., San Francisco: Sierra Club Books, pp. 53-65.

Redford, Kent H., 1990. "The Ecologically Noble Savange," *Orion Nature Quarterly* 9(3):24-29.

___, and John G. Robinson, 1987. "The Game of Choice: Patterns of Indian and Colonist Hunting in the Neotropics," *American Anthropologist* 89(3):650-667.

Reichel-Dolmatoff, Gerardo, 1971. *Amazonian Cosmos: The Sexual and Religious Symbolism of the Tukano Indians*. Chicago: University of Chicago Press.

___, 1976. "Cosmology as Ecological Analysis: A View from the Rain Forest," *Man* 11:307-318.

Repetto, Robert, 1990. "Deforestation in the Tropics," *Scientific American* 262(4):36-42.

Ross, Eric B., 1978. "Food Taboos, Diet, and Hunting Strategy: The Adaptation to Animals in Amazon Cultural Ecology," *Current Anthropology* 19(1):1-36.

Russell, Charles E., 1987. "Plantation Forestry: The Jari Project, Para, Brazil," In *Amazonian Rain Forests: Ecosystem Disturbance and Recovery*, Carl F. Jordan, ed., NY: Springer-Verlag, pp. 76-89.

Saffirio, Giovanni, and Richard Scaglion, 1982. "Hunting Efficiency in Acculturated and Unacculturated Yanomama Villages," *Journal of Anthropological Research* 38(3):315-327.

Salo, Jukka, *et al.*, 1986. "River Dynamics and the Diversity of Amazonian Lowland Forest," *Nature* 322:254-258.

Sanford, Robert L., Jr., *et al.*, 1985. "Amazon Rain-Forest *Science* 227:53-55.

Smith, Nigel J.H., 1981. "Colonization Lessons from a Tropical Forest," *Science* 214:755-761.

____, 1983. "Enchanted Forest," *Natural History* 92(8):14, 18-20.

Smole, William J., 1976. "Landscape Modification," In his *The Yanoama Indians: A Cultural Geography*, Austin: University of Texas Press, pp. 199-211.

Spielman, Richard S., *et al.*, 1974. "Regional Linguistic and Genetic Differences Among Yanomama Indians," *Science* 184:637-644.

Sponsel, Leslie E., 1981. *The Hunter and the Hunted in the Amazon: An Integrated Biological and Cultural Approach to the Behavior and Ecology of Human Predation.* Ithaca: Cornell University Doctoral Dissertation.

____, 1986a. "Amazon Ecology and Adaptation," *Annual Review of Anthropology* 15:67-97.

____, 1986b. "La Caceria de los Yekuana Bajo una Perspectiva Ecologica," *Montalban* 17:5-27.

____, 1987. "Cultural Ecology and Environmental Education," *Journal of Environmental Education* 19(1):31-42.

____, ed., 1992. *Human Ecology in Amazonia: Frontiers for Applied Anthropology*, in preparation.

Stearman, Allyn Maclean, 1990. "The Effects of Settler Incursion on Fish and Game Resources of the Yuqui, a Native Amazonian Society of Eastern Bolivia," *Human Organization* 49(4):373-385.

Stone, Roger D., 1985. *Dreams of Amazonia.* NY: Penguin Books.

Taylor, Kenneth I., 1974. *Sanuma fauna: Prohibitions and classification.* Caracas: Fund La Salle de Ciencias Naturales Instituto Caribe de Antropologia y Sociologia Monografia No. 18.

Tucker, Richard P., 1990. "Five Hundred Years of Tropical Forest Exploitation," In *Lessons of the Rainforest*, Suzanne Head and Robert Heinzman, eds., San Francisco: Sierra Club Books, pp. 39-52.

Uhl, Christopher, *et al.*, 1989. "Disturbance and Regeneration in Amazonia," *The Ecologist* 19(6):235-240.

Vecsey, Christopher, and Robert W. Venables, eds., 1980. *American Indian Environments: Ecological Issues in Native American History.* Syracuse: Syracuse University Press.

Wilson, Edward O., 1989. "Threats to Biodiversity," *Scientific American* 261(3):108-116.

Exploitation and Management of the Surinam Forests
1600-1975

Peter Boomgaard
Free University


In 1975, the year that Surinam became an independent nation, slightly under
150,000 of its total surface area of somewhat more than 160,000 km$^2$ was covered
by forests. Ninety percent of this forest area consisted of mesophytic tropical rain
forest (Bruijning 1977, 78-80). This makes Suriname one of the few countries in
the modern world with a surface area that is largely taken up by "primary"
(climax) forest.

In this paper, I want to trace the reasons behind this extraordinarily high proportion of forest cover, in a country thas has been a plantation economy for centuries.
As a rule, plantation economies do not have a good reputation regarding the
preservation of forests, and I want to find out why Surinam seems to be an exception to this rule. We will be looking at the usual sources of destruction of wooded
areas, such as land-clearing for subsistence agriculture and the laying out of plantations, timber felling for fuel and construction, and at export of timber and other
forest products. Government policy will be another topic. Finally, the role of
Western enterprise will be discussed.

Around 1600, the English and the Dutch began to arrive at the "wild coast". In
1650, the British founded a colony here, which lasted until 1667, when it was
conquered by the Dutch. With some short interruptions, Surinam was a Dutch
"possession" between 1667 and 1975. It is this proto-colonial and colonial period
which will be dealt with here.


**Prelude to Possession (1600-1667)**

Although a lasting Dutch presence would not be established until 1667, Dutch
traders had already been visiting the coast of Guiana for many decades. During
these early encounters, wood was mentioned as a commercial product of some
importance. When David Pietersz de Vries arrived here in 1634, he came across a
merchant from Flushing who, during a four-month stay at the mouth of the river
Commewijne, had taken in 30 tons of snakewood (*Piratinera guianensis*). De Vries
himself had arrived too early for the "new harvest" of snakewood, which would

252

not start until the onset of the rains in November. The Arawak (Indians) living near the mouth of the Marowijne river promised him a ship full if he returned after a year (Colenbrander 1911, 203/6).

This brief glimpse at one of the earlier European sources on the "proto-colonial" Surinam timber trade seems to teach us 2 things. In the first place the Indians were willing to fell the required amount of timber but they did not have large stocks at their disposal. This seems to suggest that a large-scale timber trade was not one of the customary features of Arawak commerce. In the second place, they had to wait for the rains before they could provide the foreign merchants with a fresh load of timber. As far as I can see, this can only mean that snakewood grew only in upland areas, from where it had to be floated down-river, which was only possible after the onset of the rains. This form of timber transport would remain in use up to the present.

Snakewood remained an export commodity during the 18th century, but already at the end of the century it was becoming scarce in the inhabited parts of the country. At the end the 19th century, new stands of snakewood must have been discovered because it was exported in considerable quantities. In 1975 it had again become rather rare, at least in those parts of the forests that were accessible to commercial exploitation.[1]

Other commercial products from the forests, mentioned by early Northwestern European merchants, were gums, honey, wax, beefwood (*Manilkara bidentata*), "redwood" (probably *Guilandina* spp. ["brazil"] and *Haematoxylon* spp.) and fustic (*Chlorophora tinctoria*), a dyewood. Some of these products were still being exported during the 18th century.[2] In the 18th century, *Surinam quassia* or bitterwood (*Quassia amara*) was mentioned as a "new" export commodity, employed by physicians as a febrifuge. From time to time it was exported in considerable quantities.[3]

It is highly unlikely that the gathering or cutting of any of these products did much environmental damage. This observation also applies to the conuco agriculture (slash-and-burn) and the gathering and hunting activities of the local Indians, predominantly Arawak and Carib (Watts 1987, 41-70).

Potentially much more damaging was the cutting of timber for export to Barbados in the 1660s. Around 1625, Barbados had been settled by the British, who had started to grow tobacco and cotton, but had switched to sugar shortly after 1640. The success of sugarcane cultivation led to a rapid depletion of the forested areas, and fairly soon timber had to be imported for the construction of houses in Bridgetown (Watts 1987, 156-205). As Surinam also had been recently settled by the British, its proximity to Barbados and its virtually untapped resources of excellent timber led to an incipient export trade in timber. After 1667, when the Dutch took over Surinam, this trade came to an end, and henceforth Barbados had to import its timber from New England, which removed a potential threat to the forests of Surinam. From then on, developments inside the colony itself largely determined the fate of the forests. We will now turn to a description of these developments.

## A Buoyant Plantation Economy (1667-1775)

As in Barbados, sugar took root in Surinam after a short experimental tobacco phase. The technology of sugar plantations was imported from Brazil and Barbados, where Dutch, Portuguese Jewish, and British settlers had experimented with this form of production. The first settlers founded their plantations upstream along the rivers of Northeast Surinam. The downstream areas, although of course nearer to the coast and therefore better situated for export to Europe, were too wet to be settled without major technological improvements in the laying out of a plantation. Fear of British, French, and Spanish naval raids was another reason for plantation owners to avoid the coast.

After 1700, however, some planters started to establish themselves in the downstream areas along the rivers. The required technology for the permanent drainage of these wet areas was imported from Holland. There, agriculture on "polders", or areas surrounded by dikes and permanently drained by windmills, had been a feature of the landscape for centuries. With some adaptations (e.g. no windmills, and a more elaborate system of drainage ditches), this polder model could be used for the plantations in the swampy areas. Between 1700 and 1740, some plantations were still being laid out in the inland areas, but after 1740 all new plantations - now mostly coffee instead of sugar - were established downstream. Between 1740 and 1800, some of the old plantations were abandoned.

The gradual shift from inland to coastal areas can be attributed to various motives. In the first place, the inland soils were less fertile than the original settlers had expected, and certainly less fertile than the downstream soils. In the second place, the threat, posed by European naval powers to the coastal establishments of the plantation economy turned out to be less formidable than the dangers from the hinterland. Inland plantations were raided by Indians and, as time went by, also by maroons (runaway slaves), whose numbers increased in proportion to the growth of the plantation area (Stipriaan 1991).

Planters, working on tiny islands of clearings in a sea of forests, had very few incentives to use their wooded environment sparingly. Large areas were cleared for sugar and coffee, and only some of the timber felled in this way would be used for the buildings of the plantation. In Surinam, as in most countries with rain forest vegetation, the variety of species per hectare is high and only a small number of these species was used for construction purposes. Around 1900, it was calculated that 1 hectare of forest did not contain more than 80 trees of over 40 cm in diameter, belonging to more than 40 different species, of which perhaps not more than 4 had any commercial value. In terms of volume, therefore, a hectare would not yield more than 10 m$^3$ squared timber with commercial value, compared to 100 m$^3$ in Germany around the same time (Berkhout 1903, 19-21).[4]

In publications and archival sources dating from the 18th century, particularly in probate inventories, we come across a fairly limited number of species which were used for construction purposes. Often mentioned were basralocus (*Dicorynia guianensis* sv *paraensis*), brownheart (*Vouacapoua americana*), bullet-tree wood or beefwood (*Manilkara bidentata*), cedar (*Cedrela odorata*), greenheart (*Tabebuia serratifolia*), kopie or kabokali/kabukalli (*Goupia glabra*), kwarie or iteballi (*Vochysia* spp.), locus(t) or kawanari (*Hymenaea courbaril*), purpleheart (*Peltogyne* spp.), wane or determa (*Ocotea rubra*), and wassie wassie (*Qualea* spp.).[5] The pina palmtree (*Euterpe oleracea*) was used for the construction of dwellings for the slaves.[6] It might surprise the modern silviculturist that some of these trees were used for construction purposes, but one has to remember that planters often had to make do with locally available material.

Although some of these species of timber could resist Surinam's wet and hot climate and the concomitant insects for a long time, all plantation buildings, mills, vessels and carts had to be renewed regularly. This, of course, meant that new patches of forest had to be cleared from time to time. Only a limited amount of non-timber obtained by clearing could be set aside for fuel, employed for the production of sugar. The remainder was left to be burned (e.g. Herlein 1718, 67; Blom 1786, 21).

Sugarcane could be harvested from the same cane stool for a number of years (ratooning). In Surinam in the 18th century, ratooning up to and including the fourth crop was considered to be optimal, and only ca. 20 percent of all crops consisted of fifth ratoons and over (Stipriaan 1989, 100). This practice implies that plantation owners had to clear part of their still forested areas for additional cane fields after 4 years. The original cane fields were then left to lie fallow. From then on, the cane could rotate between fields of which the original forest cover had already been cleared. Nevertheless, many a planter had to resort to additional clearings, because the fallow period had been too short owing to exhaustion of the soil. This may have been largely the result of the insufficient application of manure during the cane-growing period.

Moreover, if the plantation owner could dispose of sufficient capital or credit, he would be able to buy more slaves, and could, therefore, expand the area at any one time under cane. In times of high and increasing sugar prices most owners would be tempted to do so. Taken together, rotation and expansion could exhaust the plantation's timber supply within a few decades. If we add the recurrent demand for fuel and timber for maintenance mentioned above, it will be clear that fairly soon after the laying out of their estates, most planters had to look for sources of wood outside their plantations. There is no evidence that planters set aside part of their estates for sustainable wood production, as was done in Saint-Domingue (Haiti) in the 18th century (Debien 1952, 43). For the 19th century we have an explicit denial, by Van Sijpesteyn, that planters used part of their arable lands to plant trees (Sijpesteyn 1851, 18).

## Production of Timber and Firewood

The quest for timber and fuel from other sources than the plantation itself led to the acquisition by planters of additional lands in the neighbourhood of their estates. These timber estates[7] (Dutch: outgronden) had only 1 purpose, namely the production of timber and firewood for the export crop plantations. If no suitable lands could be found in the immediate neighbourhood, planters acquired pieces of forest further away, preferably along the same river. These timber estates functioned as annexes of the "mother"-plantation, and had no economic life of their own (Bruijning 1973, 481).

Soon, however, some people started to specialize in the production of timber and firewood on independent timber estates. Some of these estates were converted export crop plantations, particularly sugar plantations in the upland areas that had stopped producing cane when the downstream plantations with their higher yields per hectare came into production.

A good example is the plantation Hanover, located upstream along the Para River. The plantation was mentioned for the first time in 1737, with a surface area of 2,200 akkers or 940 hectares, but it is probably older. Hanover started out as a sugar plantation, but switched to the production of timber ca. 1750, during a period of low sugar prices. Around 1760, the estate had been almost entirely stripped of its forest cover, and the owners had to acquire an additional 1,000 akkers in the neighbourhood. In an inventory dated 1784, Hanover was reported to have a total surface area of 5,422 akkers (2,320 ha.), so that apparently yet another expansion had taken place, probably for the same reasons. There are several indications that it was not a profitable enterprise. From time to time, the owners tried to diversify their production and experimented with various export crops, such as coffee in 1772, cotton in 1784, indigo in 1829, and tobacco in 1845. None of these experiments seem to have been successful. Somewhere in the late 19th century, Hanover was left by its owners, and taken over by its former slaves. In 1903, it was still being exploited by them as a timber estate.[8]

Other timber estates, particularly the ones in the downstream areas, started out as such. Much more archival research will have to be done before we can establish the ratio between converted and original timber estates. It is also unclear when the timber estates of any description began to appear in considerable numbers, although it can be said that it is highly unlikely that this occurred before 1750. Pistorius mentions the existence of 10 to 12 houtzagerijen, which in all probability should be read as timber estates, in 1761 (Pistorius 1763, 26). At the end of the 18th century, when the total number of plantations reached a peak of ca. 600 (the timber estates included), there were somewhat under a 100 timber estates.[9]

As to the technology of timber production we can be brief, owing to a lack of detailed information. We do know that sawmills were introduced at an early stage, the first one probably in 1677. In 1716, a sawmill was erected on the government owned estate Andresa (Oudschans Dentz 1949, 13, 20). Oddly enough, Blom seems to suggest that all timber from the estates was sawn by hand (Blom 1786, 340). So either the saw-mills mentioned had remained a rarity or they had gone out of business.

There are no indications that timber estates ever planted trees when they had felled the original stands (Blom 1786, 345; Sijpesteyn 1851, 18). As was the case with Hanover, they just added another piece of virgin soil to the estate, which then was stripped of its forest cover. If no expansion was possible or desirable, the estate was abandoned.

The timber estates were not the only ones involved in lumbering. They had competition from the maroons, who in Surinam are called bosnegers (bush-negros). During the second half of the 18th century they numbered a few thousand, probably between 5,000 and 10,000 at the end of the century. Felling timber for sale to the plantations or to Paramaribo was a sideline for many of them, after the conclusion of peace treaties with the colonial government (1760, 1762, 1767). Among the annual gifts to the bush-negros, specified by the peace treaties, we encounter steel axes and carpenter's tools, which facilitated the felling of trees and enabled them to produce squared beams. The treaties guaranteed them free passage to the coast, and henceforth they floated their timber down the river on rafts.[10]

How much damage was being done to the forests by the maroons, the timber estates, the plantations, and the town population? Exact figures are lacking, but we can safely assume that the environmental impact of their activities was limited. A few thousand maroons, a few thousand Indians, 60,000 slaves and under 5,000 Europeans, free blacks and mulatto's, and about 600 plantations comprising some 250,000 hectares cannot have had all that much influence.[11] Large forest fires, lasting for a month or more, such as happened in 1746, 1769, 1779, 1797 and 1825, may have destroyed more forest vegetation than did the people living there (Oudschans Dentz 1949, 24/7; Stipriaan 1991, Ch. 3).

## The Long 19th-Century Retreat (1775-1900)

Around 1775, the plantation economy of Surinam had been at the peak of it's prosperity. Thereafter, high debts forced many a planter into bankruptcy. After the turn of the century, the number of plantations declined, as did the total surface area under plantation agriculture. This development was speeded up by the abolition of the slave trade in 1806. As the slave population did not reproduce itself sufficiently, the number of slaves decreased gradually. This is not to say that all plantations fared badly. Cotton - a newcomer - and sugar did rather well from time to time, and although the number of sugar plantations declined, their average surface area increased, which must be interpreted as a sign of economic vigour (Stipriaan 1991). Nevertheless, as a whole, the plantation area was shrinking continuously, as was the population. The final blow to the system came with the abolition of slavery, in 1863, followed by an "apprenticeship" period, when slaves were still under the obligation to work on the plantations albeit now for wages, ending in 1873.

The timber estates shared in the general downward trend. The ca. 90 estates in existence around 1800 had declined to 64 in 1827, dwindling to ca. 20 in the 1850s and 60s (Koloniaal Verslag). This development may have boosted the lumbering activities of the bush-negros, who had come to depend more on their timber trade anyway, because in 1849 government had stopped sending the annual gifts which had been part of the 18th-century peace treaties (Beet & Sterman 1981, 14/5). Given the restricted nature of demand, this increase in production must have been limited.

Not all demand, however, was generated locally. Throughout the 19th century, timber in various forms was being exported to the Netherlands and the Caribbean islands, most of it as round logs or squared beams. However, this never amounted to more than 2 or 4,000 m³ per year, with a monetary value that fluctuated between 0.5 and 1.5 percent of total export earnings (Koloniaal Verslag).

At the end of the century, the Surinam forests found themselves in a paradoxical situation. On the one hand, most of the forests were not accessible to commercial exploitation, owing to lack of roads or navigable rivers. Only a few thousand Indians used these forests for their fairly limited wants. In some areas, where settlers had moved away, forest cover was slowly encroaching on stretches of savanna vegetation (Berkhout 1903, 9). On the other hand, most accessible forests had been ruined by bush-negros or timber estate owners, who could not be bothered with sustainable exploitation. Surinam, therefore, had treated the fringes of its forests badly, but it had no part in the global deforestation of the 19th century (Tucker & Richards 1983).

**Government and Western Enterprise (1900-1975)**

The role of government in forest management and exploitation had been negligible up to the end of the 19th century. Until 1755, lands for (timber) estates had been given out for free, and for an annual token recognition thereafter. In 1820, these rules were changed, in view of the fact that so much land had been left. From then on, abandoned lands would revert to the status of domain lands. In 1873, at the end of the apprenticeship period, government switched from a policy of alienation to one of concessions of forest lands (Plasschaert 1910, 69/70). At the turn of the century, some 125,000 hectares had been given out in this way (Berkhout 1903, 11).

During the 19th century, various officials in the Netherlands and Surinam had attempted to stimulate the production of durable timber, to be used in the mother country for shipbuilding, sleepers for railways, etc., largely because oak and other species of hardwood were getting more expensive. In 1836, it was decided to start an experiment with government exploitation of a forest area to this end. For this purpose, a forest range was established at the government owned estate Andresa, on the Coppename River. The attempt failed because much of the timber shipped to the Netherlands left much to be desired and was, moreover, more expensive than oak. In 1842, the Navy stopped buying Andresa timber, and although other outlets were found, and a state-of-the-art steam sawmill had been installed in 1846, the forest range was given up in 1849. Throughout its 13-year existence it had been run at a loss (Kessler 1989).

Small wonder that the next attempt at government exploitation had to wait until 1904, when it was combined with the creation of an experimental Forest Service. The combination of government exploitation and government management was copied from the Java teak forests in the Netherlands Indies (now Indonesia), and the first foresters in Surinam came all from the experienced Netherlands Indies

Forest Service (Plasschaert 1910, 109). Nevertheless, government exploitation was stopped in 1910 owing to continuous financial losses. A skeleton Forest Service was, however, retained (during many years 1 forester!), in order to continue the silvicultural research that the Service had initiated, and to supervise the exploitation of the forests by private entrepreneurs. Since 1916, a number of enterprises from the USA, the Netherlands and Belgium had been given large concessions, sometimes measuring hundreds of thousands of hectares. It seemed as if the Surinam forests had been finally discovered by Western capital, and that the dawn of large-scale forest exploitation had arrived. The 1921 recession, however, put an end to these hopes. The enterprises had to limit or give up their operations, and the Forest Service was abolished in 1925 owing to budget cuts. Although better times returned in 1926, measured in terms of timber exports, the 1929 crisis followed too soon for a revival of the Forest Service (Berkhout 1917; Pfeiffer 1929, 7/8; Traa 1946, 116-122).

In the mean time, Surinam had been lucky with its forests for once, when balata--a rubber-like substance from the bullet tree--was discovered in exploitable quantities. Exploitation started ca. 1890 and picked up speed after 1907, with record export returns of over 4 million guilders in 1911 and 1913, and fairly high export levels up to and including 1931. It was insufficient to save or revive the Forest Service (Traa 1946, 128-132).

That had to wait until after World War II. In 1947, the future of Surinam's forests was determined for a long period to come by 2 events, namely the resurrection of the Forest Service and the arrival of the Dutch Bruynzeel Company, which received an initial 500,000 hectares in concession. In 1973, a total of 1.7 million hectares had been given out, with a production of 168,000 m$^3$, of which 56,337 m$^3$ was exported. This was approximately twice the quantity exported in 1950, when Bruynzeel had recently started its operations, and a quantum leap compared to the 4,000 m$^3$ of 1900 (Bruijning 1977, 80/3, 104/5). The attack on the Surinam forests had been finally launched in earnest.

## Conclusions

Summing up this short essay on the management and exploitation of the Surinam forests, it can be said that, at least until the 1950s, exploitation was limited to the accessible fringes of the forests, and management was virtually absent.

This state of affairs can be largely ascribed to a lack of people. Settlements of planters were concentrated in areas near the coast and along the rivers, and the vast hinterlands were left to the often hostile Indians and maroons. Fear for these people was not the only reason that the planters did not venture into the inland areas. Limited soil fertility and rapids in the rivers (which made transportation onerous) also caused them to stay near the coast. Even in the coastal areas not all suitable lands were taken up by plantations for lack of roads or rivers.

Lack of expertise in forestry was another factor. Knowledge of the forests was limited, and methods of wood cutting were primitive, which made for low quality timber. As Surinam was already heavily subsidized by the mother country since the beginning of the 19th century, money for a permanent professional Forest Service was not available until 1947.

## Notes

1. Snakewood is used for cabinet-making and marquetry; data are taken from Blom 1786, 340; Leon 1791, 76; Berkhout 1903, 58/9; Voort 1973, 238; Bruijning 1977, 287.

2. These references are taken from Leon 1791, 76; Berkhout 1903, 60; Harlow 1925, 135; Record & Hess 1944, 274-279; Voort 1973, 238; Bruijning 1977, 286/7.

3. Data on quassia are from Record & Hess 1944, 512; Oudschans Dentz 1949, 23/6; Voort 1973, 238.

4. These figures depend, of course, heavily on what is considered as commercially valuable wood. In the course of the 20th century, the number of species of wood that were regarded as valuable increased, which naturally led to an increase of the estimated average yield per hectare (Gonggryp & Burger 1948, 82-90). Oddly enough, in more recent research we find lower estimates, comparable to those quoted ca. 1900 (Boerboom 1965, 3). As an approximation of 18th-century ideas of what was valuable, the 1900 figures are probably nearer to the mark than the 1948 estimates. Planters, therefore, had only limited use for the trees they had to cut before they could start planting sugar or coffee.

5. Probate inventories of 18th-century plantations can be found in the Algemeen Rijks Archief (General State Archives; The Hague; ARA for short), 1st section, Bestuursarchieven Suriname, in the collection Oud Notariële Archieven (ONA for short). Details on the properties of these trees and their use in the 18th century can be found in Pistorius 1763, 51-3, Hartsinck 1770, 66-85, Fermin 1785, 218-228, and Blom 1786, 335-345. For the identification of wood species mentioned in the archives, I have used Bruijning 1977, 287, Gonggryp & Burger 1948, and Record & Hess 1944.

6. See e.g. Blom 1786, 9/10, Gonggryp & Burger 1948, 22, and Bruijning 1973, 133.

262 Changing Tropical Forests

7. I have avoided the term plantation for these lands because it carries the connotation of sustainable wood production, for which there is no evidence.

8. Data on Hanover can be found in ARA, ONA, 192, 211, 235, 703, 263, 832; ARA, 1st section, Bestuursarchieven Suriname, Onbeheerde Boedels & Wezen 1828-1876, 1949; Gemeente Archief Amsterdam (Municipal Archives Amsterdam), Particuliere Archieven 600, 544; Quintus Bosz 1954, 429; Hostmann 1850; Berkhout 1903, 9.

9. Of the 591 plantations in 1791, 139 estates were timber estates and foodcrop estates (Dutch: kostgronden) (Leon 1791, 5). In 1827, there were 64 timber estates and 36 foodcrop estates (together 100), when the total number of plantations was 443 (ARA, Archief Ministerie van Koloniën 1813-1849, 791: Verbaal 25.3.1831, No. 41). If the number of timber estates had dropped in proportion to the total number of plantations, there must have been ca. 90 timber estates in 1791.

10. On maroons, peace treaties, and lumbering see Hartsinck 1770, 780-806; Beet & Sterman 1981, Price 1983, Hoogbergen 1985, and Groot 1986.

11. Probably, I have overstated the number of plantations and hectares somewhat. Stipriaan, who has carefully scanned the archival records for data on plantations, found ca. 400 sugar and coffee plantations in the period 1770-1795, with total holdings in 1770 of 161,300 hectares (Stipriaan 1991, tables 1 D and 2 C). However, Leon (1791, 5) reports 591 plantations in 1791 and Oudschans Dentz (1949, 33) 614 in 1793. These figures may include plantations that already had gone out of business. Stipriaan's figures do not include the timber and foodcrop estates, which, according to Leon, totalled 139 in 1791. If their average holdings were similar to those of the sugar and coffee plantations, we may add ca. 55,000 ha. to Stipriaan's total, which leads to a grand total of ca. 215,000 ha. For population figures see Lier 1971, 22/3; Stipriaan 1991, Ch. 9.

# References

Beet, C. de, & M. Sterman, *People in between; the Matawai maroons of Suriname*. Unpublished Ph.D. thesis, Rijks Universiteit, Utrecht, 1981

Berkhout, A.H., *Rapport over de Surinaamsche bosschen*. 's- Gravenhage: Mouton & Co., 1903

Berkhout, A.H., *Rapport over het Surinaamsche Boschwezen (...)*. 's-Gravenhage: Algemeene Landsdrukkerij, 1917

Blom, A., *Verhandeling over den landbouw in de colonie Suriname*. Haarlem: Van der Aa, 1786

Boerboom, J.H.A., *De natuurlijke regeneratie van het Surinaamse mesofytische bos na uitkap*. Wageningen, 1965

Bruijning, C.F.A., J. Voorhoeve & W. Gordijn (eds.), *Encyclopedie van Suriname*. Amsterdam/Brussel: Elsevier, 1977

Colenbrander, H.T. (ed.), *Korte historiael ende journaels aenteyckeninge van (...) David Pietersz de Vries*. (Werken Linschoten Vereeniging, No. 3) 's-Gravenhage: Nijhoff, 1911

Debien, G., *Une plantation de Saint-Domingue; la sucrerie Galbaud du Fort (1690-1802)*. (Notes d'Histoire Coloniale, I) s.l., s.a. [1952]

Fermin, P., *Nieuwe algemeene, historische, geografische, en natuurkundige beschryving van de colonie van Surinaame (...)*. Vol. I. Amsterdam: Roos & Zn, 1785

Gonggryp, J.W., & D. Burger Hzn., *Bosbouwkundige studiën over Suriname*. Wageningen: Veenman, 1948

Groot, S.W. de, A comparison between the history of maroon communities in Surinam and Jamaica, in G. Heuman (ed.), *Out of the house of bondage*. London, 1986, pp. 173-184

Harlow, V.T. (ed.), *Colonising expeditions to the West Indies and Guiana, 1623-1667*. (Works Hakluyt Society, 2nd series, No. LVI) London, 1925

Hartsinck, J.J., *Beschryving van Guiana of de wilde kust in Zuid-Amerika (...)*. Vol. I. Amsterdam: Tielenburg, 1770

Herlein, J.D., *Beschryvinge van de volk-plantinge Zuriname*. Leeuwarden, 1718

Hoogbergen, W.S.M., *De Boni-oorlogen, 1757-1860; marronage en guerilla [sic] in Oost-Suriname*. Unpublished Ph.D. thesis, Rijks Universiteit, Utrecht, 1985

Hostmann, F.W., *Over de beschaving van negers in Amerika door kolonisatie met Europeanen (...)*. II Volumes. Amsterdam, 1850

Kessler, G., De gouvernementshoutvelling aan de Coppename 1836-1849. Een onderzoek naar de bosontginning in het negentiende-eeuwse Suriname, unpublished paper, Vrije Universiteit, Amsterdam, 1989

*Koloniaal Verslag*. 's-Gravenhage, 1849-1930 (the official annual colonial report of the minister of colonies to the second chamber of parliament)

Leon, M.P. de, S.H. de la Parra, J. de la Parra, D. de I. Cohen Nassy & S.W. Brandon, *Geschiedenis der kolonie van Suriname (...)*. Amsterdam/Harlingen, 1791 (1st ed., in French, 1788)

Lier, R. van, *Samenleving in een grensgebied. Een sociaal-historische studie van Suriname*. Deventer: Van Loghum Slaterus, 1971

Oudschans Dentz, F., *Geschiedkundige tijdtafel van Suriname*. Amsterdam: De Bussy, 1949

Pfeiffer, J.Ph., *Bevordering van den houtuitvoer van Suriname (...)*. 's-Gravenhage: Algemeene Landsdrukkerij, 1929

Pistorius, T., *Korte en zakelyke beschryvinge van de colonie van Zuriname (...)*. Amsterdam: Crajenschot, 1763

264    *Changing Tropical Forests*

Plasschaert, E.K., *Der Forstbetrieb in Surinam*. München: Kastner & Callwey, s.a. [c. 1910]

Price, R., *To slay the Hydra; Dutch colonial perspectives on the Saramaka wars*. Ann Arbor: Michigan U.P., 1983

Quintus Bosz, A.J.A., *Drie eeuwen grondpolitiek in Suriname. Een historische studie van de achtergrond en de ontwikkeling van de Surinaamse rechten op de grond*. Assen: Van Gorcum & Co., 1954

Record, S.J., & R.W.Hess, *Timbers of the New World*. New Haven, etc.: Yale University Press, 1944 (first published 1943)

Stipriaan, A. van, The Surinam rat race: labour and technology on sugar plantations, 1750-1900, *Nieuwe West-Indische Gids/New West Indian Guide*, 63, 1/2 (1989), 94-117

Stipriaan, A. van, *Overleven in een plantage economie, Suriname 1750-1863*. Unpublished Ph.D thesis, Vrije Universiteit, Amsterdam, 1991 (forthcoming)

Sijpesteyn, C.A. van, *Het Surinaamsche hout, bruikbaar en voordeelig bij den aanleg van spoorwegen (...)*. Breda, 1851

Traa, A. van, *Suriname 1900-1940*. Deventer: Van Hoeve, 1946

Tucker, R.P., & J.F. Richards (eds.), *Global deforestation and the nineteenth-century world economy*. Durham: Duke U.P., 1983

Voort, J.P. van de, *De Westindische plantages van 1720 tot 1795, financiën en handel*. Unpublished Ph.D. thesis, Katholieke Universiteit, Nijmegen, 1973

Watts, D., *The West Indies: patterns of development, culture and environmental change since 1492*. Cambridge, etc.: Cambridge University Press, 1987

Foreign Investors, Timber Extraction, and Forest Depletion
in Central America Before 1941

Richard P. Tucker
Oakland University and University of Michigan

The forest cover of Central America has receded dangerously in the twentieth century, and especially since 1950. The major cause has been the expansion of agriculture: plantation crops for export markets, beef for both domestic and foreign consumption, and mixed cropping for the subsistence needs of a rapidly expanding rural population, peasants many of whom have been dispossessed from their traditional homes.

However, commercial timber operations have also encroached on the region's forest resources, turning trees into commodities for both domestic and international trade. This element of resource extraction was first linked with the world economy in the early sixteenth century, in the first years after the Spanish conquest. Thereafter, until well after 1950, nearly all commercial logging was merely high-grading, especially in the moist lowland forests, degrading both their biotic diversity and their silvicultural quality. In the pine forests a very different pattern evolved, but it was equally wasteful of the resource, for both local use and export.

## Hardwood Logging and Forest Degradation

### The Colonial Dyewood Trade

On the Bahia coast of Brazil, the first Portuguese intruders in the early sixteenth century began exporting Brazil wood to Europe; its red dye was highly valued for Western Europe's rapidly expanding woolen cloth industry.[1] By the end of the century, selective felling of these species reduced them to insignificance along coasts and riverways. But constrained by the primitive timber transport methods then available, loggers rarely penetrated far from waterways, and the full diversity of the deeper forest remained.

During the same era European dyewood hunters in the Caribbean discovered logwood, an equally valuable source of red dyes. A hardwood of the lowland moist forest, logwood grew prolifically along river banks from Campeche and Yucatan in Mexico through coastal Belize to the Miskito Coast of Honduras and Nicaragua.

British privateers in the seventeenth century, despite the Spanish navy's attempts to interdict them, generated a large-scale export of logwood for the clothing mills of Europe, an export which lasted 300 years.[2] The logwood harvesters' anarchic ways served them well in the turbulent political conditions which prevailed until 1670, when Spain and Britain agreed by treaty that Belize would become the British possession of British Honduras.[3]

The work was itinerant, difficult and very crude. Logs could be floated to the coast only in the rainy season. And since logwood was desired for the dye extracted rather than for lumber, the size of the tree did not greatly matter. "Baymen," or British pirates-turned-loggers, cut all trees in their working area, large and small. One early traveller observed, "During the wet season, the land where the logwood grows is so overflowed, that they step from their beds into water perhaps 2 feet deep, and continue standing in the wet all day, till they go to bed again; but nevertheless account it the best season in the year for doing a good day's labour in. ... When a tree is so thick that after it is logged, it remains still too great a burthen for 1 man, we blow it up with gunpowder."[4] Under that regime, easily accessible stands of logwood quickly declined.

After the treaty of 1670 stabilized the political situation, the Baymen began using oxen for hauling logs to the rivers, as well as African slaves from the British Caribbean islands for labor, thereby expanding the reach of their operations. These were the only changes in extraction techniques over a 300-year logwood era, but they enabled the loggers to respond to expanding demand in Europe in the nineteenth century. Central America's logwood exports rose from 700 tons in 1800 to 35,000 tons in 1896. But by 1900 chemical dyes came to dominate the market throughout Europe, and the pressure on Caribbean logwood stands vanished. By 1913 only 3,600 tons were cut, and thereafter amounts were negligible.[5]

**Brown Gold: the Mahogany Trade**

In the early seventeenth century, several species of cabinetwoods from the Caribbean coastal lowlands came into demand on world markets. Mahogany soon dominated the trade, but a few other species, notably Spanish cedar, were also selectively cut.[6] In the seventeenth century, mahogany was used largely for shipbuilding. But in the early eighteenth century mahogany came into major demand throughout Europe for furniture and cabinetry. By 1800 the scale of mahogany extraction became far greater than the dyewood trade had ever been.

Baymen searched the entire coast from Campeche onward for cabinetwoods, competing often violently with each other. In a setting of colonial rivalry between

British and Spanish, they worked the forests without bothering to establish formal rights.[7] Their methods were primitive, and they could not operate far inland from the short coastal rivers. Waterside stands of mahogany and cedar became severely depleted over the years. By the mid-nineteenth century the mahogany business, which had long dominated the economy of British Honduras, went into severe recession. Forests at any greater distance from water remained largely untouched.[8] Only a technological transformation in timber transport and milling could extend the exploitation of those forests; that would come in the 1920s, in conjunction with the beginning of formal silviculture.

The mainland was not the only mahogany exporter to Europe. Through the nineteenth century the larger Caribbean islands were significant competitors in the trade. Cuba's mahogany forests were cleared for the expansion of plantation sugar. Then from about 1900 hardwoods from West Africa moved onto European markets, as British, French, and German firms turned to other sources of cabinet woods in an increasingly complex international marketing system, and North American firms rapidly expanded their control of Caribbean basin mahogany outside the small British colonies.

U.S. timber companies first became a major force on the Caribbean coast during the last half of the nineteenth century. Their operations became the major pressure on the lowland forests outside the British possessions, in an intricate interweaving of the U.S. and Central American timber economies. Even in British Honduras they made their presence felt after 1900. American firms entered that market just after 1900. Their arrival resulted in a short-lived boom for Belize, whose exports rose from 6 million board feet in 1899 to over 16 million in 1914, with the additional factor of steadily rising prices.[9]

Until the 1920s the Yankee hardwood purchasers were for the most part not lumbermen but shippers, middlemen between the forest and furniture manufacturers from New Orleans to Boston. Their network of investment and trade traced back to the colonial New England entrepreneurs in the Caribbean who had financed and marketed sugar and other Caribbean commodities. Throughout the nineteenth century they maintained their offices in Havana and other ports on the islands. They then began opening branch offices on the Central American mainland after the U.S. took control of Texas and California in mid-century.[10]

Until about 1890 no U.S. logging firm actually operated on Latin soil.[11] Thereafter the leading firm was the George D. Emery Company of Boston, which in 1894 bought the only 2 existing timber concessions of coastal Nicaragua. Casa Emery exported about 1,000 mahogany and Spanish cedar logs monthly to Boston. Their work force of 1,300 was mostly Miskito, but some 100 Americans were imported for the more highly skilled functions.[12] The Emery-Williams operation was a microcosm of the operations of U.S. commercial power in the Caribbean basin, revealing the link between U.S. capital and tropical natural resources exploitation. Two major New York banks, Brown Brothers and J. & W. Seligman, became involved in its financing, in an extension of their penetration of Cuban sugar in those years. In 1909 the new President Zelaya of Nicaragua annulled Emery's claim, and Brown Brothers bought the concession. The U.S. government, by then deeply embroiled in Nicaraguan politics, took up the claim; in September 1909 the Nicaraguan government agreed to pay Emery $600,000 in 5 installments. But in October Zelaya was overthrown. In 1910 Brown bought Emery's U.S. and Nicaraguan interests, falling heir to $550,000 in remaining payments. A year later the bank sold the operation to the Ichabod Williams family of New York, which specialized in tropical hardwoods throughout its long career from 1838 to 1966. Williams continued the lumbering operation for some years longer.[13]

After the war, the 1920s saw a vast expansion in international commodity trade, including tropical timber. 1927 was the year of greatest mahogany exports from Belize. Through the 1920s in its northern mahogany forests, according to the leading authorities of the time, "logging. . .is generally carried on in a haphazard manner. . . . The recent strong market demand [has] led to wholesale slaughter of immature timber."[14] The American forester Tom Gill corroborated this. Surveying the coastal forests in the late 1920s, he reported that there was little regeneration of mahogany, and that local foresters (such as they were) seemed passively complacent about the situation.

> Today. . .logging is a thoroughly primitive, unsatisfactory, and haphazard process. Methods are slow and wasteful, and because of the highly selective character of logging, the costs vary from merely high to prohibitive. . . . There is no steady flow of ties or logs or any other product. For that reason industry has been very loath to consider seriously so uncertain a source of supply.[15]

By 1930 the Great Depression hit, and world markets shrank severely. Timber cutting went into a global decline. In other words, the Depression temporarily saved the mahogany forests from desperate depletion before the capacity to manage them professionally could evolve; that was to begin in the wake of yet another world war. In Belize, in fact, the timber export depression lasted through the 1950s, until the forest service was able to develop a more advanced system of silviculture.[16]

**Technological Change in the Early Twentieth Century**

The evolution of timber harvesting in these decades also reflected several technical and managerial factors, which can only be alluded to in this paper. One was the beginnings of modern timber technology and silviculture. In the aftermath of World War I, the impact of the automotive industry was first felt in many forests of the world, including Central America. After the severe but brief immediate postwar depression, British and American firms made considerable new logging investment in the 1920s, centering on access to more remote forests. Motor-driven tractors appeared in the 1920s, along with skidders and log-wagons which were able to haul logs several miles from cutting sites to floatable rivers, and on hillier terrain than oxen were able to work. In 1933 the most progressive firm, Belize Estate and Produce Company, built the country's first modern sawmill, with major government financial support, in Belize City.[17] Finally, in the wake of World War II, power saws and heavy trucks and bulldozers appeared in force.

Another important dimension of the region's forest products industry was the beginning of local processing of timber. Mexico in the 1920s was the first Central American country to require exporters to process timber in the country before exporting it, thereby bringing more modern sawmilling and veneer-cutting technology and more jobs into the country and allowing less of the value added to escape. A plywood plant in Yucatan built in 1946 was perhaps the first of its kind in the region.[18] Local operators in other areas too began to produce mahogany furniture, but their efforts were largely stymied by the financial and organizational power of U.S. and European industry, in a classic case of dependent development. As Standley & Record noted in 1936, "there is a small but steadily increasing local industry producing mahogany lumber, mainly from inferior material unsuitable for export, but inefficiency of plant and limited shipping facilities have hitherto precluded it from competing to any material extent with the highly organized lumber manufacturing industry in the United States."[19]

The 1920s also saw the beginnings of silvicultural research, at the United Fruit Company's research center at Tela, Honduras, which specialized in the technical aspects of fruit culture. The laboratory included commercial tree plantations as well. These were largely the work of UFC's Director of Tropical Research, Dr. Vining Dunlap, who took personal interest in tree plantations. Its first plantings were made in 1927, but they were poorly maintained. A second round of plantations, established in 1942-43, were better maintained. By 1949 the nursery there had experimental intensive plantations of mahogany, Spanish cedar, primavera, and 1 exotic species, teak.[20] By 1960 UFC's Honduran teak plantations covered 2,000 acres, and it had begun similar teak plantations on the Pacific coast of Costa

Rica. But these were commercial ventures; they could hardly be expected to restore the forest biota which had been severely degraded throughout the Caribbean Basin.

**The Pine Forests of the Caribbean Littoral**

A very different dimension of timber extraction evolved in the pine forests which grew on drier land and poorer soils in various parts of Central America, both along the Miskito coast and in the interior highlands.[21] Long-established peasant communities, some dating back to pre-Columbian times, survived in close symbiosis with the pine forests. Any commercial pine exploitation would involve much more intricate relations with the local populace than in the thinly populated lowland rain forest.

Little is known about relations between subsistence life and commercial pine extraction until very recently, though the issue must have been significant as far back as early colonial times. The Spanish were actively interested in the pine belt, for several species of pine produced not only low-cost building materials for local use, but naval stores, just as did the pine forests of the Atlantic coast of the United States. The colonial headquarters of Peru, in dry coastal Lima and its shipbuilding center of Guayaquil, relied extensively on the Central American pine forests for shipbuilding and naval maintenance.[22] By the nineteenth century many stands of pine were depleted. Regrowth was inhibited by annual fires (the worldwide regulator and scourge of monospecies pine forests). And finally, in an ecological cycle entirely different from the mahogany cycle, these drier forests were gradually turned into grazing grounds for cattle, a frequent sign of biotic decline.[23]

Some pine forests were protected from commercial exploitation by their remoteness from transport arteries. Methods of logging were still primitive and wasteful. There is little evidence of export pine lumber production until well into this century. Belize again is a case in point. Despite its considerable pine reserves, little was logged, except for a bit exported westward. Most construction lumber used in Belize was yellow pine imported from the American Southeast, which gave a better finish and was available in convenient sizes. Rosin and turpentine might be marketable; but no one tried it commercially until recently.[24]

**Southern Pine Exports from the U.S.**

The exchange of primary products between Yankees and Ladinos--food and lumber moving southward, plantation crops moving northward--was a major feature of the dependency relationships of the region. As in other aspects of the linked international economy and its ecological consequences, we must take into account conditions of resource extraction and industrial economics in the United States.

In the aftermath of the U.S. Civil War's massive dislocation of economy and land use in the southern states, the southern lumber industry gradually rebuilt and expanded its scale. The forests of the New England and Great Lakes states were severely depleted by 1880, and the Mississippi Valley was entering an era of industrialization from Minneapolis and Chicago to New Orleans. From the cutover northern states, lumber firms moved into the southern pine and cypress belt which stretches 1,500 miles from Virginia to east Texas, in a burst of land speculation equalling anything happening in the western states in those years.[25] Southern pine lumber production reached one peak in 1909, when 20 billion board feet were cut. The insatiable appetites of World War I further accelerated cutting. The ecological results were devastating. A 1919 survey declared that nearly half of the region's 200 million acres of pine forest were cutover, and most of them deserted, not fit even for crops or cattle. But the loggers' profits were high, and markets both domestic and international were eager for the strong, handsome lumber of the longleaf or yellow pine.

With help from the Southern Pine Association, the trade association which the loggers had assembled in the 1890s, these companies had already found Latin American markets for up to 10 percent of their total production by 1900.[26] The scale of Southern pine imports was greatest in the larger Caribbean islands, which had lost most of their pine reserves.[27] But with the rapid decline of southern pine forests, expansion of exports into the Caribbean was slow. Indeed, by 1920 southern firms began actively studying Caribbean Basin forests as potential sources of future supplies.[28]

The first U.S. capital invested in financing local timber operators occurred during the 1920s.[29] But this trend did not last long. Yankee loggers could see the great difficulty in setting up financially viable operations where no transport infrastructure existed except for maddeningly shallow rivers or the banana companies' railroads. And they faced the legendary political turbulence of the region, which only companies the size of the fruit giants could hope to influence or withstand. In Nicaragua a few small entrepreneurs were operating by 1930, and when Augusto Sandino's revolutionary forces turned on *gringo* companies, the small loggers joined the chorus of U.S. interests calling on their government's Marines to protect them from local revolutionary nationalism.[30] Moreover, by the late 1920s the southern pine forests had begun a remarkable recovery, under new techniques of fire control and forestry management. The pressure to look beyond U.S. borders for additional supplies gradually diminished. Nonetheless, that brief trend indicates how closely the extension of U.S. power into the Caribbean basin has reflected domestic circumstances.

In sum, yellow pine provided an important source of supply for countries with no more than a fledgling construction lumber industry of their own. But conversely, the foreign competition made it extremely difficult to develop profitable, efficient indigenous logging operations. Beyond this, it increasingly drained scarce foreign exchange (or provided additional pressure to expand export cropping) in payment.

**Timber and Bananas: Bragman's Bluff Lumber Company**

U.S.-based tropical fruit companies began clearing the wet lowland forests of the Caribbean coast on a large scale in about 1900. In the early years of their operations, they simply removed the primary forest, making no effort to harvest timber for any commercial profit. On the lumbering side, small firms from Louisiana and elsewhere had negotiated sometimes conflicting concessions with unstable Nicaraguan governments.

In 1923 the Vaccaro Brothers, soon to found the Standard Fruit Company, became the first to unite the 2 operations.[31] Three years earlier, at the depths of the pinelands scarcity in the southeastern U.S., a Louisiana lumbering family had purchased a concession giving it the rights to 80,000 acres of pine forest behind Bragman's Bluff on the Miskito coast of Nicaragua. It began with a modest declared value of $50,000 and the intention of exporting pine lumber to New Orleans. The Vaccaros' banana empire was well established in the adjacent lowlands of Honduras. Looking for land rights in Nicaragua, they purchased the small firm in 1923 and rechristened it Bragman's Bluff Lumber Company. They renegotiated the concession with Managua, gaining the customary privilege of duty-free import of machinery; in return the company agreed to pay a small export tax on lumber.

The new owners quickly invested $5 million to construct the port of Puerto Cabezas and 100 miles of private railroad to serve both lumber and banana operations. They imported a modern sawmill from Louisiana, to supplement the 3 existing small mills along the coast, and began exporting pine boards in 1925. This was the first step in its long-range ambition to operate in Cuba, Puerto Rico, and other Caribbean sites.

A few figures present a surprising picture for Nicaraguan exports in the late 1920s. In 1926 its primary export, coffee, was valued at $8 million. Next to this, timber exports virtually equalled bananas. That year it exported $1.5 million in mahogany and pine, and only $1.5 - 2 million in bananas, based on Standard Fruit and Bragman's Bluff's joint investment which then amounted to $8 million.[32] That investment totally dominated Nicaragua's Caribbean coast in the 1920s. Harold Denny, the *New York Times* correspondent, who travelled throughout Nicaragua in 1929 investigating the civil war there and the American corporate involvement in it, described the 2 main ports on the Miskito coast, Bluefields, and Puerto Cabezas, as Yankee enclaves. "Puerto Cabezas is even more American than Bluefields. It is an industrial village of some 1,200 population, situated on a broad, flat plain overlooking the Caribbean. It looks and is precisely like a lumber mill village in some southern state in North America. . . . [Standard Fruit / Bragman's Bluff] owns the town and everything in it. The inhabitants, American families from American villages, live in quantity production wooden houses rented from the company, buy their clothing and groceries from the company's store, and find their relaxations in a club built by the company. ..."[33] It was extremely difficult to maintain orderly operations in the conditions of endemic civil war in Nicaragua in the following years. The Caribbean coast was frequently in revolt against the remote capital. Augusto Sandino, from his remote base along the Honduran border, periodically attacked the fruit companies and their subsidiaries from 1926 onward, until 1934, when he was trapped and assassinated by Somoza's men in Managua.

Further difficulties arose over the duration and territorial extent of the company's concession. Neither the company nor the government was careful about local smallholders' prior ownership claims. Some years previously the Miskito Indians in the area had been confirmed in their title to the lands they tilled, in one of their periodic negotiations with the distant government. Reacting to the fruit company's new claim, the Miskitos angrily reminded the government that it had casually given away their ancestral lands to the company. They were, of course, ignored. This was only one instance of the frequent conflict between Indians and foreign capitalists, and between Indians and national governments, in locations where detailed cadastral surveys had never been done, and land records were consequently chaotic.

A more serious blow to the banana operations, one that probably reduced the pressures on the Miskitos, was the attack of Panama Disease and Sigatoka on Standard's Caribbean coast banana plantations, which crippled their Nicaraguan operations by the mid-1930s. In 1935 Standard Fruit began preparing to move from Nicaragua to Mexico and the Pacific lowlands. They exported their last Nicaraguan bananas in 1942, and in the same year liquidated Bragman's Bluff entirely. Standard kept control of the port facilities, but leased the pine forests to a New Orleans lumber firm, which was later reorganized as the Nicaragua Longleaf

Pine Lumber Company or NIPCO, and constructed a network of permanent logging roads.[34]

## Conclusions

Efficient sustained-yield forestry management is a long-term venture demanding political stability and a well developed transport infrastructure. That stability did not exist in most countries of the Caribbean Basin. This is one key to the inefficiency of lumbering techniques, the low level of capital investment in the forest economy, and the resulting degradation of the forests. It also helps explain the inability of timber products to compete with agriculture as an economically viable alternative for the management of marginal lands including many mountain watersheds.

Ultimately a combination of political turmoil, extractive investments from the wider world, and the pressure of subsistence needs all contributed to the degradation of the natural forests of Central America before World War II, and set the stage for accelerated forest extraction in the postwar era.

## Notes

1. F. W. O. Morton, "The Royal Timber in Late Colonial Bahia," *Hispanic American Historical Review*, 58:1 (1978), pp. 41-59.

2. For this trade from the perspective of Europe, see A. M. Wilson, "The Logwood Trade in the Seventeenth and Eighteenth Centuries." in D. C. McKay, ed., *Essays in the History of Modern Europe*. New York: 1936.

3. S. L. Caiger, *British Honduras, Past and Present*. London: Allen and Unwin, 1951, pp. 50-59.

4. Narda Dobson, *A History of Belize*. Trinidad and Jamaica: Longman Caribbean Ltd., 1973, p. 55. See A. R. Gregg, *British Honduras*. London: Her Majesty's Stationery Office, 1968, chap. 8, for a description of the logging camps.

5. Caiger, pp. 135, 139.

6. See F. Bruce Lamb, *Mahogany of Tropical America: Its Ecology and Management*. Ann Arbor: University of Michigan Press, 1966.

7. Dobson, pp. 141-144.

8. W. A. Miller, "Mahogany Logging in British Honduras," *Caribbean Forester*, 2:2 (January 1941), pp. 67-72; Edward Benya, "Forestry in Belize, Part I: Beginnings of Modern Forestry and Agriculture, 1921 to 1954," *Belizean Studies*, 7:1 (1979), pp. 16-28; and Edward Benya, "Forestry in Belize, Part II: Modern Times and Transition," *Belizean Studies*, 7:2 (1979), pp. 13-28; N. S. Stevenson, "Forestry in British Honduras," *Caribbean Forester*, 1:1 (October 1939), pp. 1-3.

9. Caiger, p. 140.

10. See Craig L. Dozier, *Indigenous Tropical Agriculture in Central America: Land Use, Systems and Problems.* Washington: National Research Council--National Academy of Science, 1958, for a detailed survey of the British and American eras on the Miskito coast, and Bernard Nietschmann, *Between Land and Water: The Subsistence Ecology of the Miskito Indians, Eastern Nicaragua.* New York and London: Seminar Press, 1973, for indigenous life and its transformations.

11. James J. Parsons, "The Miskito Pine Savanna of Nicaragua and Honduras," *Annals of the Association of American Geographers*, 45:1 (1955), pp. 36-63, notes that State Department archives contain extensive materials on small U.S. timber merchants in the Caribbean Basin from the early nineteenth century. These should be investigated more thoroughly.

12. Parsons, p. 55.

13. Harold N. Denny, *Dollars for Bullets: The Story of American Rule in Nicaragua.* Westport, Conn.: Greenwood Press, 1980, reprint of 1st ed., 1929, pp. 144, 166-167; John C. Callahan, "The Mahogany Empire of Ichabod T. Williams & Sons, 1838-1973," *Journal of Forest History*, 29:3 (July 1985), pp. 120-130.

14. P. C. Standley and S. J. Record. *The Forests and Flora of British Honduras.* Chicago: Field Museum of Natural History, Botanical Series, Vol. XII,, 1936, pp. 26-28, 31.

15. Tom Gill, *Tropical Forests of the Caribbean.* Washington, D.C.: Tropical Plant Research Foundation, 1931, p. 18.

16. Benya I, p. 20. By 1959 sugar plus citrus exports passed timber as the biggest export earner in Belize, ushering the little country into its contemporary era. Dobson, p. 265.

17. Dobson, pp. 261-264.

18. Tom Gill, "Tom Gill Looks at Tropical Forestry, 1928-1971," *Forest History*, 15:1 (1971), pp.16-21, p. 19; F. Bruce Lamb, "Status of Forestry in Tropical America," *Journal of Forestry*, (1948), pp. 721-726.

19. Standley and Record, p. 30.

20. Hugh M. Raup, *Notes on Reforestation in Tropical America, II [Honduras].* Typescript, April 1950, 83 pp. This paralleled what the forestry profession was beginning to do in British Honduras.

21. William M. Denevan, "The Upland Pine Forests of Nicaragua: A Study in Cultural Plant Geography." *University of California Publications in Geography*, 12:4 (1961), pp. 251-330; Carl L. Johannesen, *Savannas of Interior Honduras*. Berkeley: University of California Press, 1963; and Parsons, pp. 36-63.

22. Lawrence A. Clayton, *Caulkers and Carpenters in a New World: The Shipyards of Colonial Guayaquil*. Athens, Ohio: University of Ohio Press, 1980; Denevan, p. 298.

23. Parsons, pp. 51-52.

24. Standley and Record, p. 33.

25. James E. Fickle, *The New South and the "New Competition": Trade Association Development in the Southern Pine Industry*. Urbana: University of Illinois Press, 1980; John Hebron Moore, *Andrew Brown and Cypress Lumbering in the Old Southwest*. Baton Rouge: Louisiana State University Press, n.d.

26. Fickle, pp. 199-200, 204.

27. Harold W. Wisdom, James E. Granskog, and Keith A. Blatner, *Caribbean Markets for U.S. Wood Products*. New Orleans: U.S. Department of Agriculture, Forest Service, Research Paper SO-225, July 1986, p. 10.

28. Gill 1931, p. 272.

29. F. Bruce Lamb, "Status of Forestry in Tropical America," *Journal of Forestry*, (1948), pp. 721-726.

30. Mira Wilkins, *The Maturing of Multinational Enterprise: American Business Abroad from 1914 to 1970*. Cambridge: Harvard University Press, 1974, p. 98 fn.

31. This account is taken primarily from Parsons, pp. 54-61, and the somewhat different detail in Thomas L. Karnes, *Tropical Enterprise: The Standard Fruit and Steamship Company in Latin America*. Baton Rouge: Louisiana State University Press, 1978.

32. Denny, p. 61.

33. Denny, p. 64.

34. Karnes, ch. 10.

USDA Forest Service Involvement in Post World War II International Forestry

Terry West
USDA Forest Service

(The views expressed in this paper are solely those of the author and do not represent any official position of the USDA Forest Service. The essay is not intended to be a complete account of the International Forestry program of the Forest Service nor of the many people who worked in tropical forestry in Latin America. Instead, it is a more modest attempt to outline the steps by which the program grew and where it is now.)

The role of the USDA Forest Service in international forestry is a neglected area of agency history (Steen 1977). The aim of this essay is to briefly explore the story of Forest Service involvement in international forestry from 1939 until today. It is during this period that the Forest Service international forestry program originated and grew into a new deputy area of the agency. The essay will feature agency work in Latin America because global concern over tropical deforestation has made the region a priority of forestry foreign aid programs funded by the U.S. Government.

It may be said that Forest Service involvement with foreign forestry followed the flag after the Spanish-American War of 1898. Captain George P. Ahern of the U.S. Army organized the Philippines Bureau of Forestry in 1900 and invited USDA Bureau of Forestry director Gifford Pinchot to visit and offer advice in 1902. Creation of the Luquillo (now Caribbean National Forest) forest reserve in Puerto Rico in 1903 further involved the agency in tropical forestry (Cristobol and Lugo 1991). The Forest Products Laboratory (Madison, WI) began a program of tropical wood research shortly after being founded in 1910, with employee Eloise Gerry writing the first of a series of research reports on South American forests and woods of commerce in 1918 (Forest Products Laboratory 1991). Later, in 1940, the USDA Forest Service established the Tropical Research Experiment Station in Rio Piedras, Puerto Rico. Outside of this work, however, the Forest Service was not formally involved in Latin America or Caribbean forestry because it had no authority to spend money on foreign forestry until 1939.

**Tropical Forestry Research**

The scant knowledge that existed on Latin American and Caribbean tropical forests prior to this period was largely the result of Tropical Plant Research Foundation survey work financed by the Charles Lathrop Pack Forestry Trust (Gill 1931). The Pack Foundation was concerned with hardwood depletion in the United States and considered Latin America as a potential new source of wood. In 1927 the Foundation loaned their forester, Tom Gill, (who had resigned from the Forest Service to take the job) to the Tropical Plant Research Foundation for a 3-year period to do forest inventories and market studies. Tom Gill (1931:12) and his assistant William Barbour (1930) decried the general ignorance of forest conditions in Latin America. The onset of the economic depression and later World War II limited further work by the Foundation in the region. As late as 1939, when the U.S. Congress authorized what became the USDA Forest Service Institute of Tropical Forestry (ITF) in Puerto Rico, the situation had not improved. The goal of the ITF was to conduct and disseminate research. "The work we were doing was almost unique because there was so little money in Latin America for tropical forestry research" according to early employee Frank Wadsworth (Ebenreck 1988:53). The staff of the Institute in the early years was filled with a mission. Arthur Bevan, the first director, in a *Journal of Forestry* article (1942:169), exclaimed: "Forestry in the tropics of the Western Hemisphere is standing at the crossroads as we stood in 1900 when Gifford Pinchot and his men took up the challenge" (of managing the forest reserves). He ended by saying: "...any research worker willing to work in the tropics will find a field which will compensate for any disadvantages...learn Spanish because the tropical forest is going to come out of the great unknown and take its rightful place alongside the forest of the temperate zone." (Ibid:172)

The dilemma faced by those few who heeded the call for tropical forestry was the lack of employment opportunities. Tom Gill comments on the employment situation in the 1930s (in Fry 1969:15-16) "...there was little incentive to prepare for foreign forestry work because appointments were few and there was little security in employment, even when you got it. A few Americans had gone into foreign forestry fields, but not many; and there was no formal way in which men could be prepared and channeled into this foreign work." Of those few who did find work, often they were former Forest Service employees or ones on leave from the agency. For example, William T. Cox, a former employee of the Forest Service, borrowed four Forest Service forest engineers (Philip Wheeler, Montgomery Payne, Edward Hamill, and Donald Winters) in 1930 to conduct field work in Brazil when he was employed to organize the Department of Forestry by its government. In a progress report he (Cox 1930:5) made a common complaint of field workers on overseas assignments:

In conducting this important work we have been greatly handicapped by having very little money (in some cases none) made available to the engineers to defray the cost of hiring local men, mules and canoes as needed. Even with practice of the greatest economy it has been impossible for our men to make certain trips that would have given us valuable information.

## The Origin of International Forestry

It was the onset of World War II that set the basis for increased U.S. involvement in international forestry. Passage of Public Law 63 (in 1939) authorized funds for U.S. technical assistance to Latin America, Liberia, and the Philippines. The next year the Second Deficiency Appropriations Act (of 1940) provided the first funds for the direct financing of agricultural and forestry experts in all foreign countries. These acts led to agreements with several Latin American nations for agricultural projects with forestry being included. A forestry program was started in Paraguay in 1943 with John Camp, recruited from the Forest Service, to head the project (Winters 1980:2).

During the War, defense needs led the U.S. government to foster studies of forest conditions in selected Latin America nations (Merker 1943, Holdridge 1947). Wartime need for quinine to treat malaria led to teams of foresters being dispatched to South America to locate sources of cinchona bark. Forest Service employees Leslie Holdridge and Donald Winters explored Columbia, and Gordon Fox and Earl Rodgers worked in Peru (Winters 1977). But the onset of foreign aid projects, including forestry, after World War II by the U.S. Government was the genesis of international forestry in the Forest Service. This remained a marginal activity, however, because infatuated with the success of the Marshall Plan in Western Europe, developmental agencies emphasized highly visible, quick result projects such as airports and bridges, rather than long-term ones such as forestry (Gill 1960:29). Two organizations that did involve U.S. foresters, including many from the Forest Service, in forestry projects during this period were the U.S. Agency for International Development (AID) and the United Nations Food and Agricultural Organization (FAO).

## FAO

The FAO was born in 1943 when President Franklin D. Roosevelt convened a conference to consider ways to organize international cooperation in agriculture. The model for the FAO was the International Institute of Agriculture, based in Rome (Italy), that had been organized in 1905 (Winters 1971). Forestry was excluded from the FAO agenda until a group led by the Forest Service in Washington, DC, managed to get it added in 1945 during the first FAO conference. "It was on the forestry panel of this Interim Commission of FAO (chaired by former Forest Service Chief Henry Graves, with Tom Gill representing Latin America) that American participation in world forestry reached maturity" (Gill 1960:290). When the forestry division of FAO was set up, chief of the Forest Service Lyle F. Watts served as chair of its standing advisory committee.

**Latin America**

At an FAO-sponsored Conference on Forestry and Forest Products held in Brazil in 1948, a Forestry Working Group for Latin America was established with an office in Rio.  FAO forest development work in Latin America had two objectives: opening up little developed forests to stimulate economic and social development and bringing all forests under proper management to avoid waste and check erosion (USDA Forest Service Information Digest, February 18, 1949).

Participation by Forest Service personnel in FAO programs in Latin America operated at 2 levels--as delegates to conferences or as project staff.  Raymond Graver, of the Division of Forest Economics, had a 3-month assignment to the FAO Agricultural mission to Nicaragua.  In his 1950 report he stressed that forestry should receive its proper place among other types of land use.  The struggle to persuade developmental agencies that forestry was a critical element in land use planning was to be waged by foresters for years.  The basic problem was that most of these agencies were primarily concerned with agricultural production to feed a growing world population.

It was left to the Forest Service to promote forestry wherever its staff could find a forum.  Assistant Chief Kotok in 1950 attended the 4th InterAmerican Conference on Agriculture (held in Montevideo, Uruguay), where as the U.S. delegate he presented a paper  "Forest Policy Development in the United States of America." Institute of Tropical Forestry (ITF) director Henry Bosworth and his assistant Frank Wadsworth also attended, and later all went to a meeting of the Forest and Forest Products Commission of the FAO in Santiago, Chile (Information Digest November 1950).  In 1955, at the Fifth Session of the Latin American Forestry Commission in Caracas (Venezuela), Assistant Chief Verne Harper headed the U.S. delegation.  A major agenda item was a plan for establishing the Central Institute for Forestry Research and Training at the Forest School of the University of the Andes (Merida, Venezuela).  The same year (1955), a FAO-sponsored short course in tropical forestry was conducted at the Institute of Tropical Forestry. Half of the 25 trainees were from Eastern nations, especially India, the remainder from Latin American nations. It was this role that led the State Department to recognize the ITF as the primary center for training Latin American forestry technicians (Gill 1960).

## USAID

In addition to FAO-sponsored activities, other international forestry opportunities for the Forest Service originated with the work of the International Cooperation Administration (ICA), a semi-autonomous agency with the U.S. Department of State. Born out of the Point IV Program of 1950, ICA evolved from an earlier series of offices having the common purpose of aiding the "free nations in the interest of international peace and security." Related to Cold War security interests, ICA aimed to help nations raise living standards through material (jeeps, sawmill and logging equipment) donations and technical-scientific expertise. "Designed originally to secure rapid and impressive results, ICA in its early years provided a poor climate for forestry" (Gill 1960:291). Early ICA forestry work was small-scale, with one individual being assigned to a nation. For example, in the early 1950s Forest Service employee Eugene Reichard served as forester in Colombia and Bolivia (Mayer 1989). Nonetheless, it was a primary conduit for Forest Service participation in international forestry, with the agency working with ICA since 1952 by helping to recruit foresters for overseas assignments (Gill 1960:291). ICA became the Agency for International Development (AID) on 4 November 1961 by an Act of Congress. In 1959, The Forest Service managed 200 foreigners who arrived to study U.S. forest conditions and practices. The majority were there via ICA programs, the remainder were affiliated with other State Department programs or FAO projects, a few were financed by their own government or private funds. Tom Gill wrote (1960:292) that "Perhaps it was inevitable that the great bulk of America's activity in foreign forestry has been channeled through the Federal government." Not all of this foreign forestry work involved the Forest Service.

ICA had a forestry contract with Purdue University in 1960 to advise the Brazilian government on setting up the first forestry school in Brazil. Douglas Knudson, a Colorado State forestry school graduate, was given the task and recalls there were only eight foresters in Brazil at the time, all foreign educated or foreigners living in the country. The first federal forestry school (1960) was sited at the state university now known as the Federal University of Vicosa in Minas Gerais.

Two decades later in the 1980s Knudson was employed by AID, the agency that replaced ICA, to conduct forestry work in the Dominican Republic. Created by an act of Congress on November 4, 1961, AID is the U.S. State Department agency that administers U.S. foreign aid.

Knudson found AID committed to forestry, however, its employee performance standards can act as an impediment at the field level:    "The agency does support forestry but the problem is the career development of the AID mission staff, assigned to a post for only 2 to 3 years in which to prove their success, their natural reaction is forestry is too slow to demonstrate results for their resume" (personal communication, 5-31-91).  The result is that the AID field officer may not be as enthusiastic as the consulting forester about the prospects of a proposed forestry project.


**Forest Service Participation**

If over the years the USDA Forest Service has shared the field of international forestry with foresters based outside the agency, it nonetheless remained a primary source of experienced professionals to serve as technical advisors.  As already noted, individual members of the agency were active following the birth of the FAO forestry division in 1946, with a further impetus in 1950 when President Truman announced the Point-4 program of bilateral technical assistance to the newly independent countries and to other developing nations.  The Forest Service was called upon to provide two kinds of help: (1) recruiting foresters and technical leaders for assignment overseas; and (2) receiving foreign nationals sent to the United States for academic studies or on-the-job training in forestry and related areas (LaMois 1975).

Over the next two decades (1950 to 1970) the Forest Service furnished over 150 professionals for long-term assignments or short-term details to technical assistant programs overseas; in the same period over 2,500 foreign nationals went through Forest Service training programs.

The first mention of foreign forestry in the agency directory lists William Sparhawk as the sole staff member, with his section part of the Division of Forest Economics in 1948.  This division in turn was a subordinate part of Research, the third "leg" of the stool (Forest Service) with the other two being State and Private and National Forest Administration.  These three "legs" represented the historical mission of the Forest Service with international forestry at that time being a very minor and new responsibility.  By 1951 Albert Cline was head of this foreign forestry staff, which totaled six employees.  Placement of international forestry in research is attributed to the strong personality of Verne Harper, who out of personal interest in the program, asked Forest Service Chief Lyle Watts that it be assigned to research. Harper left the directorship of the Northeastern Forest Experiment Station to become Deputy Chief for Research in 1952. By 1953, Harper assigned Albert Cline to his personal staff in the capacity of Technical Assistance Specialist, Foreign Forestry.

## IUFRO

In 1956 the Forest Service became a member of the International Union of Forestry Research Organizations (IUFRO), with Dr. Harper playing an active role in the organization. Meanwhile, Forest Service activity in FAO units was growing, and preparation for the 5th World Forestry Congress was also time-demanding, with the result that in 1958 Cline was assisted by two others, Hubert L. Pearson and Robert K. Winters, and the unit became known as the Foreign Forestry Service in the Office of the Deputy Chief for Research. A.C. Cline was designated director of Foreign Forestry in 1959, and in 1961 two new sections were added to Harper's office staff: technical support of foreign programs (Hubert Person) and training of foreign nationals (B.J. Huckenpahler and Carl Olsen). A more staid version of why international forestry was placed in research was later supplied by an internal report (1974:2) that stated: "This is probably a natural development in that the PL-480 Special Foreign Currency Research Program, the International Union of Forestry Research Organizations, and the bilateral science exchanges are all predominantly research oriented."

## CIA Contract

A contract in 1949 with the Central Intelligence Agency (CIA) led to formation of a World Forest Resource subgroup in the Division of Forest Economics Research. This unit was created to provide intelligence on forestry and the forest products situation in selected countries, a task first assigned to William Sparhawk, who had earlier worked with Raphael Zon on the global forest situation (LaMois 1975:2). The workload was later transferred in 1961 to the International Forestry Staff. The contract was terminated on June 30, 1971, but during the contract years "a considerable library was built up of publications and statistics on forest resources and industries for most of the nations of the world" (International Forestry report 1974:7).

**Forest Service Paradigm**

Multiple-use was the theme of the Fifth World Forestry Congress, hosted by the
U.S. in Seattle (Washington) in 1960. Although developed by the Forest Service
in the 1950s, the integrated land use concept was new to many nations. In fact, it
is argued by Winters (in Roth 1986:26) that it had an impact on the policies of a
large number of nations. Almost a decade earlier Earl Loveridge (1952), assistant
chief of the Forest Service, had pointed out to the Venezuelan Ministry of Agriculture the need for central management of natural resources.

In the 1960s the Forest Service underwent a period of growth in numbers and
occupational diversity of its employees. A result was that not only foresters went
overseas as advisors but specialists in range, wildlife, and watershed management
as well. By 1968 at least 60 U.S. foresters were on leave, many from the Forest
Service, in order to work overseas (Winters 1977:167). A problem for the agency
became finding enough suitable candidates to serve overseas. As early as 1962,
Chief Edgar Cliff mentioned this concern in a memo to Forest Service Regional
Foresters and Directors. Because FAO and AID preferred to recruit experts with
years of experience he wrote: "Our problem, therefore is to cultivate the interest
of enough of the more promising of the younger men to meet the expected future
foreign forestry needs, and to give them the experience... for a career in foreign
forestry." He mentioned that one way of obtaining experience was Peace Corps
service. A major factor why Forest Service personnel were hesitant to accept long
term overseas assignments was concern over its impact on their career advancement in the agency.

The formal program in International Forestry (IF) began in 1958 with creation of
the Foreign Forestry unit in Research, and by 1968 it was upgraded to a Division.
This came as the result of a 1967 workload analysis of the IF staff, which noted
that with decreased funding by the CIA and AID, its two main outside sources of
funding, increased support by the Forest Service was necessary to continue the
current workload (International Forestry 1968:5).

At that time the budget for the Division of International Forestry totalled
$372,363, with 22 percent reimbursed by the CIA, 8 percent by Forest Service
Research, and the remainder by AID, the latter mostly for technical consultation
and training costs. In FY 1967 252 "foreign nationals" were in the United States
for training. It was one of the objectives of this training to prepare a member of
the host nation to replace the U.S. specialist after his tour of duty was over in
order to ensure project continuation (Winters, in Roth 1986:40). Over 800
requests for technical consultation from 50 countries were filled in 1967, ranging
from: (1) advice on technical forestry problems, (2) procurement of seed,

(3) technical publications and training films, (4) procurement of specialized equipment and supplies, and (5) testing of wood samples. The same year (FY 1967) 35 Forest Service personnel were serving on 1-year assignments in 20 foreign nations, with 8 others on short-term projects rendering technical assistance in recreational planning, range management, land use planning, forest industries, nursery development, etc. (World Forestry Resources processed 206 requests for information in 1967). By 1967, the de facto division of International Forestry had 11 staff and 8 clerical positions.

Clark E. Holscher, director of International Forestry in 1968, in a letter (August 20, 1968 on file WO History Unit) to his staff reported agency approval of the 1967 workload study: "We now have full responsibility for leadership in IF programs with only broad direction review and counsel from the Deputy Chief (of Research)." International Forestry was evolving into a major program in the Forest Service.

The future did not turn out so bright, since shifts in direction by outside funding agencies in subsequent years meant reduced levels of IF activity in the agency. FAO project assignments peaked during the five-year period 1965-69 (Winters 1980:10), while AID employment of Forest Service personnel in overseas projects peaked in the mid-1960s after growing during the decade 1955-65 (Winters 1977, 1980:11). Bilateral forestry assistance program grants peaked in the 1950s and were eliminated in 1977 (Winters 1980:10). The IF staff decreased from 19 in 1968 to 6 in 1974. Changes in AID program emphasis in this period no doubt reflected shifts in U.S. foreign policy interests related to the Vietnam war. Nor was the Forest Service interested in funding international forestry given its internal concerns with environmental conflicts at home such as clear-cutting (Roth and Harmon 1989).

**Peace Corps**

By 1961, with the creation of the Peace Corps, there was a new source of labor to work on natural resource projects overseas. From 1962 to 1983, 5,000 forestry-related and natural resource management specialists completed Peace Corps service (Smith 1984:553). Although the majority of these volunteers did not continue to work in international forestry after their two-year stint, those who have made a career of international development work have had a tremendous influence in shaping policy and programs in agencies such as AID, the Peace Corps, and the Forest Service--the three main government organizations involved with international forestry today. The Peace Corps clearly dominates the field in terms of numbers of people (past and present) active in developmental forestry or natural resources development projects. To quote Bruce Burwell, a former volunteer in Chile and now serving on the staff of the Peace Crops, "Peace Corps, forestry-wise, since World War had more people in direct contact with Latin American foresters than any other U.S. government agency. These former volunteers now staff positions in the Forest Service and AID and maintain their ties with old friends, often their host country counterpart (personal communication 7-31-91)." In Chile, and elsewhere, the Peace Corps forestry program evolved from individual volunteers being assigned as agricultural extension agents who ended up doing local forestry projects such as reforestation. From this basic forestry work, programs were developed involving volunteers with in regional and national forest planning, wildlife and park management. It is this experience that led them to a career in international forestry and to join the cadre of mid-level managers at AID and Forest Service International Forestry.

"I estimate that the Forest Service Forestry Support Program (a part of International Forestry) has 80 percent of its staff filled by former Peace Corps volunteers," says Burwell. The linkage of the two agencies exists in more than just people but in programs as well. The impetus for mutual cooperation stems from a revival of interest in forestry on the part of AID. This support for international forestry has grown at a rapid pace since 1980 (Smith 1984:552).

**Rebirth of IF**

The year 1980 began a decade-long surge in public interest in international forestry following increased publicity over the environmental impact of tropical deforestation. In a cover letter to his field supervisors that accompanied a copy of his testimony to Congress (May 7, 1980) on tropical deforestation, Chief R. Max Peterson wrote of "our increasing need for involvement in forestry problems beyond our own domestic programs." The movement accelerated with a flurry of publications (Office of Technology Assessment 1983, 1987; World Resources Institute 1985; Padoch, Schmink and Stone 1987) calling for action. AID acted early with its Forest Resources Management Project in 1980 that led to the Forestry Support Program (FSP) in the Forest Service and a joint AID/Peace Corps Initiative. Jointly managed by the Forest Service and the Office of International Cooperation and Development, the FSP helps AID identify, design, execute, and evaluate forestry and natural resource conservation projects in tropical countries. The program now has a roster of over 2,500 resource specialists available for overseas projects (USAID 1988-89 Report to Congress). David Harcharik, IF director, notes that the return of AID to supporting forestry beginning in 1980 "may reflect the influence of the World Bank and the FAO...but much credit goes to Dan Deely of AID because the FSP was his brainchild" (personal communication, 7-31-91).

A decade later the 101st Congress passed legislation--the Global Climate Change Prevention Act and the International Forestry Cooperation Act--that greatly expanded the role of the Forest Service in international resource management (Jesch 1991). The Global Climate Change Prevention Act directed the secretary of agriculture to establish an Office of International Forestry under a new and separate deputy chief in the Forest Service. Jeff Sirmon, the newly selected deputy chief of International Forestry believes that the Congress intent was to prod the Forest Service into taking a more aggressive leadership role in IF for a long time (Davis 1991:18). The bias of the agency toward its traditional three stool "legs" (research, national forest administration, and state and private forestry) has relegated it to the rather passive role of filling requests for technical help made by AID field staff. And while Forest Service funding for IF is increasing it still is small ($5-6 million) compared to the $77 million AID expects to spend on tropical forestry.

Another sign of growing interest in international forestry in the United States is that the number of forestry and natural resource volunteers in the Peace Corps (over 500) reached an all-time high in 1990 (Durst and Norris 1990). Tropical Forestry Program director Sam Kunkle (personal communication on 12-26-90) cautions, however, that real job opportunities are limited despite the expanding number of candidates. The field of international forestry--with a total of 753 people estimated to be employed in 1983--is still not large enough to merit formal programs in forestry school (Smith 1984). The two largest sources of employment were the Peace Corps (250) and AID (250) with the Forest Service, Private Voluntary Organizations (PVOs), private firms, and Multi-Lateral Agencies (World Bank, FAO) employing smaller numbers (ibid:552). The Forest Service International Forestry program, for example, employs only about 10 full time people overseas in a year, plus several dozen on short-term details. Thus the intense competition that exists for the few available jobs makes a career in tropical forestry as risky as risk Tom Gill noted in the 1930s. Job seekers may have cause to hope, however, given the increased attention by Congress to international forestry. The Washington Office of IF currently has 38 employees and approval for 49 in FY 1992, with part of the funding for positions (FSP) from AID. Congress funded nearly $4 million for a Tropical Forestry Program within the Forest Service in FY 1990, with priority for work in tropical Latin America, and islands of the tropical Pacific Ocean and the Caribbean. Eight project areas were selected for the Tropical Forestry Program: agroforestry and fuelwood, social forestry, forest and watershed management, environmental education, forest planning, forest protection, forest inventory, and wood utilization.

Foreign forestry projects often fail because of the U.S. experts' ignorance of local socio-economic conditions. The failure to brief specialists prior to their assignments overseas on local customs is a frequent complaint, as noted by Winters (1980:15-16) and Fortmann (1986). The importance of adapting technology transfer to native culture is that the failure to do so often leads to the lack of success of developmental projects. Thus the field of social forestry was developed by social scientists concerned with native land management practices and natural resource specialists aware of the need for gaining the support of people to ensure the success of technological projects (Messerschmidt 1988).


**Final Words**

Involvement in international forestry by U.S. foresters was marginal until after World War II. The absence of specific authority to work in overseas projects until 1940 limited Forest Service activity in the area. War related needs led to some early inventory work in the 1940s but not until the U.S. government began foreign aid projects following World War II did the agency get formally involved in international forestry programs in Latin America and elsewhere. The decades of the 1950s and 1960s were ones of increased participation by Forest Service employees, although these efforts were those of individuals recruited by AID or FAO rather than a formal agency program. The emphasis of developmental agencies in 1950s and 1960s on agricultural production to feed a growing world population relegated forestry to a minor concern. Part of the problem was that when these foreign aid programs were founded, the stress was on crash programs (Gill in Fry 1969:41), with forestry being slow to show results. Infrastructure projects such as bridges and roads were the priority of the State Department. The overall result was that (Gill in Fry 1969:29) "Except for the Forest Service itself, the U.S. Government has never had any policy for foreign forestry; through its AID, the State Department sent many foresters to work in foreign lands, but it was a haphazard, uncoordinated effort."

Yet, as small-scale as these efforts were by the standards of Tom Gill and other visionaries, these AID projects did mark the formal beginning of international forestry in the Forest Service. Based in Research the IF program expanded during the 1950s and 1960s, only to peak during the mid-1960s when the War in Vietnam began to impact foreign aid priorities. By the 1970s the Forest Service International Forestry program was scaled-back and went into a lull until the global environmentalist surge of the 1980s resulted in its rebirth.

This revival of developmental agencies, including AID, program emphasis on natural resource conservation was a response to the global alarm over the rapid rate of tropical deforestation coupled with the recognition that only through sustained yield of natural resources would less developed countries protect both jobs and the future. For example, Congress amended the Foreign Assistance Act in 1983 to direct AID to address biological diversity and tropical forests. The Forest Service was given authority to engage in international research by the Research Act of 1986. The Peace Corps has labored in forestry and related areas for decades but since 1980 enjoys greater funding and support for its natural resources program by AID. There is now a troika teamed together with the Peace Corps providing grass roots labor, Forest Service technical experts, and Agency for International Development funds and guidance.

# References

Bevan, Arthur, 1942. "Tropical Forest Research." Pp. 169-172. Vol. 40. No. 2. February. *Journal of Forestry.*

Barbour, William, 1930. "Forest Types of Tropical America." Unpublished ms. Forest History Society Archives. Tom Gill Papers. Box 4.

Cox, William, 1930a. "Report to the Minister of Agriculture, Industry, and Commerce: Progress in the Work of Organizing the Brazilian Forest Service" Unpublished ms. October 18, 1930. Box 4. Tom Gill Papers. Forest History Society.

___, 1930b. "Diversified Forestry in Tropical America." In, *Documentary Material on the Inter-American Conference on Agriculture, Forestry and Animal Industry.* Pp. 97-100. National Agricultural Cooperating Committee of the United States. Washington, DC: Government Printing Office.

Cristobol, Carlos and Ariel Lugo, 1991. "The Origins and Management of the Caribbean National Forest: First in the System." Paper presented at the National Forest History and Interpretation symposium and workshop. Missoula, MT. June 20-22, 1991.

Davis, Norah Deakin, 1991. "International Forestry: The Forest Service's Fourth Leg." Pp. 17-20. Vol. 97. Nos. 7 & 8. *American Forests.*

Durst, Patrick and Marcie J. Norris, 1990. "A Profile of U.S. Tropical Foresters." Vol. 88. No. 2. February. Pp. 17-20. *Journal of Forestry.*

Ebenreck, Sara, 1988. Frank Wadsworth: Tropical Forester." Vol. 94. No. 10-11. Nov.Dec. *American Forests.*

Fry, Amelia (Interviewer), 1969. A Summery of the Career of Tom Gill, International Oral History Office. Brancroft Library. Berkeley: University of California.

Figueroa, Julio C., Wadsworth, Frank, and Susan Branham, eds., 1987. *Management of the Forests of Tropical America: Prospects and Technologies*. An Institute of Tropical Forestry Publication. USDA Forest Service. Southern Forest Experiment Station. Government Printing Office.

Forest Products Laboratory, 1991. *Bibliography of Forest Products Laboratory Tropical Forest Utilization Research: 1910-1989*. General Technical Report. FPL-GTR-66. Forest Products

Fortmann, Louise, 1986. "Linking Forestry with People: The Changing Face of International Forestry." *Proceedings* of the Society of American Foresters National Convention. Pp. 44-50. Bethesda, Maryland.

Gill, Tom, 1928. "Tropical Forests and Tomorrow." Pps. 859-864. Vol. XXVI. No. 7. November. *Journal of Forestry*.

___, 1931. *Tropical Forests of the Caribbean*. Tropical Plant Research Foundation. Baltimore:Read-Taylor Co.

___, 1960. "America and World Forestry." In, *American Forestry: Six Decades of Growth*. Edited by Henry Clepper and Arthur B. Meyer. Pp. 282-294. Washington, DC:Society American Foresters

Hamre, Robert, 1981. *The Forest Service Role in International Forestry*. USDA Forest Service. Washington, DC: Government Printing Office.

Holdridge, L., *et al.*, 1947. *The Forests of Western and Central Ecuador*. USDA Forest Service. Washington, DC.

International Forestry, 1974. Workload Analysis. Internal Memo to M.B. Dickerman, Deputy Chief. Dated July 8, 1974. Copy on file at WO History Unit of USDA Service.

___, 1968. Organizational Study International Forestry Staff. Prepared by John F. Prokop, Administrative Management and Marshall Spencer, Research. July. Unpublished report copy on file WO History Unit. Washington, DC.

Jesch, Katherine A., 1991. "An Overview of International Forest Activities in the Forest Service." Briefing Paper, USDA Forest Service. Unpublished ms. on file WO History Unit. Washington, DC.

LaMois, Lloyd, 1975. "Division of International Forestry, Forest Service." USDA Forest Service. Unpublished ms. on file WO History Unit. Washington, DC.

Loveridge, Earl W., 1952. *A Reconnaissance Study of Organization* and *Administration of the Forestry Department of the Venezuelan Ministry of Agriculture*. Caracas: Consejo de Bienestar Rural.

Lugo, Ariel and John Ewel, Susanna Hecht, Peter Murphy, 1987. Christine Padoch, Marianne Schmink, and Donald Stone. *People and the Tropical Forest*: A Research Report from the United States Man and the Biosphere Program. United States Department of State. Washington, DC: Government Printing Office.

Messerschmidt, Donald A., 1988. "Notes for a Social Science of Development Forestry: Approaches to Interactive Research Linkages." USDA Forest Service Forestry Support Program. Unpublished ms. USAID/Science & Technology Bureau/Rural Development. Washington, DC.

Mayer, Karl, 1990. "A Tribute to Gene Reichard." pp.8-9. Dixie Ranger. Vol. XX. No. 3. August. Southern Forest Retirees Association.

Merker, C.A., Barbour, William, Scholten, John, and William Dayton, 1943. *A General Report on the Forest Resources of Costa Rica.* USDA Forest Service. Washington, DC.

Office of Technology Assessment, 1983. *Sustaining Tropical Forest Resources: Reforestation of Degraded Lands.* Background Paper #1. Congress of the United States. Office of Technology Assessment. Washington, DC: Government Printing Office.

___, 1987. *Technologies to Sustain Tropical Forest Resources.* Congress of the United States. Office of Technology Assessment. Washington, DC: Government Printing Office.

Roth, Dennis (Interviewer), 1986. Interview with Robert K. Winters, Former Director, International Forestry. USDA Forest Service. Unpublished transcript on file WO History Unit. Washington, DC

Roth, Dennis and Frank Harmon. "The Forest Service in the Environmental Era." Unpublished ms. USDA Forest Service, History Unit. Washington, DC.

Smith, Christopher, 1984. "Education and Careers in International Forestry." *Proceedings* of the 1983 Convention of the Society of American Foresters. Pp. 551-555. Bethesda, Maryland.

Stahelin, Rudolph and William P. Everard, 1964. *Forests and Forest Industries of Brasil.* Forest Resource Report No. 16. USDA Forest Service. Foreign Forestry Services, Research. Washington, DC.

Steen, Harold K., 1977. *U.S. Forest Service: A History.* Seattle. (Second Edition) University of Washington Press

Winters, Robert K., 1971. "How Forestry Became Part of FAO." pp. 574-577. Vol. 69. No. 9. September. *Journal of Forestry.*

___, 1977. "U.S. Participation in International Forestry. March. *Journal of Forestry.*

___, 1980. International Forestry in the U.S. Department of Agriculture. Staff Report: National Economics Division. Economics, Statistic and Cooperative Service. USDA. Washington, DC.

___, 1983. "International Forestry" In, *Encyclopedia of American Forest and Conservation History.* Vol.1. pp. 313-315. Edited by Richard Davis. New York: Macmillan Publishing Company.

World Resource Institute, 1985. *The World's Tropical Forests:* A Call for Accelerated Action. World Resources Institute/The World Bank and the United Nations Development Programme. Report of an International Task Force. (Draft Copy on file at WO History Unit). Washington, DC.

Latin America and the Creation of a Global Forest Industry

M. Patricia Marchak
University of British Columbia


This paper is about capital invested in pulpwood forests and mills, more than about trees. It is also about history, but mainly the history of the past 2 decades. My objective in studying the movement of capital is to develop a global context for understanding what is happening to tropical and temperate forests, and to the indigenous peoples and communities affected by the massive changes in the industry. So far my research has focussed on Canada, Japan, and Indonesia, and I am only now turning attention toward Latin America, so these comments are still at a preliminary stage. I should add that the title of the overall project is, "For Whom The Tree Falls."


**Depletion of Northern Forests**

In the period before 1960, the pulpwood forest industry was virtually confined to northern softwood regions. Pulp was sold on a world market, but it was produced primarily in the northern hemisphere.

The northern countries, and in particular Canada and the United States, severely depleted the forest and failed to replant it. Softwood forests take up to 80 years to reach commercial size. The industry invested in biotechnology research and experimentation with potential new fibre sources, and in new pulping technologies rather than in reforestation of northern regions. These investments have resulted in pulping techniques that can transform pine and hardwoods into high grade kraft pulp. The pines can be grown on plantations in 15-30 year cycles. The hardwoods, including eucalyptus, can be grown in tropical climates in 3 to 10 year cycles. Thus the companies are moving more of their capital to southern climates, rather than into reforestation of northern softwood forests.


**New Fibre Sources and New Producing Regions**

The new producing regions include northern New Zealand, Tasmania in Australia, Indonesia, Malaysia, Thailand, and some other parts of South-East Asia, parts of Africa, Portugal, Spain, Brazil, and Chile, with some potential in Argentina and

292

Venezuela. The southern United States have become a major new region, and the boreal forests of Canada are just beginning to be exploited.

Companies in New Zealand, Australia, and the southern United States developed the radiata pine during the 1960s and 1970s. In New Zealand, native forests have been burned and bulldozed to make way for radiata pine plantations. These can be grown in relatively short cycles (about 30 years) and have high productivity per hectare. Chile is the most promising source of new radiata pine fibres, with superior growing conditions and extensive afforestation projects already underway, and a reported 100,000 hectares planted by the late 1970s. Caribbean pine is becoming a major new fibre source in Central and South America. Ellioti pine is being grown in South Africa and Brazil. In Canada, and now in Sweden, lodgepole or jack pine is being harvested. Pine is the major plantation species in the southern United States, where, according to Sedjo, 1 million hectares per year were planted during the 1980s.[1]

Tropical hardwoods have become major sources of wood fibre for pulp, paper, and paperboard only since the mid-1970s. In 1971, when the Paper Industry Corporation of the Philippines (PICOP) was constructed in Mindanao,[2] the utilization of local hardwoods was largely experimental. Since then, with genetic experimentation and extensive trials on all tropical woods in the Philippines, it has been established that local woods, either alone or in combination with softwoods sourced elsewhere, are viable sources for high grade papers and paper-boards. Similar experiments have been conducted in other tropical regions including Brazil, and plantations have been established. Most of these trees can be grown in under 10 years, and the yield per hectare is greater than for softwoods. Yields from intensively managed plantations in the tropics are estimated to be up to 10 times greater than those in temperate zones.[3] Further, in some regions - notably northern Brazil - the plantation acreage is enormous; in the Aracruz project it is reported to be 1.6 million hectares (an area half the size of Belgium).

Eucalypts have more recently become the major plantation fibre. Pulping technology changed from sulfite to sulfate methods over the 1960s, and then the range of options increased to include several chemical and thermal mechanical methods in the 1970s and 1980s. The new techniques allowed companies to utilize a wider range of fibres. Eucalypts were found to be ideal sources. Their advantages included resistance to insects and fungi, and growth cycles of under 10 years. They produce over twice as much pulp wood as pine, and are easier to cut and transport. Several harvests can be obtained from each stump. Companies introduced plantations in warm climates well served by fresh water.[4] Eucalyptus plantations have been established in Chile, Morocco, Spain, and Brazil. The *eucalyptus grandis* can be harvested in 7 years for pulpwood, and the stumps then sprout new stems to be harvested again on 7-year cycles.[5] *Eucalyptus globlus* is now touted by industry experts as the best fibre source for computer papers.

In Brazil, productivity per hectare per year for the Aracruz pulpmill is about 15 times greater than in Scandinavia and northern North America. Brazil's eucalyptus plantations and land are already the base for 3 major pulpmills, with more in construction and planning. Half the output goes to the export market, and Brazilian manufacturers anticipate continued growth because, they say, "eucalyptus is now a desired fiber worldwide," and:

> "The forestry resources [have] a rapid growing rate when compared with the Northern Hemisphere, offer low wood cost, high productivity per unit area, great land spaces ready for reforestation, and biomass resources as an energy source. The pulp demand in the international markets along with increasing prices and low labor cost are promoting a leading technology for eucalyptus pulp and paper manufacture as well as more efficient mills. ...There is the possibility to transform a part of the foreign debt loans to risk capital, and the participation of interested groups in existing and new projects."[6]

These plantations are not tropical forests in the fashion designed by nature. Those forests are expensive to log because of their density and the extraordinarily rich mixture of woods, where many are unsuitable for pulping. More economical from the perspective of large logging operations are plantation forests, where the jungle variety is eliminated and only those trees which can be transformed into high grade pulps are seeded. The view of the vice president of Indah Kiat, a major pulpmill in southern Sumatra, provides a succinct expression of this:

> "We have 65,000 ha now; we are now in the process of getting concessions for another 65,000. Basically we are looking for forest which can be clear cut and replaced with eucalyptus and acacia."[7]

**The Investors**

Japan demonstrated that it was possible to create a huge pulp and paper industry with relatively little domestic fibre, by sourcing supplies elsewhere. The growth of the Japanese paper industry after 1960 relied on the availability of logs, wood-chips, and raw pulp sourced in nearby Asian countries, New Zealand, Russia, and North America, plus recycled newsprint from domestic sources. As supplies dwindled from traditional sources, Japanese investors began to seek new sources, including plantation fibres in Latin America.

Cenibra, established in the Minas Gerais state of Brazil in 1977, is a joint venture between the state-owned Brazilian mining, minerals, and industrial giant, Compania Vale Do Rio Doce, and a consortium called Japan-Brazil Pulp and Paper Resources Development (JBP) made up of 18 Japanese papermakers and administered by the Japanese engineering and sales company, C. Itoh, together with the Japanese government agency, the Overseas Economic Cooperation Fund.[8] Despite the shaky Brazilian economy and escalating interest rates, Cenibra, with cheap wood resources, low costs, and growing demand for eucalyptus pulp, is expanding its share of export markets. The Japanese companies take half the output of the mills, manufacturing it into high grade papers in Japan. As well, C. Itoh has started expanding sales from this mill to China.

Mitsubishi formed a joint-venture with Forestal Colcurra of Chile in 1987 to export chips to Japanese producers, but Japanese investment in Chile lags behind American and New Zealand investments there.

While American companies remained dominant throughout the 1980s, current listing in *Pulp and Paper International* show Oji, Jujo, Sanyo-Kokusaku, Honshu, and Daishowa among the top 25, and Rengo and Taio among the next 15. Japanese activity is directed primarily toward obtaining raw material, while American activity is directed toward establishing production units elsewhere. Much of Japanese investment takes form within consortia and intricate joint ventures in developing countries. Japan has grown rapidly as a producer of paper and paperboard. It is now second only to the United States, but there are other newcomers as well, including New Zealand.

Fletcher Challenge of New Zealand is the most notable of the new companies. The original company began experimenting with exotic pines in the 1960s, and burnt off old-growth forests in order to plant radiata pine in the 1970s. Starting as a relatively small player in the forest industry, Fletcher Challenge (formally established as such in 1981) acquired several other companies in New Zealand before becoming a global player. It owned Tasman Pulp and Paper, New Zealand Forest Products, and Carter Holt Harvey by the late 1970s.[9] In 1983 FC acquired Crown Zellerbach Canada. In 1986, it purchased half of the Papeles Bio-Bio newsprint mill in Chile (later increasing its holdings to 100 percent). In 1987, it bought into BC Forest Products, and subsequently acquired 72 percent of the shares. In 1988, it acquired 50 percent of the holdings in Australia's only newsprint producer, Australian Newsprint Mills.

The move to Chile, where large radiata pine plantations were already scheduled to reach maturity at the same time as the identical plantations in New Zealand, averted competition. Both New Zealand and Chile are export-oriented and have small domestic markets, and both are aiming at the Asian markets. The purchase in Chile was made through debt swap financing. Through Carter Holt Harvey as well as in its own name, it acquired both plantations and other mills. Lumber, plywood, and pulpmills are either in construction or planned, and1of these may involve Daishowa as a partner.

In 1988, Fletcher Challenge moved into Brazil, obtaining half the shares of the Papel de Imprensa S.A. (Pisa) mill through debt swap financing.[10] Some of its B.C.- produced newsprint could be allocated to Brazil so that FC might optimise capacity in the softwood region while preparing for additional capacity in Brazil. Having supplies in diverse regions puts this company in a market-control position. It also allows it to avoid stoppages and failed shipments due to strikes, political events, natural disasters, or other impediments that affect market supplies from single-region companies.

By 1988, Fletcher Challenge had become the largest producer of market pulp, the second largest newsprint producer, the third largest lumber producer, and, with all forms of forest production combined, the fourth largest forest company in the world. It owns or has harvesting rights on 3,386,250 hectares (13,074 square miles) of land in Canada, the USA, Australia, New Zealand, Chile, and Brazil.[11] Canadian and American companies (these are virtually indistinguishable) have either been displaced, or they have, themselves, internationalized their holdings. Weyerhaeuser, International Paper, Georgia Pacific, and James River have retained their dominant positions by investing in new plantations in warm climates and linking up with capital elsewhere. They have also been active in takeovers of their weaker competitors, several of which have disappeared from the ranks over the past few years.

Of these international holdings, 2 are of particular interest for Latin America. Abitibi and Bowater entered a joint-venture with the Venezuelan government and major Venezuelan publishers (the company has 67 percent; Venezuela, 18 percent; and the publishers, 15 percent of the shares) to build a greenfield newsprint mill near Cuidad Guyana, based on Caribbean pine, with output to be marketed by the Canadian partners. Scott Paper, together with Shell and Citibank, took over the Papeles Sudamerica Nacimiento eucalyptus kraft mill in Chile in 1988.

These companies still have extensive forest harvesting rights in softwood regions of North America, and several have private lands in the United States. This base continues to be their strong suit, but their futures lie in the new plantations either off-shore or in the southern United States. In the United States, a shift from the northwestern to southern regions is clearly taking place. In Canada, there is growth based on the remaining softwood forests and in the new hardwood forests, but most of this growth is occasioned by foreign direct investment driven by a search for raw materials; there is no real growth in manufacturing capacity beyond basic pulpmills.

The Scandinavian countries have not deforested their own lands to the same extent as North America. Reforestation and afforestation projects at the turn of the century in Sweden, Norway, and Finland have continued to provide them with second-growth fibre supplies. Until the 1960s, they were peripheral staples producers, similar to Canada. However, with the unification of Europe underway, Swedish and Finnish companies began to diversify their holdings in Europe. The Finnish company, Kymmene, and the Swedish company, Stora, have established large-scale production units for printing and writing papers within the EC. Kymmene has attempted to obtain fibre supplies from southern United States to supplement Finnish sources. Stora has obtained eucalyptus plantations in Portugal, and has expanded as well to Brazil and North America. It is a major shareholder in the Aracruz mill in Brazil.

The huge mills built by these international companies are capital-intensive.[12] They rely on cheap wood raw materials, but the risk to capital in a notoriously cyclic pulp market is considerable. Some of the present companies will be taken over or drop out, and not all the new mills and plantations will succeed.

On the basis of general history of corporate activity, we might anticipate a trend toward greater concentration of ownership, but we are also seeing a trend toward investment by joint arrangements that include traditional and occasional new paper companies, banks such as Citibank of New York, indigenous pools of capital and government backing in underdeveloped countries.

**Summary**

I have described a few of the investments and companies, and hinted at the nature of the pulp market. As I said at the beginning, my objective is to provide an economic context, the global context, for local developments. If we understand the movement of capital and the market, we might be better able to create success-ful strategies for protection of the forests and the people. Some plantations are helpful to newly developing countries, and this is especially true where they are situated on land that is not forested or is otherwise degraded and unsuitable for agriculture and natural tropical forests. Other plantations are high-risk for indigenous peoples, wildlife, and more generally, the life on this planet. I do not think local action alone, though it is also essential, will save tropical forests. If we are to conserve the tropical forests and distinguish between diverse uses of land, we will have to come to grips with such companies as Oji, Daishowa, Fletcher Challenge, Abitibi, and Scott, because it is these companies - and the consumers of their paper products - for whom the tree falls.

**Table 1.** Top 40 Pulp and Paper Companies, 1988, as listed by *Pulp and Paper International* (Sept. 1989)

| Rank by sales, all products | Company | # of Countries |
|---|---|---|
| 1 | Weyerhaeuser (US) | 4 |
| 2 | International Paper (US) | 18 |
| 3 | Georgia-Pacific (US) | 3 |
| 4 | Fletcher-Challenge (NZ) | 6 |
| 5 | James River (US) | 10 |
| 6 | Stora (Sweden) | 11 |
| 7 | Kimberly-Clark (US) | 21 |
| 8 | Champion International (US) | 3 |
| 9 | Scott (US) | 21 |
| 10 | Oji Paper (Japan) | 3 |
| 11 | Mead (US) | 8 |
| 12 | Boise Cascade (US) | 2 |
| 13 | Jujo (Japan) | 1 |
| 14 | Noranda (Canada) | 2 |
| 15 | Stone Container (US) | 1 |
| 16 | Great Northern Nekoosa (US) | 1 |
| 17 | Sanyo-Kokusaku (Japan) | 1 |
| 18 | Svenska Cellulosa (Sweden) | 15 |
| 19 | Honshu (Japan) | 2 |
| 20 | MoDo (Sweden) | 6 |
| 21 | Abitibi-Price (Canada) | 2 |
| 22 | Union Camp (US) | 4 |
| 23 | MacMillan Bloedel (Canada) | 2 |
| 24 | Daishowa (Japan) | 1 |
| 25 | Kymmene (Finland) | 4 |
| 26 | Jefferson Smurfit Group (Ireland) | 12 |
| 27 | Canadian-Pacific FP (Canada) | 2 |
| 28 | Enso-gutzeit (Finland) | 6 |
| 29 | Rauma-Repola (Finland) | 2 |
| 30 | Buhrmann-Tetterode (Netherlands) | 2 |
| 31 | Feldmuhle (FRG) | 3 |
| 32 | Amcor (Australia) | 2 |
| 33 | Domtar (Canada) | 1 |
| 34 | Westvaco (US) | 2 |
| 35 | Rengo (Japan) | 1 |
| 36 | PWA (FRG) | 4 |
| 37 | Consolidated-Bathurst (Canada) | 4 |
| 38 | Wiggins Teape (UK) | 11 |
| 38 | Metsa-Seria (Finland) | 6 |
| 40 | Taio Paper (Japan) | 1 |

**Notes:** Some companies (e.g. Noranda) are not consolidated; their subsidiaries (e.g. MacMillan Bloedel) are listed seperately. Countries include only those where company has pulp, paper, and converting operations under the corporate name of the parent.

# Notes

1. Roger Sedjo, "the Expanding Role of Plantation Forestry in the Pacific Basin," paper presented to a national conference on *Prospects for Australian Plantations*, Canberra, Australia, August 21-25, 1989, p. 2; see also Richard W. Haynes, "An Analysis of the Timber Situation in the United States: 1989-2040," Part II United States Department of Agriculture, Forest Service, 1988, and Roger A. Sedjo, 1983, *The Comparative Economics of Plantation Forestry: A Global Assessment*, Resources for the Future, Johns Hopkins Press, Baltimore, MD, USA.

2. Cecil MacDonald, "Tropics could be major fiber source." *Pulp and Paper International*, vol. 25(7) July, 1983, 26-28.

3. R. C. Kellison and B. J. Zobel, "Technological Advances to Improve the Wood Supply for the Pulp and Paper Industry in Developing Countries," in FAO, *Proceedings*, 1987, 136-144.

4. Lars Kardell, Eliel Steen and Antonio Fabiao, "Eucalyptus in Portugal: a Threat or a Promise," *Ambio*, 15(1) (1986): 643-650.

5. Ed Williston, "The industry in the 1990s: where technology is headed," *World Wood*, vol 30(1), February, 1989: 20-21.

6. Alberto Fernandez Sagarra, Director FICEPA, Sao Paulo SP, Brazil, in *TAPPI*, Dec. 1988, p. 25.

7. *Pulp and Paper International*, January, 1988, p. 41.

8. *Pulp and Paper International*, April, 1987, p.42.

9. A critical company biography is given by Bruce Jesson, in *The Fletcher Challenge: Wealth and Power in New Zealand*, Pokeno, N.Z.: Jesson, 1980.

10. *Pulp and Paper International*, vol 30(8), August 1988,p. 7 provides details on the Brazilian acquisition. *Appita*, vol. 41 (5), September 1988, pp. 349-350 provides overview of all holdings.

11. Toronto Stock Exchange *Review*, March, 1989: Fletcher Challenge Limited.

12. Heikki J.W. Salonen and Pekka Niku, "The Future of the Forest Products Industry: A Worldwide Perspective," in Gerard F. Schreuder (ed.), *Global Issues and Outlook in Pulp and Paper*, Seattle: University of Washington Press, 1988:285-300. These analysts estimate that to achieve a 10 to 12 percent return on investment in a pulp mill similar to the Aracruz or Cenibra in Brazil (where Stora has holdings), would require an investment of U.S. $2,000 per annual ton and operating costs of U.S. $240 to $280 ton with a current (1988) international pulp price level.

# Author Affiliation

Balée, Dr. William
Associate Professor
Department of Anthropology
Tulane University
New Orleans, Louisiana 70115
USA

Barborak, Dr. James
Associate Professor
University for Peace
Apartado 277
3000 Heredia
COSTA RICA

Boomgaard, Dr. Peter
Vuije Universiteit
Faculty of History 7A-20
De Boelelaan 1105
1081 HV Amsterdam
THE NETHERLANDS

Brown, Dr. Larissa V.
Dept. of History
301 Morrill Hall
Michigan State University
East Lansing, MI 48824-1036
USA

Budowski, Dr. Gerardo
Head, Natural Resources
Program
University for Peace
3000 Heredia
COSTA RICA

Dargavel, Dr. John
CRES
Australian National University
GPO Box 4
Canberra, ACT 2601
AUSTRALIA

Dean, Dr. Warren
Dept. of History - NYU
19 University Place
New York, NY 10003
USA

del Amo R., Dr. Silvia
Directora General
Gestion de Ecosistemas
Asociacion Civil
AT, Postal 19 - 182
CP 03910
MEXICO DF

Graham, Dr. Elizabeth
Dept. of Anthropology
York University
4700 Keele St.
North York
Toronto, Ontario
CANADA M3J 1P3

Guillaumon, Dr. J. Régis
Instituto Florestal
Rua do Horto 1197
CP 1322
Horto Florestal
Sao Paulo
BRAZIL

Hoffman, Ms. Rhena
Sechzehneichen
0-1901
Krs. Kyritz
GERMANY


Horn, Dr. Sally P.
Dept. of Geography
University of Tennessee
Knoxville, TN  37996-1420
USA


Kengen, Dr. Sebastião
Projeto
PNUB/FAO/BRA/87/007
SAIN-AV. L4 Norte
70.800-Brasilia-DF
BRAZIL


Konrad, Dr. Herman
Dept. of History
Calgary University
Calgary, Alberta
CANADA  T2N 1N4


Lehmann, Mary Pamela
5117 8th Street
Zephyrhills, FL  33540-5162
USA


Leier, Robert D.
RK 4 Box 355B
Tyrone, PA 16686
USA


MacLeod, Dr. Murdo J.
Dept. of History
4131 Turlington Hall
University of Florida
Gainesville, FL  32611
USA


Marchak, Dr. M. Patricia
Dean, Faculty of Arts
1866 Main Mall
University of British Columbia
Vancouver, B.C.
CANADA  V6T 1W5


Melville, Dr. Elinor
History Dept.
York University
North York
Toronto, Ontario
CANADA  M3J 1P3


Pierce, Ms. Susan M.
Forest/Wood Science
Colorado State University
Fort Collins, CO  80523
USA


Sponsel, Dr. Leslie E.
Dept. of Anthropology
University of Hawaii
Honolulu, HI  96822
USA


Steen, Dr. Harold K.
Executive Director
Forest History Society
701 Vickers Avenue
Durham, N.C.  27701
USA

Tucker, Dr. Richard P.
University of Michigan
School of Natural Resources
Dana Building
Ann Arbor, MI 48109-1115
USA

West. Dr. Terry
Historian
Public Affairs Office
USDA-Forest Service
201 14th Street, S.W.
  at Independence Ave., S.W.
Auditors Building
Washington, D.C. 20250
USA